PLANT-WATER RELATIONSHIPS

EXPERIMENTAL BOTANY
An International Series of Monographs

CONSULTING EDITORS

J. F. Sutcliffe

School of Biological Sciences, University of Sussex, England

AND

P. Mahlberg

Department of Botany, Indiana University, Bloomington, Indiana, U.S.A.

PLANT-WATER RELATIONSHIPS

R. O. SLATYER

C.S.I.R.O. Division of Land Research
Canberra, Australia

1967

ACADEMIC PRESS · LONDON AND NEW YORK

ACADEMIC PRESS INC. (LONDON) LTD
Berkeley Square House
Berkeley Square
London, W.1.

U.S. *Edition published by*
ACADEMIC PRESS INC.
111 Fifth Avenue
New York, New York 10003

REPRODUCED PHOTOLITHO IN GREAT BRITAIN BY
J. W. ARROWSMITH LTD., BRISTOL 3

Preface

Some aspects of plant physiology can be studied without direct reference to the environmental conditions under which plants grow in nature. In the case of water relations, however, this is seldom possible; the water status of individual cells and tissues being determined, to a considerable degree, by the pattern of water exchange between the entire plant or plant community and the environment. Because of this phenomenon, the study of plant-water relationships is not only the domain of the plant physiologist and ecologist, but also of the soil physicist, the micro-meteorologist, and of scientists in interdisciplinary fields such as agronomy, forestry and horticulture.

In consequence, this book has been prepared with two main objectives in view. The first of these is to present an account of contemporary thinking and research concerning the distribution, movement and function of water in plant cells, tissues and organs, the development of internal water deficits and their significance to physiological processes. The second is to place these phenomena in an ecological context, with special reference to the soil factors which affect soil water status and water supply to plant roots, and the atmospheric factors which affect energy exchanges in the biosphere, particularly evaporation.

The subject matter coverage has developed from a course of lectures given to advanced graduate students at Duke University during academic year 1963–64. At this time the author was Visiting Professor in the Department of Botany and School of Forestry with the support of a National Science Foundation Fellowship.

I would like to take this opportunity of acknowledging with gratitude the assistance of the National Science Foundation in providing this generous assistance. I would also like to acknowledge the counsel of many colleagues who have read part or all of the manuscript and offered many valuable suggestions. In this regard, special mention should be made of Dr. Paul J. Kramer and Dr. K. R. Knoerr of Duke University, Dr. Sterling A. Taylor of Utah State University, Dr. P. G. Jarvis of the University of Aberdeen and my colleagues in the CSIRO Division of Land Research. Finally, my thanks are due to Miss D. Lee and Mr. O. R. Johnson who have borne the brunt of the considerable amount of effort involved in the indexing, reference checking and typing.

Permission to reproduce certain figures was kindly given by the following authors and publishers. Figures 2.1, 8.1 and 2.12 from

v

"Energy Exchange in the Biosphere" by D. M. Gates (Harper & Row, New York, 1962); Figures 1.1 and 1.3 from "Water" (A. M. Buswell and W. H. Rodebush, eds) copyright 1956 by *Scientific American, Inc.*, all rights reserved; Figure 3.1 and Table 3.1 from "Soils and Soil Fertility" by L. M. Thompson, copyright 1957, McGraw-Hill Book Company; Table 2.1 from "Radiation Biology", Vol. 3, Withrow and Withrow, copyright 1956, McGraw-Hill Book Company; Figure 2.10 from "Transpiration by Sudangrass as an Externally Controlled Process" (C. H. M. van Bavel, L. J. Fritschen and W. E. Lewis, eds), *Science* **141**, 270 (1963); Figure 6.8 from "Water Transport Across Root Cell Membranes; Effect of Alkenylsuccenic Acids" (P. J. C. Kuiper, ed.), *Science* **143**, 691 (1964); Figures 9.2 and 9.8 from "Lower Limit of Water Availability to Plants" (W. R. Gardner and R. H. Nieman, eds), *Science* **143**, 1461 (1964); Figure 2.7 from the Soil Science Society of America Proceedings, Vol. 21, 1957, page 64; Figure 2.6 from the Soil Science Society of America Proceedings, Vol. 24, 1960, page 5.

Canberra   R. O. SLATYER
January, 1967

List of Frequently Used Symbols

Symbol	Description	Main Units
A	Area	(cm^2)
D	Diffusivity, diffusion coefficient	(cm^2 sec^{-1})
E	Internal energy	(erg)
G	Gibbs free energy	(erg)
$\bar{G}$	Molal or partial molal Gibbs free energy	(erg mole^{-1})
H	Enthalpy	(erg)
I	Flow of current	(amp cm^{-2})
J	Flow in linked transfer systems	(cm sec^{-1})
K	Hydraulic conductivity of soil	(cm sec^{-1})
L	Transfer coefficient in linked transfer systems	(cm sec^{-1} bar^{-1})
M	Molecular weight, molarity	(g)
N	Mole fraction	
P	Pressure	(dyne cm^{-2})
R	Universal gas constant	(erg deg^{-1} mole^{-1})
S	Entropy	(erg deg^{-1})
$\bar{S}$	Molal or partial molal entropy	(erg mole^{-1} deg^{-1})
T	Temperature	(deg C or deg K)
V	Volume	(cm^3)
$\bar{V}$	Molal or partial molal volume	(cm^3 mole^{-1})
W	Weight of plant material	(g)
X	Force causing flow in linked transfer systems	(bars)
$\mathbf{A}$	Net photosynthetic rate	(g cm^{-2} min^{-1})
$\mathbf{E}$	Evaporation rate	(g cm^{-2} min^{-1})
$\mathbf{H}$	Sensible heat flux	(cal cm^{-2} min^{-1})
$\mathbf{I}$	Cumulative infiltration	(cm water)
$\mathbf{K}$	Atmospheric turbulent transfer coefficients	(cm^2 sec^{-1})
$\mathbf{O}$	Surface run off	(cm water)
$\mathbf{R}$	Radiation	(cal cm^{-2} min^{-1})
$\mathbf{S}$	Soil heat flux	(cal cm^{-2} min^{-1})
$\mathbf{U}$	Drainage below the root zone	(cm water)
$\mathbf{W}$	Soil water content in root zone	(cm water)
a	Chemical activity	
c	Concentration	(mole cm^{-3}, g cm^{-3})
d	Roughness length, boundary layer thickness	(cm)
e	Water vapour pressure	(mm Hg)
g	Acceleration due to gravity	(cm sec^{-2})
h	Soil water suction	(cm)
k	Permeability or transfer coefficient	(cm sec^{-1})
m	Mass	(g)
m	Metre	
m	Milli-	
n	Number of moles	
p	Partial gas pressure	(mm Hg, dyne cm^{-2})

Symbol	Description	Main Units
q	Flux	(g sec^{-1}, cm^3 sec^{-1})
r	Diffusive resistance	(sec cm^{-1})
r	Radius	(cm)
u	Wind speed	(cm sec^{-1})
v	Velocity	(cm sec^{-1})
z	Height, depth	(cm)
a	Chemical energy conversion coefficient	(cal g^{-1})
k	von Karman constant	
l	Latent heat conversion coefficient	(cal g^{-1})
r	Albedo	
v	Relative cell volume	
Δ	An increment of	
$\boldsymbol{\Delta}$	Slope of the saturation vapour pressure/temperature curve	(mm Hg deg^{-1})
Φ	Total potential	(bars)
Ψ	Water potential	(bars)
α	Absorption coefficient for incident radiation	
β	Bowen ratio	
γ	Psychrometric constant	(mm Hg °C^{-1})
ε	Elastic modulus	(bars)
η	Viscosity	(dyne sec cm^{-2})
θ	Volumetric soil water content	(cm^3 water cm^{-3} soil)
μ	Chemical potential	(erg mole^{-1})
μ	Micron	
π	Osmotic pressure (osmotic suction)	(bars)
π	Pi	
ρ	Density	(g cm^{-3})
σ	Reflection coefficient	
σ	Surface tension	(dyne cm^{-1})
τ	Matric pressure (matric suction)	(bars)
ϕ	Chemical activity coefficient	
χ	Mixing ratio	

Main superscripts

i	Inwards, inside
l	Lower
o	Outwards, outside
0	Reference value

Main subscripts

a	Air
i, j	ith, jth components in a sytsem
l	Leaf
s	Solute
w	Water
o	Reference value

Contents

Chapter 9

CHAPTER 1

Some Properties of Water and Aqueous Solutions

In common with other organisms, life in plants takes place in an aqueous medium. Water is essential for the structural integrity of biological molecules and hence for the integrity of cells, tissues and the organism as a whole. It also performs a vital role as a solvent, mineral nutrients and other foodstuffs being translocated in solution throughout the plant body. Furthermore, and in contrast to the situation in terrestrial animals, in all actively growing plants there is liquid phase continuity from the water in the soil through the plant to the liquid–gas interface at the evaporation sites in the leaves. The proliferation of roots in the soil provides an extensive absorbing surface across which passes virtually all the water and mineral nutrients utilized by the plant.

The following discussion first deals briefly with the composition, structure and related properties of water in the solid, liquid and vapour phases and then with some of the colligative properties of aqueous solutions of particular importance to plant water relations. Additional useful reference material can be found in Pauling (1960), Buswell and Rodebush (1956), Penman (1955), Némethy and Scheraga (1962a, 1962b), List (1963), Kavanau (1964), Bernal (1965), and in physical chemistry textbooks.

1. STRUCTURE OF WATER

Although almost all water occurs as simple H_2O, there are three isotopes of hydrogen, H^1, H^2 and H^3 and three of oxygen, O^{16} O^{17} and C^{18}, which may be combined in eighteen different ways. Tritium (H^3) and O^{17} are extremely rare, Deuterium (H^2) occurs at a concentration of about 200 p.p.m. in ordinary water and O^{18} at a concentration of about 1000 p.p.m.

The hydrogen atom consists of a positively charged proton and a negatively charged electron. The oxygen atom has a proton and eight electrons, six of which are arranged in an outer shell. Because the hydrogen shell has room for one more electron and the oxygen outer shell has room for two more electrons, the atoms have an affinity for each other. In the water molecule, therefore, two O-H bonds develop and, from x-ray diffraction and other studies (Pauling, 1960) the arrangement of the atoms in the ice molecule is now known to consist of two hydrogens bonded to an oxygen atom at an angle very close to the tetrahedral angle of 109°, with the internuclear O-H distance being 0·96 Å. In this arrangement, the positively charged proton of each hydrogen atom has an attraction for the negatively charged electrons of an adjacent water molecule.

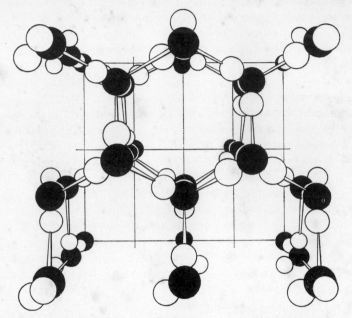

Fig. 1.1. Diagrammatic sketch of ice-type structure, showing oxygen atoms in black and hydrogen atoms in white (after Buswell and Rodebush, 1956).

In each molecule there are four regions where the density of the outer electrons is maximal. Two of these regions are associated with the O-H bonds and coincide with the positions of the protons and the other two are associated with lone pairs of electrons. These are located above and below the plane of the atomic nuclei on the opposite side of the oxygen nucleus from the protons. In consequence, the net charge distribution is similar to a tetrahedron with two positive and two negative corners and there is electrostatic attraction between a positive tetrahedral corner of one molecule and a negative tetrahedral corner of another. The resultant bonds are termed hydrogen bonds since the proton of the hydrogen atoms is involved. Clearly, each molecule can form four hydrogen bonds and the resultant structure of an assemblage of molecules is one in which each molecule tends to be hydrogen bonded to four others which surround it in a tetrahedron. The average distance between the centre of one oxygen atom and the next is approximately 2·75 Å. The H atom is located about 1·00 Å from one O atom and 1·76 Å from the other (Pauling, 1960).

This type of structure is illustrated diagramatically in Fig. 1.1. From the angle at which the structure is viewed in the figure a hexagon pattern is revealed. The similarity of this pattern to that exhibited on a macroscopic scale by snowflakes is immediately apparent.

The structure of ice appears to be a regular tetrahedral array (Pauling, 1960) with the forces of attraction between the molecules producing a structure rather like an archway under strong downward pressure. When the temperature rises to 0°, the thermal agitation of the molecules is sufficient to bend or stretch the hydrogen bonds and cause the ice structure to break down, the water becoming fluid. In the process the molecules move further apart, the distance between the internuclear-oxygen distance increasing to an average value of 2·9 Å. This would be expected to make water less dense than ice, but close packing of the molecules in groups more compact than the ice tetrahedra causes the opposite effect. Thus, in liquid water near 0°, there are slightly more than four nearest neighbour molecules for each water molecule, even though the average internuclear distance has increased.

Despite this change in density, the structure of liquid water is still similar to that of ice and has been referred to by Morgan and Warren (1938) as having a "broken down" ice structure. This can be appreciated by reference to the relative values for the heat of sublimation and heat of fusion of ice (Tanford, 1963). The heat of sublimation at 0° is 12,200 cal mole $^{-1}$, of which about 1400 cal mole $^{-1}$ represents the heat required to give water molecules the random motion they possess in the gaseous state. The remaining 10,800 cal mole $^{-1}$ represents the energy required to break the bonds which hold the crystal together. The heat of fusion of ice at 0° is only 1440 cal mole $^{-1}$, about 15 % of the total energy needed to break all the hydrogen bonds in the crystal. It is clear therefore that liquid water must retain a high degree of structure, probably containing more than 85% of the hydrogen bonds present in ice, since the bonds will be no stronger than in ice and may be significantly less so. (If they are 15% weaker than in ice, it can be appreciated that the number of bonds would be about the same in both ice and liquid water).

No single theory at present accounts for all aspects of the structure and properties of liquid water but several fairly closely related views appear to combine most of the important features. All envisage that liquid water has a somewhat broken down ice structure but that extensive hydrogen bonding still exists. Perhaps the most popular model at the present time is the "flickering cluster" model of Frank and Wen (1957) (see also Frank, 1958, 1963). This postulates that the formation of hydrogen bonds in liquid water is predominantly a cooperative phenomenon caused by the partially covalent nature of the hydrogen bond. The covalency arises from the displacement of the lone-pair electrons of one bonded molecule toward the proton of the other as a result of their mutual attraction. This displacement of electrons increases the polarity of the molecule and increases the

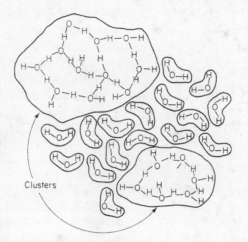

Fig. 1.2. Schematic representation of hydrogen bonded "flickering clusters" of water molecules and unbonded molecules (after Némethy and Scheraga, 1962a).

possibility of bond formation with a second molecule and so on. Thus hydrogen bond formation is regarded as a cooperative phenomenon, in the sense that the bonds are not made and broken singly but several at a time, producing short-lived clusters of highly bonded ice-like molecules surrounded by non-bonded molecules (Fig. 1.2). The clusters have rigidity because the covalent hydrogen bonds are capable of relatively little bending. When the bonds are ruptured, electrostatic interaction remains so that the non-bonded molecules are not random in the distribution and the energy changes involved when a cluster melts are small and localized. The clusters are assumed to appear or disappear when local energy fluctuations form cold or warm regions.

The flickering cluster theory can account quantitatively for most properties of water and has been used by Némethy and Scheraga (1962a, 1962b) to give quantitative predictions of thermodynamic and

TABLE 1.I

Some important physical properties of liquid water†

Temperature (°C)	Density (g cm⁻³)	Surface tension (g sec⁻²)	Dynamic viscosity (g cm⁻¹ sec⁻¹) × 10⁻²	Heat of vaporization (cal g⁻¹)	Specific heat (cal g⁻¹ deg⁻¹)	Thermal conductivity (cal cm⁻¹ sec⁻¹ deg⁻¹) × 10⁻³
−10	0·99794	—	—	603·0	1·02	—
−5	0·99918	76·4	—	—	—	—
0	0·99987	75·6	1·7921	597·3	1·0074	1·34
4	1·00000	—	—	—	—	—
5	0·99999	74·8	1·5188	594·5	1·0037	1·37
10	0·99973	74·2	1·3077	591·7	1·0013	1·40
15	0·99913	73·4	1·1404	588·9	0·9998	1·42
20	0·99823	72·7	1·0050	586·0	0·9988	1·44
25	0·99708	71·9	0·8937	583·2	0·9983	1·46
30	0·99568	71·1	0·8007	580·4	0·9980	1·48
35	0·99406	70·3	0·7225	577·6	0·9979	1·50
40	0·99225	69·5	0·6560	574·7	0·9980	1·51
45	0·99024	68·7	0·5988	571·9	0·9982	1·53
50	0·98807	67·9	0·5494	569·0	0·9985	1·54

† based on van Wijk (1963).

density data. However, some of the other properties of water are not adequately accounted for by the theory in its present form. For a detailed discussion of the relative merits of other models for water structure, and of the flickering cluster theory, the reader is referred to Kavenau (1964).

Important physical properties of water are tabulated in Table 1.I (after van Wijk, 1963).

2. STRUCTURE OF AQUEOUS SOLUTIONS

A. Solutions of Electrolytes

From the foregoing account it can be appreciated that the distinctive properties of water can be explained, to a considerable degree, in terms of bond angle combined with electrostatic forces arising from the charge distribution of the water molecule. Since simple ions have dimensions and carry charges comparable to those of the water molecule, it is only to be expected that ions will attract water molecules, leading to the formation of ion-dipole bonds, and that the structure of water will be considerably modified in ionic solutions.

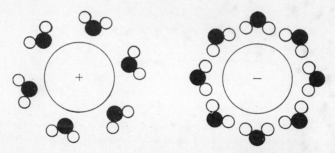

FIG. 1.3. Diagrammatic sketch of the manner in which ions (shown as large circles with + and − signs) are kept apart in water because the water molecules surrounding them become polarized. Around the cation, the water molecules are oriented with their oxygen atoms (small black circles) closest to the ion; the hydrogens (small white circles) are closest to the anion (after Buswell and Rodebush, 1956).

In pure water, as discussed above, two of the four molecules arranged around a central molecule are oriented with their proton corners toward the central molecule and the other two are oriented with their lone—pair electron corners towards the central molecule and their proton corners away from it. When an ion is introduced, a different situation develops depending on whether an anion or cation is concerned. Around a cation all the water molecules have their effective electronic centres directed inwards whereas around an anion all proton corners are directed inward

(Fig. 1.3). The net effect is that even if an ion is of appropriate size to fit into the space normally occupied by a water molecule, the water of hydration does not match-up with the surrounding water and the normal water structure is disrupted. The disruption is greater if the ion differs in size from a water molecule and, in general, increases as ion size increases.

It is interesting to note that the cations and anions are separated in solution because the high dielectric constant of water effectively reduces the force of attraction between oppositely charged ions to a small fraction of its original value. When a substance is dissolved in water, the negatively charged oxygen atoms and the positive ions are mutually attracted and the positively charged hydrogen atoms and the negative ions are mutually attracted. Consequently, the structure already described develops with water molecules surrounding a positive ion becoming oriented with their oxygens nearest to the ion and the molecules around a negative ion with their hydrogens nearest to the ion. Thus the water molecules act as cages which separate and neutralize the ions.

Frank and Wen (1957) have proposed a model for ion-water interaction which envisages each ion to be surrounded by three regions. In the nearest, the water molecules are strongly oriented in the intense electric field of the ion, have little kinetic energy, and are effectively immobilized. In the next region the water structure is somewhat broken down, being more random and less like that of ice than pure water. In the outermost region the water structure is normal but is polarized by the ionic field which, at this distance, is relatively weak.

The cause of the structural breakdown in the intermediate region is attributed to competition between the normal structural requirements of neighbouring water molecules and the orienting influences of the spherically symmetrical ionic field. Around a positive ion, for example, the water molecules would be oriented with all the hydrogens outwards. In consequence, they could not all participate in the normal tetrahedral water arrangement since this would require that two of the water molecules should have their hydrogens oriented inwards.

This model appears reasonable for very dilute solutions but, as solute concentration increases, there is an increasing degree of overlap in the regions surrounding each ion. Robinson and Stokes (1959) have calculated that, in a 1 molar solution of a 1:1 electrolyte, there can be few water molecules distant by more than two or three molecular diameters from some ion, so that it is reasonable to talk of successive layers of water molecules around one particular ion only with respect to concentrations below about 0·1 molar.

B. Solutions of Non-Electrolytes

Non-polar substances have low solubility in water and it is reasonable to account for this in terms of the hydrogen bonded structure of the water molecules themselves. Since non-polar substances cannot partici-pate in hydrogen bond formation, their introduction into water leads to breaking of hydrogen bonds and the formation of "holes" in the normal water structure. Thermodynamic data (Kauzmann, 1959) indicate that the resultant water structure around the hole is "tighter" than in normal water and Frank and Evans (1945) concluded that the water existed as an ice-like cage around each solute particle.

Solutions of polar organic compounds have a more ordered structure. Tanford (1963) has suggested that the polar portion of each organic molecule may enter into the hydrogen bonded structure of water, while a partial cage of highly structured water molecules is developed about the non-polar portion. Alternatively, small micelles may be formed, in which the non-polar groups form a small droplet, the surface of which contains the polar portions.

C. Structure of Water on Colloid-water Interfaces

On hydrophilic colloidal surfaces such as clays or proteins, water may be very strongly oriented to form ice-cages which sheath the solid surfaces (Low, 1961; Bernal, 1965). As distinct from the situation with single ions, Bernal (1965) considers that, on extended protein surfaces, a high degree of ice-like structure may extend for 10–20 Å. He considers that ions in the bathing solution cannot disturb the structure of the ice-cage, and that the water so oriented may represent about 30% of the weight of the hydrated protein molecule. Beyond this zone, up to 100 Å or so, the water molecules are still considered to be oriented to some degree, so that, although they may contain ions, their possibilities of movement are restricted. Perutz (1946) drew similar conclusions as to the quantity of water which was strongly oriented around protein surfaces and the degree to which ion entry into the ice cage was possible.

Changes in the character or degree of hydration can markedly affect the properties and behaviour of the material under consideration. Aspects of clay-water interaction are considered in Chapter 3 and are also considered in detail by Low (1961). Aspects of cell and protein hydration and their effect on cell and tissue function are discussed in Chapters 5, 6 and 9 and are considered by, amongst others, Klotz (1958) and Bernal (1965).

3. PROPERTIES OF WATER VAPOUR

A. *Vapour Pressure*

A pure substance may exist as a solid, liquid or vapour. In going from solid to liquid, as has already been described, heat is absorbed on melting. In the same way heat is absorbed during vaporization so that the enthalpy increases in passing from one phase to the next. According to kinetic theory, there is a continuous flow of molecules from a liquid surface into the space outside it and, at the same time, a continuous return flow towards the liquid surface. The relative ratios depend on the concentration of the vapour and a condition of equilibrium is established when the two flows are equal. At this point, the vapour is said to be saturated and the pressure exerted by the vapour in equilibrium within the liquid is known as the saturation vapour pressure. The saturation vapour pressure is dependent on temperature but is independent of the relative or absolute amounts of liquid and vapour.

The vapour pressure increases as temperature increases since heat is absorbed during vaporization and the system at equilibrium shifts so that the vaporization of liquid just absorbs the applied heat. As the temperature increases, the density of the liquid decreases and that of the vapour increases, until, at the critical temperature, these densities become equal and above this temperature no liquid phase can exist as long as the pressure remains constant. Solids also have a vapour pressure which depends on temperature, and sublimation occurs if the solid is directly converted into vapour.

The relationship between the pressure and temperature of water, in relation to fusion, sublimation and vaporization phenomena, is shown diagrammatically in Fig. 1.4 after Daniels and Alberty (1961). The figure is not drawn to scale because of the wide range of pressures involved. Each line represents equilibrium between the two phases it separates. Thus, at pressures and temperatures represented by points below the line AB, water vaporizes completely; above the line the vapour is completely condensed to liquid. Similarly, the line AC is the sublimation curve for ice; and the line AD, the fusion curve for ice-water equilibrium. Where the three lines meet, all phases can exist together at equilibrium. This is called the triple point. The triple point of water in the absence of air is at 0·0099°C and 4·58 mm vapour pressure (measured as mm mercury). In the presence of air at atmospheric pressure (760 mm), the three phases are in equilibrium at 0° but the partial pressure of water vapour remains at 4·58 mm (Daniels and Alberty, 1961).

The thermodynamic basis of the relationship between vapour pressure and temperature, known as the Clausius-Clapeyron equation, has been extensively used in many aspects of physical chemistry. It has been used in a complicated formulation to compute saturation vapour pressure data in the Smithsonian Meteorological Tables (List, 1963) but

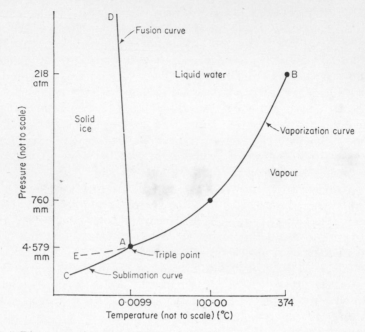

Fɪɢ. 1.4. Diagrammatic representation of the effects of pressure and temperature on fusion, sublimation and vaporization phenomena of water (after Daniels and Alberty, 1961).

the simplified form is adequate for a number of applications, if the temperature range is not too wide. It is

$$\ln e^0 = a + b/T \qquad (1.1)$$

where e^0 is the saturation vapour pressure (mm Hg), T the Kelvin temperature and a and b are constants.

Typical values for the saturation vapour pressure and density of water vapour over ice and over liquid water are given in Table 1.II.

B. Variation of the Saturation Vapour Pressure with Elevation

Since atmospheric pressure decreases with height above the earth, the partial pressures of the component gases, including water vapour,

TABLE 1.II

Physical properties of water vapour†

Temperature (°C)	Saturation vapour pressure		Density of saturated vapour		Diffusion coefficient of water vapour in air cm² sec⁻¹
	Over water (mm Hg)	Over ice (mm Hg)	Over water (g cm⁻³) (× 10⁻⁶)	Over ice (g cm⁻³) (× 10⁻⁶)	
−20	0·941	0·774	1·074	0·883	0·197
−15	1·434	1·24	1·61	1·39	
−10	2·15	1·95	2·36	2·14	0·211
− 5	3·16	3·01	3·41	3·25	
0	4·58	4·58	4·85	4·85	0·226
5	6·53		6·80		
10	9·20		9·40		0·241
15	12·78		12·85		
20	17·52		17·30		0·257
25	23·75		23·05		
30	31·82		30·38		0·273
35	42·20		39·63		
40	55·30		51·1		0·289
45	71·90		65·6		
50	92·50		83·2		

† based on van Wijk (1963).

must also decrease. This effect of elevation on total pressure can be calculated in a straight-forward manner by taking the total pressure at a height z to be P. At the height $(z + \Delta z)$, the pressure is lower than P because the weight of the air between the two levels exerts a pressure on the air below z but not on that above $(z + \Delta z)$. The mass of the air in Δz, per unit surface, is $\rho_a dz$ where ρ_a is density of the air (also referred to as the density of moist air) and its weight is $g\rho_a dz$ where g is the acceleration due to gravity. Therefore

$$\Delta P = -g\rho_a \Delta z \qquad (1.2)$$

From the ideal gas law, since the volume of one mole of air is given by M_a/ρ_a where M_a is the effective molecular weight of air,

$$\frac{PM_a}{\rho_a} = RT \qquad (1.3)$$

where R is the universal gas constant ($8\cdot314 \times 10^7$ erg mole^{-1} deg^{-1}). Then,

$$\rho_a = \frac{PM_a}{RT} \qquad (1.4)$$

Substituting in Eq. (1.2) and with infinitesimal increments of z,

$$\frac{dP}{P} = \frac{-g\, M_a\, dz}{RT} \tag{1.5}$$

Integration between the limits $z = 0$ and z gives

$$\ln P/P^0 = -M_a gz/RT \tag{1.6}$$

where P^0 is the pressure at $z = 0$.

If the surface at $z = 0$ is pure free water and the atmosphere above it is in equilibrium with the water and at the same temperature, Eq. (1.6) applies to the partial pressure of water vapour. The value at the surface is the saturation vapour pressure, e^0; at height z, it has the value e. Therefore

$$\ln e/e^0 = -M_w\, gz/RT \tag{1.7}$$

where M_w is the molecular weight of water.

Under normal conditions, the value of $M_w g/RT$ is of the order of 6×10^{-7} cm^{-1}. Hence z must be extremely large to cause e/e^0 to depart significantly from unity.

C. Influence of Liquid Pressure on the Saturation Vapour Pressure

The effect of liquid pressure on the saturation vapour pressure is commonly observed in the column of water in a capillary tube. This phenomenon can be readily evaluated by first assuming that a fine capillary tube has been inserted vertically into a beaker of water, that water has risen in the tube and equilibrium has been established.

In this situation, the vapour pressure over the liquid-air meniscus, which is at height z above the water in the beaker, must be equal to the actual vapour pressure, e, in the air outside the capillary tube, and at the same elevation. Therefore the vapour pressure over the meniscus must be less than that over a plain water surface, which would be the normal saturation vapour pressure, e^0, at that temperature and elevation.

The upward force of surface tension acting on a wetted perimeter of $2\pi\, r$ can just balance the downward force caused by the column of water in the capillary, thus

$$2\pi\, r\, \sigma = \pi\, r^2\, \rho_w gz \tag{1.8}$$

where σ is the surface tension (g sec^{-2}) and ρ_w is the density of liquid water. Since the vapour pressure at height z above a plain surface is given by Eq. (1.7), substitution of Eq. (1.8) into (1.7) gives

$$\ln e/e^0 = -\frac{2\sigma M_w}{rRT\rho_w} \qquad (1.9)$$

This relationship between relative vapour pressure and capillary pore radius is of considerable significance in considerations of sorption of water by porous materials and hence has particular significance to water retention and distribution in plants and soils. For calculation purposes, it is convenient to note that, with $\sigma = 73\cdot4$ g sec^{-2} at $15°$ $r \cong -0\cdot15/z$.

The foregoing example of liquid in a capillary tube provides an indication of the effect of liquid pressure on the saturation vapour pressure. In this case suction is caused by the weight of the water column in the tube. Since, at any height, this liquid water must be in equilibrium with the water vapour at the same height it follows from Eq. (1.7) that

$$\ln[e(\Delta P_{\text{liq}})/e^0] = -M_w gz/RT \qquad (1.10)$$

where ΔP_{liq} is the extra pressure to which the water is subjected. Since $\Delta P_{\text{liq}} = -\rho_w gz$, this gives

$$\ln[e(\Delta P_{\text{liq}})/e^0] = \Delta P_{\text{liq}} M_w/\rho_w RT \qquad (1.11)$$

This expression holds for both positive and negative values of ΔP_{liq}, hence negative and positive values of z. It follows that the vapour pressure is increased by a positive pressure exterted on the

TABLE 1.III

Relationship between relative vapour pressure, radius of curvature and capillary phenomena[†]

Relative vapour pressure at 15°C	Radius of curvature (cm)	Height of capillary rise (cm)	Equivalent tension in water (bars)
1·0000	10^{-1}	1·50	0·0015
1·0000	10^{-2}	$1\cdot50 \times 10$	0·015
0·9999	10^{-3}	$1\cdot50 \times 10^2$	0·15
0·9989	10^{-4}	$1\cdot50 \times 10^3$	1·50
0·9890	10^{-5}	$1\cdot50 \times 10^4$	$1\cdot50 \times 10$
0·8954	10^{-6}	$1\cdot50 \times 10^5$	$1\cdot50 \times 10^2$
0·3305	10^{-7}	$1\cdot50 \times 10^6$	$1\cdot50 \times 10^3$
0·000016	10^{-8}	$1\cdot50 \times 10^7$	$1\cdot50 \times 10^4$

[†] Based on van Wijk (1963).

liquid and decreased by suction. It is most important to realise that the effect is only appreciable for comparatively large absolute values of ΔP_{liq}. At $\Delta P_{liq} = 1$ bar, for example, the increase in the saturation vapour pressure is approximately $0 \cdot 1 \%$, hence the generalization that atmospheric pressure does not affect the saturation vapour pressure.

In Table 1.III values of e/e^0 for various water column heights, pressures (ΔP_{liq}) and comparable capillary radii of curvature, are given for a wide range of values of e/e^0. It is apparent that z values must exceed 10^4 cm to have significant effects on e/e^0.

D. Expressions for Water Vapour

Apart from vapour pressure, several other terms are commonly used to express the water vapour content of air. Because of the unnecessary confusion which has often existed between these expressions, the following definitions and conversion procedures are included here:

1. Absolute humidity

Absolute humidity is the mass of water vapour per unit volume of moist air. Since the density of a perfect gas at a given temperature is proportional to its partial pressure, the density can be obtained from the vapour pressure by using the ideal gas law relationship

$$\rho_y = \frac{18ke}{RT} \tag{1.12}$$

where ρ_y is the absolute humidity (g cm^{-3}), e is the water vapour pressure (mm or mb Hg), k is a conversion factor to convert e to dyne cm^{-2}, R is the universal gas constant (erg mole^{-1} deg^{-1}) and T is the Kelvin temperature.

With e in mb Hg, the numerical value of $18k/R$ is $2 \cdot 17 \times 10^{-4}$ With e in mm its value is $2 \cdot 89 \times 10^{-4}$. (1.13)

2. Specific humidity

Specific humidity is the mass of water vapour per unit mass of moist air. Thus,

$$q = \rho_y/\rho_a = \frac{M_w}{M_a} ke \tag{1.14}$$

where q is the specific humidity (g g^{-1}) ρ_a is the density of moist air (g moist air cm^{-3} moist air), k has the same value as in (1·3) and M_w and M_a are the molecular weights of water and moist air, respectively

In practice, the effective molecular weight of dry air, M_d, ($\cong 29 \cdot 0$) is used so the ratio M_w/M_a has the value $0 \cdot 622$.

3. Mixing ratio

Specific humidity is sometimes used interchangeably with the mixing ratio, which is the dimensionless ratio of the mass of water vapour per unit mass of dry air. It is therefore given by

$$\chi = \frac{\rho_v}{\rho_d} \frac{e}{(P-e)} \tag{1.15}$$

where ρ_v/ρ_d is the ratio of the densities of water vapour (g vapour cm^{-3} vapour) and dry air at the same temperature and pressure ($= M_w/M_d = 0 \cdot 622$).

If e is very small relative to P

$$\chi \cong 0 \cdot 622 \ e/P \tag{1.16}$$

4. Saturation deficit

The difference in vapour pressure between the actual vapour pressure and the saturation vapour pressure at the same temperature is termed the saturation deficit ($s.d.$). Thus

$$s.d. = (e^0 - e) \tag{1.17}$$

with units of mm or mb Hg.

5. Relative humidity

The dimensionless ratio of the actual vapour pressure to the saturation vapour pressure at the same temperature. It is generally expressed as a percentage

$$r.h. = 100e/e^0 \tag{1.18}$$

It is important to remember that relative humidity values are meaningless as humidity parameters unless additional information such as air temperature, or one of the actual vapour pressures used, is also quoted. Saturation deficit and relative humidity are simply related by the expression

$$s.d. = e^0(1 - r.h./100) \tag{1.19}$$

6. Dew point temperature

This is the temperature at which the actual vapour pressure would be the same as the saturation vapour pressure. It is obtained when a sample of air at a known temperature and vapour pressure is cooled, with no change in water vapour content, until the temperature reaches a point at which the actual water vapour content is the saturation vapour pressure at that temperature. Condensation will then commence if a surface or nucleus is available. This is therefore the dew point temperature. At temperatures below freezing it is termed the frost point temperature. It is an important, absolute and easily measured water vapour parameter.

7. *Wet bulb temperature*

Wet bulb temperature is essentially a phenomenon associated with the wet and dry bulb hygrometer, the most commonly used humidity measuring device and frequently a standard for calibration of other types of hygrometers. It is also a temperature approximated by many moist evaporating surfaces.

In a hygrometer, air is passed across two identical thermometers one of which is sheathed in a wet cloth which is supplied with pure water from a simple delivery system. Unless the moving air is saturated, water will evaporate from the wet bulb which is consequently cooled.

The theory of the wet bulb hygrometer (see, for example, Penman 1955) assumes that all the heat needed to vaporize the water is drawn from the air, that is, that the loss of sensible heat measured by the fall of temperature equals the gain of latent heat measured by the increased water content. Suppose that a volume of air at ambient temperature (T_a) and mixing ratio (χ_a) reaches the wet bulb and displaces an equal volume of air having the wet bulb temperature (T_w) and mixing ratio (χ_w), which is the saturation mixing ratio at T_w.

The decrease in sensible heat content ΔH_s (cal g^{-1} incident air) is then

$$\Delta H_s = c_p(T_a - T_w) \tag{1.20}$$

where c_p is the specific heat of air at constant pressure (cal g^{-1} deg^{-1}). The associated increase in latent heat content ΔH_l(cal g^{-1} incident air) is

$$\Delta H_l = \mathbf{l}(\chi_w - \chi_a) \tag{1.21}$$

where $\mathbf{l}$ is the latent heat of vaporization (cal g^{-1} vapour). At equilibrium the total change is zero, so $\Delta H_s = \Delta H_l$ and,

$$c_p(T_a - T_w) = \mathbf{l}(\chi_w - \chi_a) \tag{1.22}$$

since $\chi \cong 0\cdot622e/P$, the equation can be re-written

$$e = e^0 - \frac{Pc_p}{0\cdot622\mathbf{l}}(T_a - T_w) \tag{1.23}$$

Substituting normal values for P (755 mm Hg), c_p (0·242 cal g^{-1} deg^{-1}) and l (590 cal g^{-1}), the constant $Pc_p/0·622l$ has the theoretical value of 0·500 mm deg^{-1} or, if P is expressed in mb, of 0·667 mb deg^{-1}. However, this value may not be obtained, particularly if the ventilation is inadequate to completely remove the cool camp air from the wet bulb and, in consequence, the values of the constant, termed the psychrometric constant, γ, are lower for screen thermometers than with aspirated ones. Published values are given in meteorological tables [see for example Smithsonian Meteorological Tables, (List, 1963)].

4. SOME RELEVANT ASPECTS OF CHEMICAL THERMODYNAMICS

The following account deals particularly with the derivation of the chemical potential, a term of considerable significance in describing and understanding the principles of water and solute movement in biological systems.

Every attempt has been made to simplify the mathematical development without undue distortion, but the reader is referred to thermodynamic texts for more comprehensive accounts. Appropriate sources of information are Prigogine and Defay (1954), Guggenheim (1959), Lewis and Randall (1961), Babcock (1963) and Spanner (1964).

A. Derivation of the Chemical Potential

For a system containing only one component, the second law of thermodynamics can be written,

$$dE = TdS - PdV \tag{1.24}$$

where E is the internal energy in ergs, T the Kelvin temperature, S the entropy, P the pressure and V the volume.

When applied to systems in which the number of moles of material is variable, the total differential of the internal energy can be expressed as a function of S, V and n_i, where n_i is the number of moles of a given component in the system

$$dE = \left(\frac{\partial E}{\partial S}\right)_{V,\,n_i} dS + \left(\frac{\partial E}{\partial V}\right)_{S,\,n_i} dV + \left(\frac{\partial E}{\partial n_i}\right)_{S,\,V,\,n_j} dn_i \tag{1.25}$$

where n_j is the number of moles of the components not included in the summation. From Eqns (1.24) and (1.25) the first partial coefficient in Eq. (1.25) is found by dividing Eqns (1.24) and (1.25) by dS at constant

V and n_i, and equating to give

$$\left(\frac{\partial E}{\partial S}\right) = T \tag{1.26}$$

and the second partial coefficient by dividing by dP and proceeding in the same manner to give

$$\left(\frac{\partial E}{\partial V}\right) = -P \tag{1.27}$$

The "chemical potential" is now defined as

$$\mu_i = \left(\frac{\partial E}{\partial n_i}\right)_{S, V, n_j} \tag{1.28}$$

so that Eq. (1.25) becomes

$$dE = TdS - PdV + \mu_i dn_i \tag{1.29}$$

Gibbs free energy, G, is now defined as

$$G = E + PV - TS \tag{1.30}$$

This equation can be differentiated and Eq. (1.29) substituted in it to give

$$dG = -SdT + VdP + \Sigma_i \mu_i dn_i \tag{1.31}$$

Dividing by dn_i at constant T, P and n_j this yields

$$\left(\frac{\partial G}{\partial n_i}\right)_{T, P, n_j} = \bar{G}_i = \mu_i \tag{1.32}$$

where $\bar{G}_i$ is the partial molal Gibbs free energy.

Thus the chemical potential, μ_i, of a substance i, is equal to the rate of change in the Gibbs free energy of the system with n_i moles of this component, when the temperature, pressure and number of moles of all other components are held constant. Further, it is equal to the partial molal Gibbs free energy, $\bar{G}_i$, where pressure, temperature and composition are the only variables of concern. It is an intensive property, like temperature and pressure. The difference in chemical potential of a substance in two phases determines the direction in which the substance will diffuse spontaneously, just as the difference in temperature determines the direction of heat flow. It is consequently of considerable significance in considerations of water and solute movement in plant and soil systems.

An expression can now be obtained for the total differential of μ_i.

Taking T, P and n_i as independent variables, this gives

$$d\mu_i = \left(\frac{\partial \mu_i}{\partial T}\right)_{P,\,n_i} dT + \left(\frac{\partial \mu_i}{\partial P}\right)_{T,\,n_i} dP + \left(\frac{\partial \mu_i}{\partial n_i}\right)_{T,\,P,\,n_j} dn_i \qquad (1.33)$$

The first differential coefficient on the right hand side of the equation can now be evaluated in a similar way to that previously adopted by dividing Eq. (1.31) by dT at constant P and n_i to give

$$\left(\frac{\partial \bar{G}}{\partial T}\right)_{P,\,n_i} = -\bar{S} \qquad (1.34)$$

from which it follows that

$$\left(\frac{\partial \bar{G}_i}{\partial T}\right)_{P,\,n_i} = -\bar{S}_i = \left(\frac{\partial \mu_i}{\partial T}\right)_{P,\,n_i} \qquad (1.35)$$

where $\bar{S}_i$ is the partial molal entropy. Similarly, by dividing Eq. (1.31) by dP at constant T and n_i, it can be shown that

$$\left(\frac{\partial \bar{G}_i}{\partial P}\right)_{T,\,n_i} = \bar{V}_i = \left(\frac{\partial \mu_i}{\partial P}\right)_{T,\,n_i} \qquad (1.36)$$

where $\bar{V}_i$ is the partial molal volume of the ith component.

Substituting Eqns (1.35) and (1.36) in Eq. (1.33) together with the relationship $N_i = n_i/(n_i + \Sigma n_j)$, where N_i is the mole fraction of the ith component, gives

$$d\mu_i = -\bar{S}_i dT + \bar{V}_i dP + \Sigma \left(\frac{\partial \mu_i}{\partial N_i}\right)_{T,\,P,\,n_j} dN_i \qquad (1.37)$$

In both this equation and Eq. (1.33) the summation is only over the mole fraction of chemical species whose mole number changes.

Further development of this basic expression and examples of the use of the chemical potential will occur throughout the book.

B. The Chemical Potential, Water Potential and Aqueous Solutions

The thermodynamic condition for the equilibrium of a component i between the liquid and vapour phases at a given temperature and total pressure is that the chemical potential for the component be the same in the two phases, thus,

$$\mu_{i\,(\text{liq})} = \mu_{i\,(\text{vap})} \qquad (1.38)$$

Further, it is known from the ideal gas law that

$$\bar{V}_i = RT/p_i \qquad (1.39)$$

where p_i is the partial gas pressure of the ith component. Substituting

Eq. (1.39) into Eq. (1.36), with $p_i = P$, and integrating gives,

$$\int_{\mu_i^0}^{\mu_i} \mathrm{d}\mu_i = \int_{p_i^0}^{p_i} \frac{RT}{p_i} \, \mathrm{d}p_i \tag{1.40}$$

$$\mu_i - \mu_i^0 = RT \ln p_i/p_i^0 \tag{1.41}$$

where μ_i^0 and p_i^0 are the chemical potential and partial gas pressure, respectively, of the ith component in the pure phase at the reference temperature and pressure.

Thus, the chemical potential of a component may be expressed in terms of its partial gas pressure.

If the solution is ideal, Raoult's law applies which states that, for an ideal solution, the partial pressure of a component is directly proportional to its mole fraction in the liquid, and the proportionality constant is the vapour pressure of the pure substance. For present purposes this can be written

$$p_i = N_i \, p_i^0 \tag{1.42}$$

where N_i is the mole fraction of the ith component.

The chemical potential of the ith component may then be expressed in terms of its mole fraction by substituting Eq. (1.42) into (1.41) to give

$$\mu_i - \mu_i^0 = RT \ln N_i \tag{1.43}$$

In non-ideal solutions an equation of the same form can be written by using the activity coefficient, ϕ, or the activity, a, which are defined by

$$\mu_i - \mu_i^0 = RT \ln \phi_i N_i \tag{1.44}$$

and

$$\mu_i - \mu_i^0 = RT \ln a_i \tag{1.45}$$

When using electrolyte solutions it is best to use activities instead of mole fractions or concentrations, even in dilute solutions, because electrolyte solutions depart from ideality much sooner than non-electrolyte solutions. This is due to the greater distances over which the electrostatic forces between ions exert their influences compared with uncharged molecules. Also, using the law of electrostatic force, it is possible to calculate with considerable accuracy the activity coefficient for electrolyte solutions (see Guggenheim, 1959; Daniels and Alberty, 1961).

The significance of Eqns (1.41) to (1.45) for aqueous solutions is immediately apparent if the ith component, so far dealt with in general terms, is assumed to be water. In this case, the term $(\mu_w - \mu_w^0)$ is now the difference between the chemical potential of water in the system under study, be it solution or gas, and that of pure free water at the same temperature. Therefore Eq. (1.40) can be used to relate the chemical potential of water to the relative vapour pressure p/p^0. Similarly, Eqns (1.43) and (1.45) can be used to relate the chemical potential of water to the mole fraction, N_w, and activity, a_w, of water in aqueous solutions. This leads to the often quoted series of thermodynamic equivalents

$$(\mu_w - \mu_w^0) = RT \ln e/e^0 = RT \ln a_w = RT \ln \phi_w N_w \qquad (1.46)$$

It should be remembered that this expression is not an equation of state but rather an equilibrium equation relating several equivalent terms, and that $\phi_w = 1$ only when the solution is ideal.

The term $(\mu_w - \mu_w^0)$ has particular significance in studies of the plant and soil water relations since it is a measure of the capacity of the water at a point in the system under study to do work, compared with the work capacity of pure free water. Because of its significance, a term "water potential" has been introduced defined as

$$\Psi = \frac{(\mu_w - \mu_w^0)}{\bar{V}_w} \qquad (1.47)$$

where Ψ is the water potential expressed as energy per unit volume (erg cm^{-3}) and $\bar{V}_w$ is the partial molal volume of water. This provides a term with units dimensionally equivalent to pressures (erg cm^{-3} = dyne cm^{-2}) so that it can be expressed in traditional pressure units of bars or atmospheres (1 bar = 10^6 dyne cm^{-2} = 0·987 atm.). For some cases, it is preferable to retain units of erg mole^{-1}. In other cases, for example in soil physics, use is frequently made of the partial specific rather than the partial molal Gibbs free energy, with units of erg gm^{-1}. These two functions are simply related, since the partial specific free energy is given by

$$\left(\frac{\partial G}{\partial m_w}\right)_{P,\,T,\,m_j} = \frac{1}{M_w}\left(\frac{\partial G}{\partial n_w}\right)_{P,\,T,\,n_j} = \frac{1}{M_w}\mu_w \qquad (1.48)$$

where m_w, n_w and m_j, n_j, are the mass and number of moles of water and other constituents, respectively, and M_w is the molecular weight of water.

C. Osmotic Pressure and Related Phenomena

Four closely interrelated properties of dilute solutions are of particular value in studies of osmotic phenomena. They are the lowering of the water vapour pressure compared with that of pure free water, the elevation of the boiling point, depression of the freezing point and the development of osmotic pressure across differently permeable membranes. These phenomena are generally termed colligative properties. The present discussion will be restricted to dilute solutions and non-volatile solutes, and vapour pressure will not be specifically treated.

1. *Effects on the boiling and freezing points*

The effect of adding a non-volatile solute on the fusion, vaporization and sublimation curves for pure water is shown diagramatically in Fig. 1.5 (after Daniels and Alberty, 1961). The lowering of the vapour pressure means that the solution has to be heated to a higher temperature than pure water before the vapour pressure of the water in the

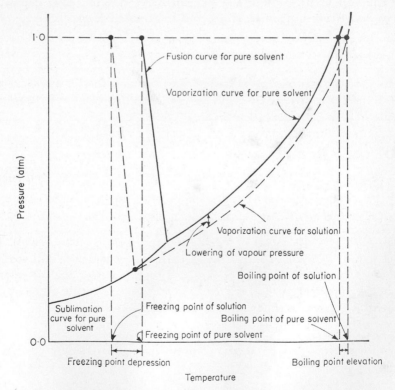

FIG. 1.5. Diagrammatic representation of the relationship between lowering of the vapour pressure, elevation of the boiling point and depression of the freezing point, caused by the addition of a non-volatile solute to pure water. Solid lines represent behaviour of pure solvent (c.f. Fig. 1.4), broken lines represent behaviour of solution (after Daniels and Alberty, 1961).

solution reaches the normal value at which pure water boils under atmospheric pressure. In similar fashion, before pure water freezes out of a solution, the temperature must be reduced below the freezing point of pure water so that the vapour pressure of the solid solvent is reduced to that of the water in the solution.

The elevation of the boiling point ΔT_b may be derived from the Clausius-Clapeyron equation or by using the chemical potential. In the latter case, it is first assumed that in order for water vapour at atmospheric pressure to be in equilibrium with the solvent in solution at a certain temperature, the chemical potential must be the same in the vapour phase (μ_{vap}) as in the liquid phase (μ_{liq}) as in Eq. (1.38).

Equation (1.43) can now be written in the form

$$\mu_w - \mu_w^0 = RT \ln N_w \tag{1.49}$$

If the vapour pressure of the solvent obeys Raoult's law, Eq. (1.49) can be substituted into Eq. (1.38) and rearranged to give

$$\frac{\mu_{vap}}{T} - \frac{\mu_w^0}{T} = R \ln N_w \tag{1.50}$$

where μ_w^0 is the chemical potential of pure water at the temperature and pressure of the solution. This equation can be differentiated with respect to temperature, at constant pressure to give

$$\left(\frac{\partial(\mu_{vap}/T)}{\partial T}\right)_P - \left(\frac{\partial(\mu_w^0/T)}{\partial T}\right)_P = \frac{R \, \partial \ln N_w}{\partial T} \tag{1.51}$$

It can be shown from the effect of temperature on the free energy that

$$\frac{\partial \ln N_w}{\partial T} = -\frac{\Delta \bar{H}_{vap}}{RT^2} \tag{1.52}$$

where $\Delta \bar{H}_{vap}$ is the molal enthalpy of vaporization (cal mole^{-1}). For small boiling point elevations $\Delta \bar{H}_{vap}$ may be considered independent of temperature, and Eq. (1.52) can be integrated over the range from the boiling point of pure water, T^0 (where $N_w = 1$, and $\ln N_w = 0$), to the boling point of a solution with mole fraction N_w

$$-\int_0^{\ln N_w} d \ln N_w = \int_{T^0}^{T} \frac{\Delta \bar{H}_{vap}}{RT^2} \, dT \tag{1.53}$$

$$-\ln N_w = \frac{\Delta \bar{H}_{vap}\,(T-T^0)}{RT\,T^0} \cong \frac{\Delta \bar{H}_{vap}\,\Delta T_b}{R\,(T^0)^2} \tag{1.54}$$

where ΔT_b is the boiling point elevation [$= (T-T^0)$].

Since, for small values of solute mole fraction, N_s, $-\ln N_w \cong N_s$,†
this gives

$$\Delta T_b \cong \frac{R\,(T^0)^2 N_s}{\Delta \bar{H}_{\mathrm{vap}}} \tag{1.55}$$

The equation for freezing point depression is derived on a similar basis assuming that pure ice separates from the solution when freezing occurs. There is then equilibrium between the chemical potential of the solid phase (μ_{sld}) and the liquid phase (μ_{liq})

$$\mu_{\mathrm{sld}} = \mu_{\mathrm{liq}} \tag{1.56}$$

and Eq. (1.49) can be substituted into Eq. (1.56) to give

$$\frac{\mu_{\mathrm{sld}}}{T} - \frac{\mu_w^0}{T} = R \ln N_w \tag{1.57}$$

Following similar procedures to those adopted previously this leads to

$$-\ln N_w = \frac{\Delta \bar{H}_{\mathrm{fus}}\,(T^0 - T)}{RT\,T^0} = \frac{\Delta \bar{H}_{\mathrm{fus}}\,\Delta T_f}{R\,(T^0)^2} \tag{1.58}$$

and

$$\Delta T_f \cong \frac{R\,(T^0)^2 N_s}{\Delta \bar{H}_{\mathrm{fus}}} \tag{1.59}$$

where $\Delta \bar{H}_{\mathrm{fus}}$ is the molal enthalpy of fusion (cal mole^{-1}), T^0 refers to the freezing point of pure water and ΔT_f is the freezing point depression $= (T^0 - T)$.

These equations provide a convenient means of relating the relative vapour pressure, and solute concentration, to relatively easily measured temperature elevations or depressions.

2. *Osmotic pressure*

When a solution is separated from pure solvent by a membrane permeable only to the solvent, there tends to be a net flux of solvent into the solution since the chemical potential of the solvent is higher in the pure phase than in the solution. This process is called osmosis and the pressure difference which must be applied to the solution to prevent a net flux of solvent is called osmotic pressure, π.

The osmotic pressure of an aqueous solution can readily be obtained using the chemical potential. In an osmometer at equilibrium the chemical potential of pure water at atmospheric pressure in one compartment (μ_w^I) equals the chemical potential of the water in the

---

† Since $N_w = (1 - N_s)$, $-\ln N_w$ can be represented by a power series in N_s thus-
$$-\ln N_w = -\ln (1 - N_s) = N_s + \tfrac{1}{2}N_s^2 + \tfrac{1}{3}N_s^3 \cdots$$
For sufficiently low concentrations of solute all terms except the first are negligible.

compressed solution in the other compartment (μ_w^{II}) since the applied pressure difference raises the chemical potential of the water in the second compartment to that of the pure solvent

$$\mu_w^{I} = \mu_w^{II} = \mu_w \qquad (1.60)$$

The effect on μ_w of adding solute and increasing the pressure is

$$d\mu_w = \left(\frac{\partial\mu_w}{\partial P}\right)_{T,\,N_s} dP + \left(\frac{\partial\mu_w}{\partial N_s}\right)_{T,\,P} dN_s \qquad (1.61)$$

Substitution of $(1-N_s)$ for N_w in Eq. (1.49) and differentiation with respect to N_s gives

$$\left(\frac{\partial\mu_w}{\partial N_s}\right)_{T,\,P} = \frac{-RT}{(1-N_s)} \qquad (1.62)$$

By substituting from Eqns (1.36) and (1.62), Eq. (1.61) can now be written

$$d\mu_w = \bar{V}_w\,dP - \frac{RT\,dN_s}{(1-N_s)} \qquad (1.63)$$

In the osmometer, the pressure is changed so that $d\mu_w = 0$. Then Eq. (1.63) becomes

$$\bar{V}_w\,dP = \frac{RT\,dN_s}{(1-N_s)} = \frac{-RT\,d\ln N_w}{N_w} \qquad (1.64)$$

Integrating from P^0 at atmospheric pressure to P^{eq}, the pressure in the osmometer at equilibrium, and from pure water ($N_w = 1$, $\ln N_w = 0$) to $\ln N_w$ in the osmometer

$$\int_{P^0}^{P^{eq}} \bar{V}_w\,dP = -RT \int_0^{\ln N_w} d\ln N_w \qquad (1.65)$$

$$\bar{V}_w(P^{eq}-P^0) = -RT\ln N_w \qquad (1.66)$$

The equilibrium pressure difference $(P^{eq}-P^0)$ is called the osmotic pressure, π, which is therefore given by

$$\pi = \frac{-RT}{\bar{V}_w}\ln N_w \qquad (1.67)$$

The osmotic pressure can also be written in terms of the relative vapour pressure, by introducing Raoult's Law, $N_w = e/e^0$ or in terms of the activity, thus

$$\pi = -\frac{RT}{\bar{V}_w}\ln e/e^0 = -\frac{RT}{\bar{V}_w}\ln a_w \qquad (1.68)$$

Equation (1.67) is frequently simplified by the approximations $-\ln N_w \cong N_s$ (see footnote p. 24) and $N_s/\overline{V}_w \cong c_s$) to give

$$\pi = RTc_s \qquad (1.69)$$

where c_s is solute concentration in units of mole cm^{-3}.

The development of osmotic pressure in plant cells is of the primary importance in determining their water relationships. In Chapter 5 these aspects will be dealt with at some length.

CHAPTER 2

Environmental Aspects of Plant Water Relationships

This chapter provides an introduction to some of the more important ecological aspects of plant water relations. A thorough treatment would require a wide-ranging excursion into micro-meteorology which is beyond the scope of this book. For such background material and for additional detail, the reader is referred to Geiger (1957), Slatyer and McIlroy (1961) and van Wijk (1963) and to the publications of the UNESCO Symposia on Climatology (UNESCO 1958a, 1958b) and on Plant Water Relationships (UNESCO 1960, 1962). For the present, attention will be confined to the primary energy and water exchanges which occur in plant communities and additional emphasis will be placed on evapotranspiration as the component of the hydrologic cycle most intimately concerned with physiological plant processes. Later, (Chapter 8) transpiration will again be considered but from a more specific point of view.

1. THE ENERGY BALANCE

Although fluctuations in the components of the water balance, particularly in precipitation, frequently appear to dominate the microenvironment, its basic characteristics are determined by the pattern and fate of the incident solar radiation. Moreover, while irrigation can modify the water regime over extensive areas, modification of the energy regime, at the present time, can only be economically achieved in special situations, such as in green-houses. Perhaps the most impressive demonstrations of the overriding significance of solar energy input are the fluctuations, diurnally and seasonally, of air temperature in all but equatorial humid regions. Although these fluctuations may be modified by the nature of the plant community, by cloudiness or precipitation, or by advective inflow of different air masses, such influences seldom act as more than modifiers, except for short periods. On occasions when they are associated with extreme microenvironmental situations, however, they can become determinants of the type of plant community which develops. The 1 in 100-year flood, the 1 in 10-year drought, the extreme freeze or heat wave belong in this category, but they are still incidental to the main long term microenvironmental controls. Considerably more detail about the energy balance can be found in the references already cited and in Lettau and Davidson (1957) and Gates (1962).

A. Solar Energy Input

Solar energy input provides two particular needs of plants, firstly the establishment of a satisfactory thermal environment for physiological function and secondly, the establishment of a satisfactory light environment for photosynthesis. While closely related, these two phenomena are not always linked in the same way.

Solar radiation reaches the outer surface of the earth's atmosphere with an almost constant intensity of about 1400 watt m^{-2} or 2·0 cal cm^{-2} min^{-1} measured perpendicular to the solar beam (Johnson, 1954). About 98% is contained in the wave-length interval 0·2–4·5μ, including about 40–45% in the 0·4–0·7μ range, and about 2% at wavelengths shorter and longer than these limits. The distribution of the incident flux with wavelenth can be regarded, for present purposes, as comparatively smooth, with few major gaps and a peak at the wavelength of green light ($\sim 0·5\mu$). Throughout the main region the distribution corresponds roughly, both as regards total energy radiated and spectral distribution, with that expected from radiation theory for a perfect absorber and emitter at a temperature of 6600 deg K, (Fig. 2.1).

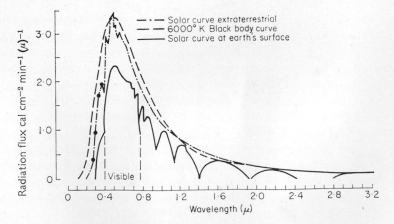

Fig. 2.1. Diagram of the extra-terrestrial solar flux and the solar flux at the earth's surface. The broken line is the curve expected from a surface temperature of 6000°K.

In contrast to solar radiation, the earth's temperature is roughly 300 deg K. The black body radiation corresponding to this temperature has its maximum spectral intensity at approximately 10μ and 98% of its energy is contained in the wavelength interval $0.5-80\mu$ (van Wijk, 1963). In consequence, the spectral range of solar radiation and terrestrial thermal radiation, although overlapping slightly, can be considered, for most purposes, as completely separate. The former, although containing some infrared radiation, is commonly called short-wave, and the latter long-wave, radiation.

In passing through the earth's atmosphere, the quantity and quality of the solar radiation is influenced in a number of ways. On the average, about one-third can be reflected back to space and smaller amounts are absorbed and scattered diffusely by various atmospheric components. In Fig. 2.1, the spectral distribution of solar radiation reaching the earth's surface is also presented, showing the depletion on the ultraviolet side of the peak by ozone and oxygen and on the infrared side by these two gases and water vapour. Absorption bands for CO_2 also exist in this region. In consequence, only about one-half of the total radiation actually reaches the earth, although this figure may be higher than 70% in arid regions with little cloud and lower than 40% in tropical rainy climates. Of course, on a heavily clouded day, the transmission may fall to a few per cent. Figure 2.2, from Geiger (1957) provides a useful illustration of the fate of incoming solar radiation.

The amount of solar radiation reaching the surface also varies with elevation, latitude and season, and time of day. It increases with elevation since there is a shorter air path to be traversed and the highest concentrations of dispersing and absorbing substances occur closest to the general ground level. This leads to the interesting effect frequently observed in elevated areas, even on quite extensive plateaux, of high radiation levels yet cool weather due to the relatively low bulk air temperatures at the particular elevation. Other effects result primarily from path length differences, although the latitude-season interaction leads to longer days with greater total radiation per day at high latitudes than at the equator, even though the peak hourly radiactive flux is less. These phenomena are illustrated in Fig. 2.3 and Table 2.I. It is interesting to note from the table that as the solar angle decreases and air mass numbers move from 1 to 5, the total radiation falls from 930 to 430 watt m^{-2} (or from 1.33 to 0.61 cal cm^{-2} min^{-1}) but the percentage of the total received in the $0.4-0.7\mu$ spectral range only drops by about 20%.

As can be seen from Fig. 2.2, the incoming short-wave radiation, as it reaches the ground surface, usually comprises a direct and a diffuse

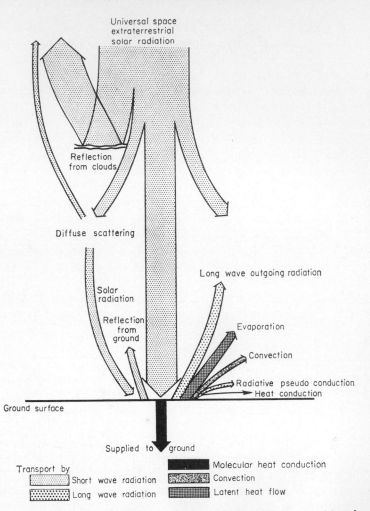

FIG. 2.2. Diagrammatic illustrations of the fate of solar radiation reaching the earth at midday on a clear day. The width of the arrows corresponds to the amounts of energy transferred (after Geiger, 1957).

component, the latter coming from the whole sky, although more intense close to the sun. Under clear sky conditions the direct beam contains more than 80% of the total, even at low solar angles, but light cloud is frequently enough to reduce the direct beam to zero. It is interesting to note that when the sun is visible through a gap in the clouds, reflection from them can sometimes increase the intensity considerably for short periods, occasionally doubling the clear sky value.

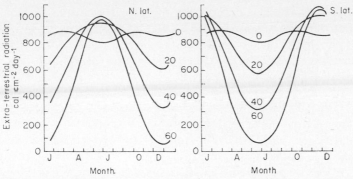

Fig. 2.3. Variation in daily totals of solar radiation received at the earth's surface at different latitudes and months of the year (after van Wijk, 1963).

TABLE 2.I

Relative solar irradiance, luminous efficiency, and colour temperature at sea level for various air-mass values†

Air mass	0	1	2	3	4	5
Solar angle		90°	30°	19·3°	14·3°	11·3°
Wave length, mμ	Percentage of total irradiance					
290– 400	7	4	3	2	1	1
400– 700	41	46	45	43	41	38
700–1100	28	33	36	38	40	42
1100–1500	12	10	9	9	9	9
1500–∞	12	7	7	8	9	10
Total	100	100	100	100	100	100
Total irradiance‡ w m^{-2}	1320§	930	740	610	510	430
lumens w^{-1}	93	105	106	103	98	93
ft-c g-cal^{-1} min^{-1} cm^{-2}	6000	6800	6850	6650	6300	6000
Colour temperature, °K	6200	5500	5100	4700	4300	4100

† From Withrow and Withrow (1956). Air mass values represent the equivalent number of atmospheric thicknesses traversed by the sun's rays.

‡ Multiply by 100 for microwatts per square centimetre and by 0·0014 for calories per minute per square centimeter.

§ Value of solar constant.

B. Energy Partition at the Plant Community Surface

On reaching the plant community surfaces, part of the already depleted radiation is reflected. The remainder is absorbed and changed into thermal energy of the absorbing materials, to be subsequently involved in other heat transfer processes.

The reflection coefficient for incoming solar radiation, referred to as the surface albedo, depends on the angle of incidence of the radiation and the physical characteristics of the surface. Therefore, the albedo varies not only with season and time of day but also with surface aspect, but these factors are generally not as important as those surface characteristics affecting reflectivity directly. For solar altitudes between 90°–20°, the albedo ranges from approximately 0·05–0·20 for clear deep water, 0·10–0·40 for most vegetated surfaces, 0·15–0·50 for soils of various colours and degrees of wetness to 0·8–0·9 for fresh snow. For the earth as a whole it is about 0·43. For any one crop or plant community, it can be expected to vary also with crop age and maturation due to changes in percentage ground cover and in the colour and reflectivity of the plant surfaces (Monteith, 1959). A number of values, collected from various sources, are given by Budyko (1956) and van Wijk (1963).

In addition to the short-wave fluxes described above, there is an appreciable long-wave exchange between earth and atmosphere. Most solid or liquid surfaces are relatively good absorbers and emitters at long wavelengths, and in this spectral region naturally-occurring surfaces generally behave as completely black bodies. Thus, irrespective of their visible colour, or the variation with wavelength of their reflectivity in the short-wave region (including the near infrared), all exposed surfaces may be considered to absorb almost completely, all long-wave radiation falling on them. They may also be considered to emit equally freely in this region, with the total energy emitted and its spectral distribution dependent solely on their absolute temperature.

On the other hand, air itself absorbs and emits long-wave radiation comparatively poorly. Except for clouds and haze, absorption and emission are almost entirely due to CO_2 and water vapour. Also, because of the normal decrease of temperature away from the ground, the radiation back from the atmosphere is generally less than from the ground, even at night with comparatively low surface temperatures. Thus there is normally a net outward long-wave flux, the only significant exceptions being with relatively warm clouds, or warm air masses over cold surfaces.

The difference between the incoming and outgoing, short- and long-wave, radiation fluxes, which represents the rate at which energy is being absorbed by the surface under study, is a most important microclimatic parameter. Usually known as the net radiation, it can be described by the expression

$$\mathbf{R}_n = \mathbf{R}_s(1 - \mathbf{r}) + \mathbf{R}_l \qquad (2.1)$$

where $\mathbf{R}_n$ is the net radiative flux, $\mathbf{R}_s(1-\mathbf{r})$ is the net short-wave flux, $\mathbf{r}$ being the surface albedo, and $\mathbf{R}_l$ is the net long-wave flux (all fluxes positive towards the surface). At night, of course, the short-wave components are cut off and net radiation is predominately negative.

With bare soil this primary energy partition takes place over a relatively simple surface but with vegetation the surface is much more complicated, some incident radiation striking the soil surface unimpeded, some being intercepted at the outer layer of leaves and the remainder at lower layers. Thus, a considerable amount of the radiative flux is transmitted by the upper canopy layers and reflected within the canopy so that diffuse components contribute to the net flux within the canopy structure.

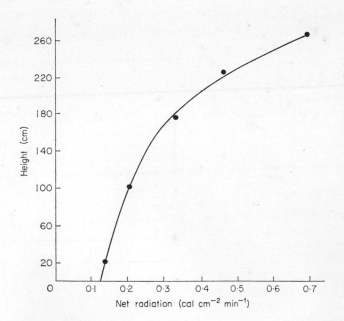

FIG. 2.4. Profile of net radiation shortly after noon in a corn crop of height 2·65 m and leaf area index 4·2 (after Allen *et al.*, 1964).

The reduction of net radiation with downward penetration into the crop is well illustrated by data replotted from Allen *et al.* (1964) (see Fig. 2.4). The data show that in a row crop of corn of leaf area index (LAI)† 4·2 and height 2·65 m, the net radiative flux at ground level

† Area of leaf per unit area land surface.

was only about 15% of that above the canopy, the larger reduction occurring in the visible portion of the spectrum.

Additional data are given in Table 2.II showing the proportion of the total, visible and near infrared spectral components which were reflected, transmitted and absorbed by the entire canopy over a whole day. It is again apparent that the photosynthetic portion of the spectrum was attenuated to a greater extent than the near infrared. Other data, obtained by Stern and Donald (1962), Begg *et al.* (1964) and other workers show almost complete light extinction at the ground surface under grass-clover swards or under plant communities with *LAI* values exceeding 6.

TABLE 2.II

Fractional disposition of total short-wave radiation in a dense corn crop†

	Total short-wave $(0\cdot3–3\cdot0\mu)$	Visible $(0\cdot3–0\cdot7\mu)$	Near-infrared $(0\cdot7–3\cdot0\mu)$
Reflected	0·170	0·035	0·135
Transmitted	0·135	0·035	0·100
Absorbed	0·695	0·460	0·235
Total	1·00	0·530	0·470

† from Lemon (1963).

The reflection and transmission of incident solar radiation by leaves has long been a subject of interest in plant physiology and many measurements have been made (see Billings and Morris, 1951; Kleschnin, 1960; Gates *et al.*, 1965). There is a considerable measure of variability between species, reflectivity depending largely on colour and surface characteristics, and transmission also depending on thickness. In general there is a tendency for both transmission and reflection to be greater in the near infrared compared with the visible part of the spectrum, as indicated in Fig. 2.5. Overall albedos for the whole spectrum range from < 10 % for dark coloured dull textured leaves to > 50 % for light coloured shiny leaves. Overall transmission factors appear to be less variable for most species, and often approximate to 20–30 %, but some very thick leaves may have zero transmission.

C. Dissipation of Energy Absorbed by Plant Communities

Of the net radiation absorbed by plant and soil surfaces through energy exchange with the sun and sky, part is dissipated as sensible heat, part as latent heat and part is involved in metabolic processes, The individual components are sensible (**H**) and latent (**lE**) heat exchange with the atmosphere, sensible. heat exchange with the vegetation itself and the soil (**G**) and the chemical energy conversions

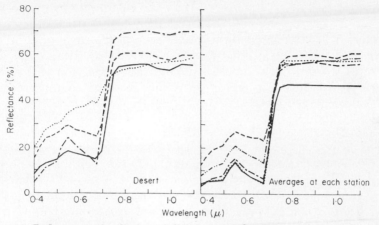

FIG. 2.5. Reflectance of radiation of different wavelengths by leaves of four desert species (left) (*Atriplex canescens* ———; *Atriplex lentiformis* — — —; *Eurotia lanata* ······; *Prunus andersonii* — · — ·) and mean values for groups of species from different environments (right) (Nevada, shaded campus ———; Nevada, open desert — — —; Pine forest, north-facing ·····; Pine forest, west-facing — · — ·; Subalpine slope, open — ·· — ··) (after Billings and Morris, 1951).

(**aA**) involved in photosynthesis and respiration of the plant community and associated organisms. These heat exchanges can therefore be written, again with flow toward the surface positive

$$\mathbf{R}_n + \mathbf{H} + \mathbf{lE} + \mathbf{G} + \mathbf{aA} = 0 \qquad (2.2)$$

where l and **a** are the latent heat of vaporization of water (~ 600 cal g^{-1}) and the chemical energy storage coefficient (~ 3600 cal g^{-1}), respectively.

In general, over time periods of the order of days the only significant components are **H** and **lE** since net changes in **G** from day to day are relatively small (inward flux by day almost balancing outward flux by night) and **aA** seldom amounts to more than a few per cent of $\mathbf{R}_n$. Over periods of hours, however, fluctuations in **G** can be significant and the flow of heat into and out of bare soil can represent a considerable proportion of $\mathbf{R}_n$ during periods of greatest upward or downward heat flow (Tanner, Peterson and Love, 1960). Heat flow into the plant community itself is generally too small to measure because the total mass and heat capacity of plant material is small compared with the mass of soil involved in heat exchange. However in dense woody plant communities, total weight of plant material may exceed 4×10^5 kg ha^{-1} (Ovington, 1963). Since this material is about one-half water, of high specific heat, it could provide substantial heat storage capacity, equivalent to the surface 5–10 cm of wet soil. Even so, heat flow into woody plant tissue is seldom as rapid as into soil because many of the main stems and branches are shaded by leaves which themselves intercept and exchange most of the incident radiation.

There are several reasons why the photosynthetic contribution seldom exceeds 2–3% of the incident radiation, except for short periods (Bonner, 1962). Firstly, only about 40–45% of the total radiation is in the $0\cdot4$–$0\cdot7\mu$ portion of the spectrum (see Table 2.I). Secondly, from the viewpoint of fundamental energy exchanges involved in the photosynthetic processes, approximately 10 quanta of light energy appear to be required at normal light intensities, for the reduction of one CO_2 molecule. While ten moles of quanta (i.e. 10 Einsteins) received in the middle of the wavelength range usefully absorbed by chlorophyll supply about 520 kg cal, 1 mole of CO_2, incorporated in plant material, only stores about 105 kg cal. Therefore, maximum photosynthetic efficiency is unlikely to exceed 20% of radiation within the visible part of the spectrum. Although experiments have demonstrated that the energetic efficiency of photosynthesis can exceed 20% of the energy supplied in the red region of the spectrum ($0\cdot6$–$0\cdot7\mu$) reaching up to 35% in some cases (Wassink, 1959), it decreases on both sides of this peak. There is a gradual decline towards the shorter wavelengths, and a rapid fall in the near infrared, practically reaching zero at $0\cdot75\mu$.

The effect of these two phenomena is to reduce the potential efficiency to 8–9% of the total incident radiation. Combined with these are other biological limitations due to such factors as light saturation of photosynthesis in any one leaf at a light intensity of perhaps only 10–20% of full sunlight; the presence of leaves which are not fully active photosynthetically as a result of age, location on the plant, or physiological disorder; or the frequently inadequate amount and orientation of leaf material to effectively intercept incident radiation, which otherwise strikes the ground. Such factors are rarely absent and substantially reduce efficiency over a whole growth cycle. Only for short periods associated with maximum growth rates do values approach theoretical levels (Watson, 1952, 1963; Black and Watson, 1960; Begg, 1965).

The remaining terms in the energy balance equation represent the net exchanges of sensible (**H**) and latent (**1E**) heat between the community surfaces and the atmosphere. Both exchanges are effected primarily by turbulent eddy movement in the lower atmosphere, the net vertical transfer being the product of the vertical temperature

$(\partial T/\partial z)$ and specific humidity $(\partial q/\partial z)$ gradients, respectively, and the eddy transfer coefficients, $\mathbf{K}_h$ and $\mathbf{K}_w$. With appropriate constants, this gives the relationships

$$\mathbf{H} = -c_p\rho_a\mathbf{K}_h \frac{\partial T}{\partial z} \qquad (2.3)$$

$$\mathbf{1E} = -1\rho_a\mathbf{K}_w \frac{\partial q}{\partial z} \qquad (2.4)$$

where c_p is the specific heat of air at constant pressure (cal g^{-1} deg^{-1}) and ρ_a is the density of moist air (g cm^{-3}); $\mathbf{K}_h$ and $\mathbf{K}_w$ have dimensions of cm^2 sec^{-1}.

If the ratio of the temperature and humidity gradients is replaced by the ratio of the differences $(\Delta T/\Delta q)$ measured over the same height interval, and $\mathbf{K}_h$ and $\mathbf{K}_w$ are assumed identical, as is the usual custom, the ratio of the two fluxes, termed the Bowen ratio, β, is given by

$$\beta = \frac{\mathbf{H}}{\mathbf{1E}} = \frac{c_p}{1} (\Delta T/\Delta q) \qquad (2.5)$$

The ratio, depending primarily on the temperature and effective wetness of the energy exchange surface, varies widely for different surfaces. From Eq. (2.5) it can be appreciated that, when the surface is wet, the specific humidity difference tends to be large and the corresponding temperature difference tends to be small. Thus a high proportion of the available energy is dissipated as latent heat, and β is low. Conversely, where the surface is dry, the temperature difference tends to be large, the specific humidity difference small, and β is high.

In vegetation well supplied with soil water, β is commonly less than 0·20 and may become negative (see below). Similar figures are found with wet bare soil. If, however, the plant stomata close, due to reductions in soil water supply or to other causes, the rate of evaporation (and hence the amount of heat used in evaporation) declines, the surface temperature rises, and more energy is dissipated as sensible heat, leading to an increase in β. Under arid conditions, $\mathbf{1E}$ tends to approach zero and β to approach infinity.

The value of β also changes during the day. This is well illustrated by Fig. 2.6 after Tanner (1960), which shows the diurnal change in the major heat balance components for vegetation well supplied with water. Averaged for the whole day, evapotranspiration ($\mathbf{1E}$) accounted for 78 % of the net radiation ($\mathbf{R}_n$), and sensible heat flow ($\mathbf{H}$) for 11 %, giving a value for β of 0·14. However around sunrise and in the afternoon, β was negative because the surface was cooler than the bulk air, resulting in a net flow of sensible heat to the surface giving positive values of **H**, while the vapour pressure gradient was still away from the surface, giving negative (but rather low) values of **1E**. The less pronounced fluctuations in soil heat flux (**G**) can also be seen.

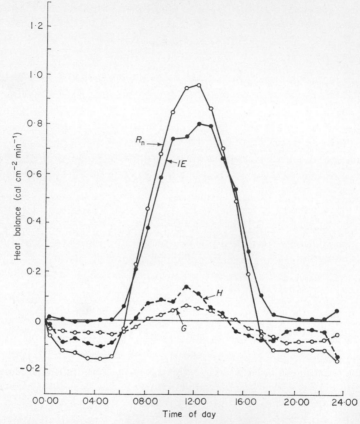

Fig. 2.6. Diurnal changes in the primary energy balance components over pasture during clear summer day conditions in Wisconsin (after Tanner, 1960).

Another phenomenon which commonly influences β and the observed values of **H** and **1E** is the contribution of horizontally advected heat and water vapour from adjacent areas. This phenomenon occurs on a major synoptic scale as air masses cross continents, but also occurs on a micro-scale within plant communities, particularly when there is spatial diversity in the plant distribution and when bare areas of dry soil alternate with areas of relatively wet plant surface. This is frequently referred to as a "clothes-line" effect. At an intermediate scale it is well demonstrated by irrigated agriculture in arid regions since the adjoining non-irrigated areas are almost invariably drier than the cultivated ones. This is often termed an "oasis" effect. Under these conditions the total heat load on an area may be well in excess of that received by direct solar radiation, and the effect may extend several miles downwind (Lemon *et al.*, 1957; Philip, 1959). A good example of this, given by Lemon *et al.*, is reproduced in Fig. 2.7.

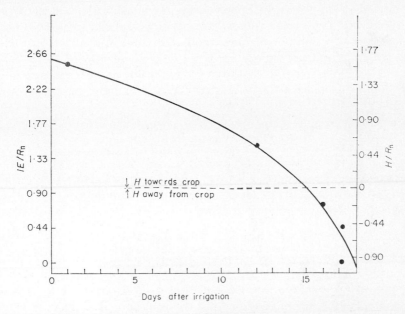

FIG. 2.7. Ratios of latent (lE) and sensible (H) heat exchange to net radiation ($\mathbf{R}_n$) in a cotton crop as a soil dried from a freshly irrigated condition to a value close to the permanent wilting percentage (after Lemon *et al.*, 1957).

   This diagram shows the proportion of the net radiation involved in sensible and latent heat transfer over an irrigated cotton field in Texas, as the top 4 feet of soil dried from a freshly irrigated condition (soil water potential, Ψ, $\cong -0.1$ bar) to a value close to the permanent wilting percentage ($\Psi \cong -15$ bars). Initially the total evaporation equalled more than two and a half times the net radiation, the large additional amount of energy being supplied from hot advected air which resulted in heat flow towards the crop. However, as the evaporating surfaces of the soil and crop dried, the latent heat flux approached zero and the surface temperature progressively increased until the outward flow of sensible heat accounted for the total direct plus advected heat load. During this period β increased from -0.66 in the extreme wet situation, to zero at the stage of no net heat flux, and to a value approaching infinity at the dry end.

Even in large areas where intermediate scale advection is of little significance, average latent heat consumption may still be found to exceed average net radiant energy received, often over months at a time if the soil water supply is not limiting evaporation. This arises from what is, in effect, large scale or upper level advection due to the movement of weather systems across the surface under study, from relatively hot, dry surfaces elsewhere. McIlroy and Angus (1964) have commented that certain regions, including those quite well provided with incoming radiant energy, may act as long-term sinks for sensible heat, due to the general pattern of circulation over them, provided that sufficient moisture is available for the amount of evaporation involved.

2. The Water Balance

Although the section just concluded has dealt with evaporation from plant and soil surfaces within a general energy balance framework, evapotranspiration also is a very significant component of the water balance, which is normally written as

$$\mathbf{P} - \mathbf{O} - \mathbf{U} - \mathbf{E} + \Delta\mathbf{W} = 0 \tag{2.6}$$

where $\Delta\mathbf{W}$ is the change in soil water storage (initial minus final) during the period under study and for the depth of measurement, and $\mathbf{P}$, $\mathbf{O}$ and $\mathbf{U}$ are the precipitation, run-off, and deep drainage, respectively. The amount of water passing beyond the root zone is defined as $\mathbf{U}$, or, for experimental purposes, the amount passing below the lowest point of measurement. Evaporation is $\mathbf{E}$ (including transpiration) from the plant and soil surfaces, as used previously. All symbols have dimensions of length e.g. cm.

Equation (2.6) can be more precisely written as

$$\int_{t_1}^{t_2} [(\mathbf{P} - \mathbf{O}) - \mathbf{E} - v_z]\, \mathrm{d}t = \int_{t_1}^{t_2} \int_0^z \frac{\partial\theta}{\partial t}\, \mathrm{d}z\mathrm{d}t \tag{2.7}$$

where $(t_2 - t_1)$ is the time interval over which the measurements are made (sec), z is the depth to the lowest point of measurement (cm), v_z is the net downward flux of water at depth z (cm sec^{-1}) and θ is the volumetric soil water content (cm^3 water cm^{-3} soil). $\mathbf{P}$, $\mathbf{O}$ and $\mathbf{E}$ now are in units of cm sec^{-1} or g cm^{-2} sec^{-1}. In the presence of a water table near the surface, v_z can be upwards.

These equations can be used on any scale, ranging from continental land masses and hydrologic catchments down to individual plants. In most cases all the elements except **E** are measured or estimated, and **E** obtained by difference.

A. Partitioning of Precipitation

Precipitation varies considerably in its amount and distribution in different geographic regions, ranging from the almost continuous precipitation characteristic of rain forest areas, where annual totals may approach 25 m, to the almost non-existent precipitation of extreme desert areas where some localities may not receive measurable quantities over an entire year. The intensity of precipitation can also vary from more than 75 cm hr^{-1} to almost zero, and its composition from snow through sleet and rain, to fog and dew (Linsley, Kohler and Paulhus, 1949).

1. Contribution of dew

It seems appropriate at this point to make a few additional comments about dewfall, not because the total amounts are large but because controversy has frequently existed as to the probable significance of this water source to plants (Duvdevani, 1953; Stone, 1957a; Slatyer, 1960c). Specific biological implications are dealt with in Chapter 7, and only its physical features will be considered here.

In the first place it should be emphasized that visible "dew" can come from three separate sources (a) true dewfall, or condensation of water vapour coming down from the atmosphere, (b) distillation, or condensation of water vapour coming up from the soil or lower plant surfaces, and (c) guttation. Since neither (b) nor (c) represent any net gain of water by the plant-soil system, and guttation is, moreover, a physiological plant process (dealt with in Chapter 6), only (a) can strictly be considered as precipitation. Alternatively it can be considered as evaporation in reverse, involving a downward rather than an upward flow of water vapour to, rather than from, an effectively wet surface.

The formation of dew is primarily a nocturnal occurrence because it depends on radiational cooling of the exposed leaf and soil surfaces to the dewpoint temperature. Continued condensation involves the maintenance of a vapour pressure gradient towards the leaf surface which is also a rare event except at night. [Downward fluxes of water vapour within plant canopies during the day have been observed by Begg et al. (1964) and Denmead (1964) but only for short periods.] In consequence, the main factors contributing to dew formation are clear skies, since cloudiness markedly reduces outgoing radiation; low wind speed, since sensible heat inflow prevents leaf temperature from falling; high bulk vapour pressure so that the amount of cooling required is small; and vegetation of low heat capacity, open-branched character, and leaf area spread over a considerable height, to aid radiation loss.

With particularly favourable conditions, dew deposition may commence before sunset and continue until after sunrise, but more generally it is restricted to the nocturnal period. Deposition seldom occurs continuously throughout the night, being more often interspersed with periods of evaporation as bulk air conditions, especially cloudiness and wind speed, fluctuate (Monteith, 1957, 1963b). The dew formed before sunrise may, in turn, persist all day (Arvidsson, 1951) although it more commonly evaporates within 1–2 hr after cessation of condensation.

Total amounts of dewfall during any one night appear unlikely to exceed about 1 mm unless there is a substantial advected component (Monteith, 1957, 1963; Slatyer and McIlroy, 1961), and reported values in the literature provide confirmation of this point (Milthorpe, 1960). However most observations comprise both distillation and dewfall so that actual dewfall amounts can be expected to be significantly less. Deposition of 1 mm of water, for example, would have to come from a considerable layer of saturated air, of the order of 100 m thick, cooling from 20° to 5°C during the overnight period, and this would require efficient transfer processes. Because transfer depends on wind speed, deposition under very calm conditions tends to be reduced and Monteith (1957) found optimum conditions within the wind speed range of 1–3 m sec^{-1}. With speeds higher than 5 m sec^{-1}, the vertical mixing is too effective and saturation is not reached, so that evaporation, rather than dewfall, occurs.

A final point about dew can be made concerning its measurement. Since the exposure and heat capacity of the surface affects the amount of deposition, it can be appreciated that the dew formed on a dew plate may bear little relation to that expected on a natural surface. Most dew measurements, therefore, while perhaps of value as relative station-to-station indicators may be meaningless in terms of dew received by plants or soils.

2. *Interception of precipitation by vegetation*

Considerable differences in the pattern of precipitation actually reaching the ground develop in many plant communities because of the interception of precipitation by the vegetation. It may subsequently be transferred to the soil by channelling down the main stems ("stem flow"), or by dripping from the branches, or may be lost by evaporation from the wet surfaces. Differences also develop between plants, due particularly to the disturbed wind structure, and most noticeable in the case of snow. The amounts of precipitation actually retained on the surfaces and the amounts channelled down the stems vary considerably with the morphological characteristics of the species concerned and the nature of the precipitation. With rain, it is usual to observe drip or stem flow after an area rain total of about 2·5 mm has been received, but with freezing rain or snow, under conditions favouring retention on the leaves and stems (low wind, temperatures only a few degrees below freezing), several times this amount may be accumulated.

Stem flow is considerably enhanced, as might be anticipated, by branches and leaves which are inclined upwards, and many arid zone plants appear to be particularly well adapted for this purpose. Hamilton and Rowe (1949) in a study of chaparral in California and Slatyer (1965) with *Acacia aneura* in central Australia, observed that up to 40% of the rainfall, incident on an area equal to the horizontal projection of the canopy, was channelled down the main stems and entered the soil with minimal out-wash at the bole. The amount of stem flow tends to decrease as wind and rainfall intensity increases. Gross morphological features of this type may be powerful ecological factors in plant distribution as, for example, in the restriction of long leaf pine (*P. australis*) to areas of southeastern U.S.A. free from regular incidence of snow and freezing rain, the weight of which can snap the main trunks.

The precipitation which remains on the leaves and stems is always depleted to some degree by evaporation and in some cases none reaches the ground. As an example of the relative amounts of precipitation partitioned into throughfall and stem flow, and lost by evaporation, the data of Eidmann (1959) are given in Table 2.III. For deciduous

TABLE 2.III

Throughfall, stem-flow and interception for evergreen (spruce) and deciduous (beech) forests†

Period	Rain (mm)	Spruce Through-fall	Stem-flow	Inter-ception	Beech Through-fall	Stem-flow	Inter-ception
Nov.–Apr.	587	465	4	118	465	97	25
May–Oct.	629	428	5	196	457	104	68
Year	1216	893	9	314	922	201	93
Year %	100	73	1	26	76	16	8

† from Eidmann (1959).

(beech) and evergreen (spruce) forests the amounts varied considerably, beech providing much more stem flow and much less interception loss. It is interesting to note that the stem flow percentage was much the same in beech in both winter and summer, even though beech is deciduous (presumably because of a greater proportion of stem flow from snow). Interception loss, however, was three times as great in the summer when the trees were in full leaf. In the spruce there was little variation in the percentage of stem flow and interception loss between summer and winter, the type of branching presumably keeping stem flow at a low value, even with precipitation as snow. Much more detailed accounts of these phenomena are given by Kittredge (1948) and Penman (1963).

3. *Surface run-off*

Of the precipitation which reaches the ground, redistribution by surface run-off further modifies the final pattern of soil water recharge. Run off occurs whenever the rate of precipitation exceeds the rate of infiltration and the accumulation exceeds the surface pondage capacity at the point of measurement. In consequence, run-off varies with amount and intensity of precipitation (and frequently with duration, since infiltration rate, initially high, tends to decrease with time, see Chapter 4) and with surface conformation, which influences the degree of which pondage can take place.

In natural situations slope is seldom constant and, while run-off tends to reduce soil water recharge at the top of a slope and increase it at the bottom, minor changes of slope generally modify the slope/run-off relationship. Very good examples of this are found where minor differences in microtopography, and resultant changes in run-off, cause spatial vegetation patterning. This occurs particularly in arid regions where relatively small increments or decrements in run-off contribution can exercise profound effects on resultant soil water storage (Slatyer, 1962a; Boaler and Hodge, 1962, 1964).

B. Drainage

In most cases determinations of evaporation from the soil water balance equation are made at times when, or in situations where, gain or loss of water, from or to the zone below maximum root depth, is considered to be negligible. However, there is evidence to suggest (Wilcox, 1960) that in many cases this term is not negligible and Rose and Stern (1965) have recently suggested a means of calculating it, from hydraulic conductivity and soil water potential profile data, under conditions where there is no lateral flow. Such conditions can be expected in most plant communities where there is a fairly uniform root distribution, as distinct from situations where isolated trees or shrubs may be separated by large areas of grass or bare soil.

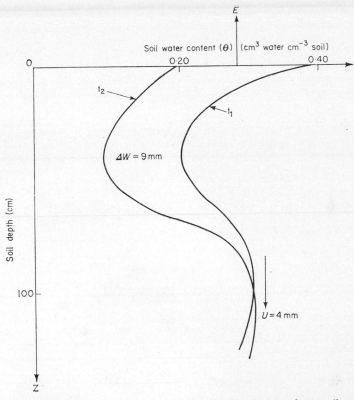

FIG. 2:8. Soil water content profiles, measured 24 hr apart, on loam soil cropped to cotton. Of the total $\Delta W = 9$ mm, 4 mm was attributable to drainage. For symbol meanings, see Eqns (2.6) and (2.7) (after Rose and Stern, 1965).

Anticipating parts of Chapter 3 and 4, Rose and Stern (1965) have written the normal equation for vertical flow of water, v_z in the form

$$v_z = K + K(\partial h/\partial z) \tag{2.8}$$

where v_z is the rate of downward flow in cm sec^{-1}, K is the hydraulic conductivity (cm sec^{-1}) and $(\partial h/\partial z)$ is the rate of change of soil water suction (h cm) with depth (z cm). Soil water suction in cm of water is obtained from soil water potential, Ψ, in dyne cm^{-2} by the relationship $h = -\Psi/\rho_w g$ where ρ_w is the density of water and g the acceleration due to gravity (cf. Eqns 3.2 and 4.3). Unless h is very small, $(\partial h / \partial z)$ is usually much greater than unity, so that the K term in Eq. (2.8) is often negligible. The through drainage term, U, of the water balance equation (see Eq. 2.6) is then given by

$$U = \int_{t_1}^{t_2} v_z dt \qquad (2.9)$$

where (t_2-t_1) is the time period between observations.

In Fig. 2.8 an example of the effects of through drainage on the soil water balance is given for a loam soil cropped to cotton. The importance of including the drainage term can be appreciated from the data, which show that through drainage accounted for more than 40% of the total change in soil water content in the surface metre of soil.

In other situations, there can be a net upward flux of soil water into the root zone from wetter underlying soil. This is particularly pronounced where there is a water table near the surface, as discussed in Chapter 4. Rose and Stern (1965) have also considered the effects of overburden pressure, and the special problems of swelling soils.

C. Evapotranspiration

Since run-off and rapid downward drainage are closely linked with precipitation and soil water recharge, the water balance Eq. (2.6) reduces to $E = \Delta W$ during intervening periods of dry weather (or between irrigations). In reality, redistribution of water in the soil continues indefinitely and even in the absence of a water table there can be detectable upward water movement under some conditions, (Rose and Stern, 1965). For many practical purposes, however, these effects can be neglected.

Evaporation from plant community surfaces is frequently termed evapotranspiration, to signify the combined plant-soil nature of the evaporating surfaces. The two words are used interchangeably in this book in most applications when free water surface evaporation is not involved. Transpiration, however, is used only to describe evaporation from plant surfaces. Evaporation from soil and transpiration from plants involve basically similar processes and both soil and plants may be thought of as alternative paths through which water flows on its way to the surface, from which it finally diffuses into the bulk air. In the case of vegetation, the heterogeneity of the effective surface, and the biologically controlled variability of the resistance to internal flow, are only modifications to this pattern, albeit important ones.

Evaporation is best thought of in terms of three inter-dependent influences. The first of these is the availability of energy at the evaporating surface, to supply the latent heat demand. Possible sources are radiation from the sun, sky, and clouds and sensible heat transferred from the adjacent air and soil. Excepting the sun, all of these can also act as energy sinks, competing with evaporation for whatever energy is being received at the surface from the remaining sources. The other two influences are those which determine the vapour pressure gradient (or difference) between the water at the evaporating surface and the bulk air, and those which contribute to resistances in the water vapour pathway.

It must be emphasized that of all the above variables only solar radiation can be regarded as at all independent of the others. It can be misleading to think of a gradient as externally controlled and itself controlling the associated flux. It is equally wrong to take the flux as given, and the gradients are thereby controlled. In fact each interacts continuously with the other, and with other relevant factors.

Under steady state conditions, all three groups of factors will adjust themselves to produce a particular rate of evaporation. A change in any one of them will not necessarily produce a proportional change in evaporation, but rather will be associated with a change in the other factors, and finally a new balance will be established.

As an example of these interactions, consider the effect of sudden widespread clouding-over on a previously bright day with a high rate of evaporation. There will be an immediate drop in the net radiant energy received by the surface. However the fall in the evaporation rate which follows, although rapid, will be more like an exponential decay. The temperature at, and near, the surface will commence to fall, and the vapour pressure likewise. A change in the gradients will therefore spread outwards from the surface, altering the rate of flow of both heat and water vapour, until finally a new energy balance is established with a reduced rate of latent heat consumption and an even more reduced flow of sensible heat.

Similarly, the consequence of a change of wind speed will be a progressive re-adjustment, not only of evaporation rate, but also of all vertical gradients and of heat flow, until a new equilibrium is reached— provided other bulk conditions remain steady.

1. *Characteristics of vapour transport*

Detailed accounts of the evaporation process are given by Sutton (1953), Slatyer and McIlroy (1961) and Rijtema (1965). The following notes briefly outline some of the factors involved.

Evaporation involves the net upward transport of water vapour, from the liquid–air interfaces of the plant community surfaces to the bulk air, by molecular diffusion and turbulent eddy movement. Initially transport is by molecular diffusion across the thin boundary layer which sheathes all natural surfaces like a continuous skin. If the liquid–air interfaces are beneath the natural surfaces (as in dry soil or plants), rather than at the natural surfaces (as in wet soil or water), the diffusion zone is extended by the extra path-length so developed. Beyond the boundary layer there is a transitional region in which both molecular diffusion and turbulent eddy movement are involved. Subsequently this merges into the zone where turbulent transport mechanisms are of overriding importance.

For simple surfaces a close relationship between the diffusive resistance external to the natural surface, usually termed r_a, and boundary layer characteristics, may be found (see Chapter 8, also Sutton, 1953 and Gates, 1962) and the fraction of the total diffusive resistance within the natural surface can be evaluated. For surfaces as complex as plant communities, however, no simple relationship exists and efforts to partition the total resistance into external and internal components are dependent on the degree to which a meaningful "effective crop surface" can be defined.

Even so, overall aspects of the net water vapour transfer can be idealized, without specifying the nature of the transport mechanisms, in expressions of the form

$$E = \frac{(c_n - c_a)}{(r_n + r_a)} = \frac{(c_n - c_o)}{r_n} = \frac{(c_o - c_a)}{r_a} \qquad (2.10)$$

where c_n, c_o and c_a (g cm^{-3}) are the water vapour concentrations (= absolute humidities) at the internal liquid–air interfaces, at the effective crop surface, and in the bulk air, respectively; and r_n and r_a (sec cm^{-1}) are the effective resistances to vapour transfer beneath and beyond the effective crop surface, respectively.

For single leaves, c_o is located at the natural leaf surface so that the internal resistance, r_n (referred to as r_l in Ch. 8) is wholly within the leaf and the external resistance, r_a, is outside the leaf. For crops, however, c_o is the vapour concentration at an effective crop surface, and a specific morphological boundary cannot be indicated. Even so, the two resistances r_n and r_a can be loosely described as a surface or crop resistance, and an external or atmospherical resistance, respectively.

Monteith (1963a, 1965) has attempted to define an effective crop surface using profile data of wind speed, temperature and vapour pressure. The form of the wind profile above a crop can be described in an equation of the form

$$\bar{u} = \frac{u_*}{k} \ln \left[(z-d)/z_o\right] \tag{2.11}$$

where $\bar{u}$ (cm sec^{-1}) is the mean wind velocity at height z (cm); u_* (cm sec^{-1}) is a velocity characterizing the particularly turbulent regime, known as the friction velocity; z_o (cm) is the roughness length, which varies with the surface characteristics, d (cm) is the zero plane displacement, and k ($\simeq 0.40$) is a dimensionless turbulence parameter, termed the von Karman constant.

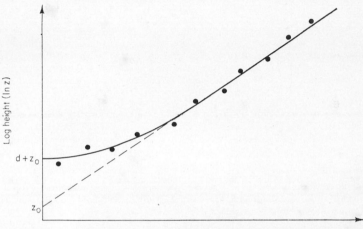

FIG. 2.9. Typical wind velocity profile above a rough surface, plotted against the logarithm of height above the ground, z. The roughness length is z_0 and d the zero plane displacement. For explanation see text (after Gates, 1962).

A typical profile of wind velocity, plotted against the logarithm of height, is given in Fig. 2.9. Assuming that the profile has logarithmic form, as indicated by Eq. (2.11), z_0 is seen to be the extrapolated height at which $u = 0$. If the actual profile is extrapolated, the intercept is at $(z_0 + d)$. Values of u_* are about 10 % of the values for u at 2 metres above the surface; and of z_0 are approximately 10 % of the average height of the vegetation (Sutton, 1953).

Where the profiles of temperature, T, vapour pressure, e, and wind speed, u, measured at several heights above a crop, are the same shape, Monteith's procedure has been to plot T, and e, against u to give straight lines intercepting the axis $u = 0$ at $T = T_o$ and $e = e_o$, respectively. Assuming that the transfer coefficients for water vapour and momentum are equal under all conditions of stability (Deacon and Swinbank, 1958), and that the vertical fluxes are constant with height, Penman and Long (1960) showed that evaporation rate could be expressed as

$$E = \frac{u\mathbf{k}^2(e_0-e_a)c_p\rho_a}{\gamma\mathbf{l}\{\ln\,[(z-d)/z_0]\}^2} = \frac{u\mathbf{k}^2(c_0-c_a)}{\ln\,\{[(z-d)/z_0]\}^2} \qquad (2.12)$$

where u is the wind speed (cm sec^{-1}), e_a (or c_a) the vapour pressure or concentration at the same height z (cm) and the other symbols have

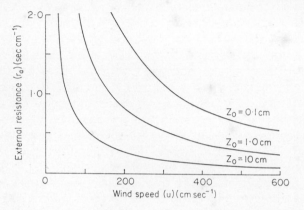

FIG. 2.10. Examples of the effect of wind speed on the calculated crop external resistance at various roughness length, values, z_0, with wind speed measured 2 m above the zero plane (after Monteith, 1965).

the same values as before. As can be appreciated from Fig. 2.9, the zero plane displacement is chosen to make wind speed proportional to $\ln\,(z-d)$. Monteith (1965) then provides the following expression for r_a

$$r_a = \frac{(c_0-c_a)}{E} = \frac{\{\ln\,[(z-d)/z_0]\}^2}{u\mathbf{k}^2} \qquad (2.13)$$

Thus for a given level of wind speed at an arbitrary height above the zero plane, it is considered that r_a can be related either to z_0 or to a dimensionless drag coefficient given by $\mathbf{k}^2/\{\ln\,[(z-d)/z_0]\}^2$. Typical examples of the expected influence of wind speed on r_a at different values of z_0 have been given by Monteith (1965) and are reproduced in Fig. 2.10.

Assuming that the water vapour concentration, c_n, at the actual evaporating surfaces is the saturation vapour concentration, $c^0(T_0)$, at the effective surface temperature, T_0, Monteith then writes the equation for crop r_n in a form analogous to Eq. (2.10):

$$r_n = \frac{(c^0(T_o) - c_o)}{E} \qquad (2.14)$$

Although it is seldom possible to identify T_o and e_o with conditions at a specific level within the canopy, Monteith (1965) considered that crop r_n and r_a can be regarded as realistic parameters, if r_n is a valid internal resistance, unaffected by wind speed. Tanner (1963) has criticized this view on the basis that the source and sink distributions for heat, water vapour and momentum may not be the same. Recent experimental and theoretical evidence has further supported this criticism (Begg et al., 1964; Denmead, 1964; Philip, 1964a), thus casting doubt on the proposition that information about internal resistances can be inferred from above-canopy profile data. Even so, the general concept of an internal crop resistance, combined with an external resistance in series, is a useful one for many purposes. For single leaf studies, where appropriate boundaries for the internal and external phases can be assigned, the concept provides a good basis for understanding the relative effects of physiological and environmental factors on leaf transpiration. These are considered in detail in Chapter 8.

From the foregoing it can be appreciated that evaporation can be regarded as a transport process in which the driving force is the surface–air water vapour concentration difference, and the rate of flow is proportional to the driving force divided by the total resistance to flow. The effect of energy supply on evaporation is primarily through changes in c_n, because of the strong dependence of surface vapour pressure (and hence vapour concentration) on surface temperature (see Table 1.III). Thus variations in c_n are generally of much greater significance to evaporation than variations in c_a. Variations in both r_n and r_a can also be quite marked. With vegetation well supplied with soil water, crop r_n is usually relatively small, but when the stomata close, or in dry soil, overall transport may be dominated by r_n. The following notes briefly outline some of the pathway characteristics in soils and vegetation which affect r_n and c_n.

2. Pathway characteristics in soils

Under freshly irrigated conditions, or following rain, the soil surface is effectively wet and evaporation proceeds from a bare soil at a rate not dissimilar to that from an exposed water surface. This is because, in general, only small differences exist, between the two types of surfaces, in net energy supply and surface vapour pressure (determined by surface temperature), and the difference in resistance to vapour flow, r_a, also is small. The presence of waves on water surfaces, or of furrows and other projections on soil surfaces, however, can influence boundary layer characteristics and the turbulent mixing above it.

As soon as the rate of evaporation exceeds the rate of flow of water to the soil surface the evaporating surface retreats into the soil. This may increase the actual evaporating surface (by exposing a wet internal surface greater than the overlying land area) or, occasionally, reduce it (by the presence of dry "islands" of soil particles among wet internal surfaces) but the primary reason for the sudden drop in evaporation,

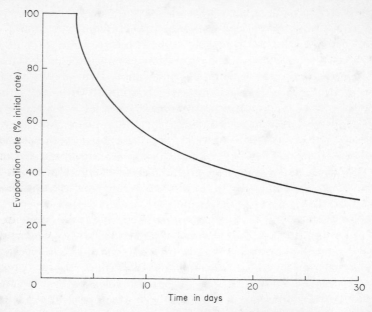

Fig. 2.11. Typical pattern of evaporation from initially saturated soil under constant meteorological conditions (after Philip, 1957a).

frequently observed (see for example, Fig. 2.11) is that the need for the evaporating vapour to diffuse to the soil surface lengthens the effective vapour diffusion pathway, and hence increases its resistance to vapour flow, which now includes an r_n, as well as an r_a, term.

Because flows of water and heat, within the soil, are closely linked, diurnal and seasonal fluctuations in soil heat flux can strongly influence the rate of flow of soil water to the surface, and hence influence the stage at which the water surface retreats into the soil, at any given evaporation rate. Philip (1957b) has shown the paradox which develops when, with high net radiation and rapid heat (and water) flow from the soil surface into the soil mass, the soil surface rapidly becomes dry, evaporation rate drops, and total evaporation over a period of days can be less than under conditions of much lower rates of energy supply. Diurnal fluctuations in soil heat flux can be largely responsible for the common observation that after rain, even though the soil mass is wet, the surface frequently dries out each day and becomes wet again at night. Seasonally, too, there is a tendency for a net upward heat flux from the soil in autumn which can lead to drying out of the soil to significant depths, with the reverse tending to occur in the spring.

As evaporation proceeds and the soil water content is reduced, the length and tortuosity of the vapour pathway progressively increases and evaporation declines. Some compensatory effect follows the associated increase in soil surface temperature, which can markedly increase the surface vapour concentration c_n, and hence steepen Δc, but this is generally small compared with the increases in r_n. The effect of increasing temperature on c_n is itself reduced by the decline in the relative vapour pressure, e/e^0, of the soil water, associated with the progressive fall in soil water potential (see Eq. 1.46). The reduction in water content, and hence in vapour pressure at the evaporating surfaces, is accentuated by absence of a water table and by water extraction by plant roots (but, as shown by Table 1.III, the soil must be much drier than the permanent wilting percentage before e/e^0 departs significantly from unity).

In consequence, with external conditions constant and in the absence of water table effects, evaporation from soil can be expected to decrease continually but at a slower and slower rate following the initial retreat of the wet surface, until effective vapour pressure equilibrium exists between soil and air. Because external conditions are not constant, this final situation is probably not achieved under natural conditions except in the layers of soil closest to the surface.

3. *Pathway characteristics in vegetation*

The evaporating surfaces of plants, except when the leaves are actually wet, are located within the leaves, primarily in the walls of the mesophyll cells exposed to intercellular air spaces and also in the external walls of the epidermal cells. This introduces an internal resistance to water vapour transfer, r_n, between these surfaces and the natural leaf surface. However, the area of actual evaporating surface may be much greater than the equivalent land area, since internal leaf area generally exceeds external leaf surface area by about an order of magnitude (Esau, 1965) and leaf area frequently exceeds land area.

As a result, total evapotranspiration from a plant community, expressed per unit land area, may exceed that from a similar area of

bare wet soil or free water, the greater surface more than compensating
for the additional resistance in the water vapour pathway. Evapotranspiration in excess of free water evaporation is accentuated in
strongly advective situations and by structural features of the vegetation which lead to enhanced aerodynamic roughness. These features
are well illustrated in Fig. 2.12, from van Bavel *et al.* 1963 which

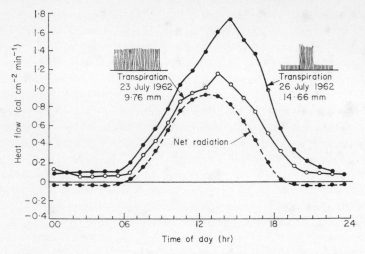

Fig. 2.12. Transpiration from Sudan grass in a closed sward (open circles) and with
the surrounding area mowed (closed circles) in relation to net radiation under summer
conditions at Tempe, Arizona (after van Bavel *et al.*, 1963).

shows evapotranspiration in excess of net radiation from a sward of
Sudan Grass at Tempe, Arizona under strongly advective arid conditions. The figure also shows how the discrepancy can be enhanced
if both roughness and advection are increased by mowing the area
surrounding the test location. It is apparent from this figure and from
Fig. 2.7 that advective heat loads can be dissipated as latent heat more
readily by the extensive leaf array than by a wet soil surface.

It is important to note that, while the foregoing applies to plants
which are well supplied with soil water, internal water deficits rapidly
lead to reductions in the aperture of the leaf stomata. Under such
conditions, the stomata progressively close and the internal diffusive
resistance, r_n, can become so high that it can dominate vapour flow,
resulting in a reduction of transpiration to a small fraction of the open
stomata value (see Chapter 8).

The vapour pressure at the liquid-air interface is often considered
to be the saturation vapour pressure appropriate to the temperature

of the evaporating surface. However, as in the case of soil water, water in the plant is usually neither chemically pure nor free of other constraints and the actual vapour pressure is, instead, a function of the water potential, Ψ, at the evaporating surface (see Chapter 1). In most cases Ψ_{leaf} does not fall below about -25 bars, equivalent to a relative vapour pressure of about 0·98 at normal temperatures (see Table 1.III), so little error is introduced to $(c_n - c_a)$ by assuming that the relative vapour pressure is unity. However, in extremely desiccated plants, Ψ_{leaf} has been observed to be about -140 bars (Slatyer, 1960a) and some lower values have been reported (Stone, Went and Young, 1950; Whiteman and Koller, 1964). This is equivalent to a relative vapour pressure of almost 0·90 and could possibly cause a significant reduction in the vapour concentration difference $(c_n - c_a)$. It must be remembered, though, that a small increase in surface temperature could readily increase c_n to compensate for this effect.

For the present, therefore, it need only be emphasized that, as leaf water content (and hence Ψ_{leaf}) falls, due to an inadequate supply of soil water to the evaporation sites to satisfy the transpiration demand, observed reductions in crop transpiration appear to be due primarily to increased diffusive resistance in the path of the evaporating water (affecting r_n) rather than to a reduction in the vapour pressure at the wet surfaces (affecting c_n). As described in Chapters 4 and 7, the soil

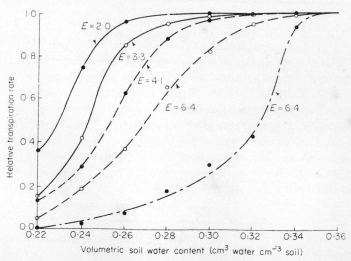

Fig. 2.13. Variation in relative transpiration rate (expressed as the ratio of actual transpiration to the maximum observed values) with soil water content under different evaporative conditions (expressed as potential transpiration and indicated on diagram as daily potential transpiration in mm of water) (after Denmead and Shaw, 1962).

water content at which observed reductions in transpiration commence
will depend mainly on the evaporative demand itself, on the lowest
value of Ψ'_{leaf} which can be tolerated without stomatal closure and on
the water transmission and retention characteristics of the soil and
plant (themselves determined largely by levels of Ψ' in the soil and
plant and by root distribution). Figure 2.13 provides a good example
of the interaction between evaporative demand and soil water content
which occurs under field conditions. For a corn crop grown in Iowa, it
can be seen that, at a volumetric water content of, say, 0·26, the relative
transpiration rate varied from about 10 to 100 % of the potential value,
depending on the evaporative demand.

3. Determination of Evapotranspiration

From the foregoing discussion it follows logically that the three
main techniques for measuring evaporation from natural surfaces
involve determinations of (1) the evaporation term in the water balance
(2) the latent heat term in the energy balance and (3) determinations
of the net upward flow of water vapour in the air layers near the
ground. In addition, methods have been developed involving combinations of (2) and (3) and a number of procedures have been developed
for empirically calculating **E** from readily measured meteorological
variables. Some of these are discussed below.

A. Water Balance Methods

As outlined previously [see Eq. (2.6)], $\mathbf{E} = \Delta\mathbf{W}$ during periods of
dry weather between rain or irrigation, neglecting, or otherwise
measuring **U**. Evapotranspiration can consequently be measured during
such periods by determining changes in soil water storage under the
plant community being studied. An alternative procedure, frequently
applied to catchments or to entire drainage systems is to calculate **E**
over long periods, between occasions of similar wetness or dryness,
when $\Delta\mathbf{W}$ is assumed to be zero, **P** and **O** are measured, the latter
across a weir or other suitable device, and **U** is again neglected or
allowed for. In this case $\mathbf{E} = \mathbf{P} - \mathbf{O}$.

For most plant communities and for operations at either the agronomic or ecological level, the former procedure is commonly adopted.
It is the most accurately conducted by the use of weighing (or hydraulic)
lysimeters provided they are properly designed and sited. However
lysimeters cannot be used when the mixed nature of the species
composition, the spatial distribution of the vegetation, the depth and

ramification of the root system or other factors make it impossible to simulate the natural environment inside the lysimeter itself. Typical examples of such limitations are found in truly arid plant communities and in natural multispecific forests.

In such cases, determinations of changes in soil water storage at different points in the community provide the only technique for evaluating **W**. Even so, integration of the observations is difficult and the presence of a water table near the surface can limit the application of the method, unless fluctuations of water table depth can themselves be used (Holmes, 1960). Typical examples of soil water storage determination by measurement of soil water content can be found in papers by Butler and Prescott (1955), Slatyer (1955, 1962a), Army and Ostle (1957) and Specht (1957). Details of the actual soil water determinations can be found in Chapter 3.

One of the major advantages of the soil water balance approach is that it demonstrates, quite vividly at times, the relative rates at which extraction is taking place from different soil zones. This information is of particular value in understanding root distribution, problems of competition between species, and reasons for species persistence or failure under extreme environmental conditions.

However, when lysimetry can be used, it can provide a more accurate means of evaporation measurement. Lysimeters are usually most effective for agronomic studies, particularly in irrigation areas, where large well-watered areas of homogeneous vegetation are available. Lysimeters vary in size from small laboratory size pots to large containers of several metres diameter inserted into agriculture fields. Typical of the larger installations are those described by Harrold and Dreibelbis (1951), Morris (1959), Pruitt and Angus (1960), McIlroy and Sumner (1961) and Rose, Byrne and Begg (1966).

The performance of any lysimeter depends on its size, construction, and location. The two main requirements for successful performance are to minimize disturbances resulting from the lysimeter itself, and to ensure the representativeness of the local conditions being sampled. Thus the contents of the lysimeter must behave in as similar a manner as possible to that in their natural unenclosed state, and this behaviour must be typical of the situation being investigated. This depends not only on establishment of a suitable crop cover in the lysimeter itself, but also on the characteristics and treatment of the surrounding area.

B. Energy Balance Methods

From Eq. (2.2) and the associated discussion, it will be apparent that if the components of the surface energy balance other than that of the latent heat term, 1E, can be measured. 1E can be obtained from the equation by difference.

In practice $\mathbf{R}_n$ can be measured with net radiometers (Gier and Dunkle, 1951; Fritschen and van Wijk, 1959; Funk, 1959) which consist essentially of a horizontally exposed flat plate, blackened on both sides and either force-ventilated or screened with transparent material to eliminate the variable ventilation effect of natural air movement. The temperature difference between the upper and lower faces (measured by embedded thermocouples or resistance thermometers), is then directly proportional to the difference in their received radiation. At heights small enough to neglect differential absorption and emission in the intervening air layer, this is equal to the net radiation received by the underlying surface.

Soil heat flux, G, can also be measured directly using soil heat flux plates (Deacon, 1950; Monteith, 1958; Philip, 1961). These are like net radiometer elements in construction, but are buried flat in the ground near the surface, and the temperature differences across their faces are taken as a measure of the heat flow perpendicular to them.

Alternatively G can be found from changes in the temperature profile of the soil (or water) combined with measurements or estimates of its heat capacity. A typical procedure employs standard soil thermometers for the profile measurement, and laboratory tests on soil samples for thermal capacity. Where continuous measurements are desired, electrical recording thermometers can be used (de Vries, 1958; van Wijk, 1963). In many cases, as mentioned previously, G can be neglected, particularly if the measurements are medium or long term in nature.

The residual terms in the equation then represent the sensible and latent heat fluxes, H and 1E. These fluxes are usually separated by evaluating the Bowen ratio, β, for the surface being studied, in terms of the corresponding vertical gradients (or differences) of temperature and vapour pressure. This involves the assumptions, stated previously, that the transfer coefficients for water vapour and sensible heat are identical and that H and 1E are constant throughout the height range used for measurement (see Eq. 2.5).

Both assumptions appear reasonable provided the measurements are made as close to the surface as possible, to minimize the effects of buoyancy on the diffusivity coefficients, and of advection on the constancy of H and 1E. This technique has been effectively used on various plant communities under a range of climatic conditions (see for example, Tanner, 1960; Lemon, 1963; Begg *et al.*, 1964 and Denmead 1964). It can be modified using an expression developed by

I. C. McIlroy which incorporates a slowly varying, temperature-dependent, weighting function. With this procedure only wet and dry bulb temperature measurements are required and the calculations are considerably simplified (see Slatyer and McIlroy, 1961).

For many purposes the energy balance approach provides a most attractive means of evaporation measurement although it is restricted to situations where the vegetation is reasonably homogeneous so that effective averaging is achieved by the measurements. This tends to preclude it from use with spatially heterogeneous communities, but King, Tanner and Suomi (1956) have described a rotating boom sampler, and Gates (1962) and Begg *et al.* (1964) horizontal line sampling procedures which assist in giving effective spatial averaging. Advection, when present, as is frequently the case, does not invalidate this approach but accentuates the importance of correctly sited instruments with measurements sufficiently close to the surface.

C. Vapour Flow Measurements

The most direct means of determining evaporation is the actual measurement of the vertical vapour flux over the surface being studied. Under natural conditions this requires the integration of the instantaneous product of vertical air flow and water vapour content at a point and, although the theory and instrumentation for this purpose has been outlined for some time (Swinbank, 1951), both these quantities fluctuate rapidly and markedly, and delicate instrumentation is required. In consequence few measurements of this type have, as yet, been reported (Dyer, 1961).

Alternatively the vertical vapour flux can be determined indirectly, and with less accuracy, from the corresponding gradient of humidity, measured as described previously, and an appropriate transfer coefficient, obtained either from empirical considerations (as in the Dalton equation) or inferred from turbulence theory (as in the aerodynamic method).

The Dalton method utilizes a relationship which considers the overall transfer process from surface to bulk air, and assumes that the corresponding transfer coefficient is a function of mean wind speed ($\bar{u}$) alone, thus

$$\mathbf{E} = f(\bar{u})(e_0 - e_a) \qquad (2.15)$$

where $\mathbf{E}$ is evaporation rate in units depending on the dimensions of $f(\bar{u})$, and $(e_0 - e_a)$ is the vapour pressure difference between the effective surface and the bulk air. Penman (1948) has modified the expression, using wind speed (miles day $^{-1}$) at a height of 2 m, to obtain

$$E = 0.4\ (1+0.17u)(e_0-e_a) \tag{2.16}$$

Expressions such as these have given satisfactory estimates of evaporation over free water surfaces (Harbeck *et al.*, 1958; Webb, 1960) once a set of empirical constants have been fitted. However, when applied to surfaces other than water, different constants are required and the assumption that e_0 is the saturation vapour pressure at the surface temperature is no longer always valid, as has been shown earlier in this chapter.

The aerodynamic, or eddy profile, approach is similar to the Dalton method but is based on a more complete understanding of the physical principles involved. It involves using the water vapour flux relationship of Eq. (2.10) written in the form

$$E = -\rho_a \mathbf{K}_w \overline{(\partial q/\partial z)} \tag{2.17}$$

where the $f(\bar{u})$ of Eq. (2.15) is replaced by the product of air density ρ_a and the eddy transfer coefficient $\mathbf{K}_w$, and (e_0-e_a) by the average gradient of specific humidity $\overline{(\partial q/\partial z)}$. Advection effects are not considered, thereby placing a limitation of application of the method.

The vapour pressure gradient can again be measured with wet and dry bulb thermometers but, because the differences are generally smaller, more sensitive elements are required, and platinum resistance thermometers, thermistors or thermocouples are frequently employed (see Slatyer and McIlroy, 1961). Generally, $\mathbf{K}_w$ is assumed identical with $\mathbf{K}_m$, the eddy transfer coefficient for momentum which in turn is, in principle, determinable from measurements of change in wind speed with height (see Eqns 2.3 and 2.11). Thornthwaite and Holzman (1939) assumed a logarithmic form for the wind speed profile and derived an expression for evaporation involving wind speed and humidity determinations at two heights (requiring, in addition to the above equipment, accurate and sensitive anemometers). This expression, subsequently modified by Pasquill (1949) can be written in the Dalton form

$$E = \alpha u_2(e_1-e_2) \tag{2.18}$$

where

$$\alpha = \frac{\mathbf{k}^2 M_w}{RT}\ (1-u_1/u_2)/\ln\ (z_2/z_1)^2$$

where e_1, e_2 and u_1, u_2 are the vapour pressure and wind speed measured at heights z_1 and z_2, $\mathbf{k}$ is von Karman's constant ($\cong 0.40$), R is the gas constant, T the air temperature (deg K) and M_w the molecular weight of water.

Although α can be fairly constant for any one surface, $\mathbf{K}_m$, and hence $\mathbf{K}_w$, increases with convective activity and decreases when there are conditions of temperature inversion. Also, the height of the vegetation, when more than a few inches, requires adjustment to account for zero plane displacement. This can be particularly important in a growing crop (Rider, 1954). In addition to these factors, the simple theory breaks down under non-neutral conditions, where the logarithmic law of wind speed with height no longer applies (Deacon, 1949) and the expression for $\mathbf{K}_m$ becomes more complicated. Because of these and other factors, the method is, in many respects, only suitable for short term use at the present time even though it has been extensively developed from the instrumental standpoint and successfully used by House, Rider and Tugwell (1960) and Rider (1960).

A final approach warranting description, which can also be regarded as a vapour flow method, is the determination of changes in water vapour content of a measured air-stream passing through a large enclosure mounted over the section of plant community to be studied. The measurement of air flow rate and of changes in humidity present little difficulty and the method has been used by several workers (Musgrave and Moss, 1961; Decker, Gaylor and Cole, 1962). However it must be realized that evaporation inside the enclosure may bear little relationship to that outside unless adequate precautions are taken to ensure that the environmental conditions inside adequately simulate natural conditions. It is probable that this can at best, only be approximated, since the problems of simulating, for example, wind structure are considerable. It is apparent that, even if rate of energy supply is reproduced, itself a difficult problem, differences in r_a and surface temperatures due to the altered wind structure will provide a different partitioning of the incident energy, and hence different evaporation rates. In many respects it seems that the method can only be used for evaluating relative evaporation rates, and then with possible error, since evaporation measured when the same experimental conditions are imposed over dry and wet surfaces will not necessarily be in the same ratio as their natural evaporation rates (see also Chapter 8).

In consequence, while this procedure can be used successfully for studies of the evaporation process under controlled environmental conditions, its use for measurement of natural evaporation should be undertaken with caution. Its most useful applications would appear to be the approximate evaluation of relative evaporation rates from different plant communities and perhaps the assessment of natural evaporation rates from droughted vegetation where internal plant and soil resistances to vapour diffusion dominate the pattern of evaporation and marked changes in r_a, for example, would not be expected to influence unduly the observed results.

D. Formulae Used for Calculation of Evapotranspiration

A number of more or less empirical procedures have been developed over the years to provide estimates of evaporation from open water surfaces. Among the better known are those of Thornthwaite (1948,. 1954), Penman (1948, 1949, 1956), Blaney and Criddle (1950), and Prescott (1958). Some of them have also been applied to plant community surfaces and to bare soils. Most have been developed with the objective of obtaining from commonly measured meteorological elements (such as bulk air temperature and humidity, wind speed, cloudiness, length of daylight) estimates of evapotranspiration which can be used to calculate the water balance of various plant communities and to predict irrigation schedules, drought frequency, stream flow, etc. Accounts of most of them have been given by Dzerdzeevskii (1958) and Penman (1963). The following brief account describes only the Penman method which is the most soundly based and has had the most successful application to biological and hydrological problems.

The method combines an aerodynamic and energy balance approach and, although more complicated than most other evaporation formulae, still requires only standard meteorological data. Essentially intended to give an expression for free water evaporation the final generalized equation is:

$$E = [\Delta R_n + \gamma E_a]/[\Delta + \gamma] \qquad (2.19)$$

where Δ is the slope of the saturation vapour pressure curve (de^0/dT) at the bulk air temperature T_a (mm Hg deg^{-1}), R_n is the net radiation for the surface under investigation, γ is the psychrometric constant (in the same units as Δ) and $E_a = \{f(u)[e^0(T_a) - e_a]\}$ with $(e^0(T_a) - e_a)$ being the saturation deficit at the bulk air at temperature $T_a[e^0(T_a)$ and e_a are the saturation vapour pressure and actual vapour pressure, respectively, at the air temperature, $T_a]$. The form of $f(u)$ has been changed from time to time, but for free water Penman (1956) has used $f(u) = 0 \cdot 35 \ (0 \cdot 5 + u/100)$ when u is wind run in miles day^{-1}. When R_n is not directly measured, Penman (1948) has developed a procedure for calculating it on the basis of latitude (solar radiation outside the atmosphere), probable surface albedo, cloudiness, air temperature and humidity.

The potential evapotranspiration from an extensive area of a dense, physiologically active, short green crop, well supplied with water, (E_t), and in the absence of advection, was originally related to the free water evaporation (E_w) by an expression of the form (Penman, 1956)

$$Et = f\mathbf{E}_w \tag{2.20}$$

where f ranges from 0·6–0·8 depending on length of day and season. Other refinements have been incorporated to allow for availability of soil water (Penman, 1949, 1956) and the expression has proved extremely useful for a wide range of studies (see for example, van Bavel and Verlinden, 1956; Slatyer, 1960b).

Although the method breaks down in the presence of advection, it was not intended for use in such situations. It has been compared with other techniques, generally to its advantage, by Tanner and Pelton (1960), Fitzgerald and Rickard (1960), Stanhill (1961) and others.

Less empirical relationships than Eq. (2.20) have now been developed, based on the crop resistance considerations discussed earlier, and these give a more valid means of estimating evapotranspiration (e.g. Slatyer and McIlroy, 1961; Monteith, 1965).

Monteith (1965) first modified the basic Penman expression to give

$$1E = \frac{\Delta R_n + \rho_a c_p [e^0(T) - e]/r_a}{\Delta + \gamma} \tag{2.21}$$

replacing the $\mathbf{E}_a$ term of Eq. (2.19) with $\rho_a c_p [e^0(T) - e]/\gamma r_a$. This expression is still only valid for wet surfaces. For surfaces such as plants and soils, where the evaporating surfaces are located within the natural surfaces, an internal resistance component is included using the relationship

$$\frac{\gamma 1E}{\rho_a c_p} = \frac{[e^0(T_0) - e_0]}{r_n} = \frac{(e_0 - e_a)}{r_a} \tag{2.22}$$

[cf. Eqns (2.13) and (2.14)]. This leads to

$$[e^0(T_0) - e_a] = (1 + r_n/r_a)(e_0 - e_a) \tag{2.23}$$

and permits an expression analogous to Eq. (2.20) to be written for plant community evapotranspiration

$$1E_{(crop)} = \frac{\Delta R_n + \rho_a c_p [e^0(T) - e_a]/r_a}{\Delta + \gamma(1 + r_n/r_a)} \tag{2.24}$$

From Eqns (2.21) and (2.23), an equation analogous to Eq. (2.20) can now be written

$$E_{(crop)}/E^0 = \frac{\Delta + \gamma}{\Delta + \gamma(1 + r_n/r_a)} \tag{2.25}$$

where E^0 is the evaporation from the crop with all surfaces thoroughly wetted.

When field crops are adequately supplied with soil water, Monteith (1965) considers that the minimum ratio of r_n to r_a is probably 1–2 and the corresponding ratio of $E_{(crop)}/E^0$ is between 0·6–0·8, similar in general terms to the empirical values adopted by Penman (1956). For short grass, however, McIlroy and Angus (1964) were unable to detect an increase in evapotranspiration on spraying with water, suggesting that r_n/r_a was probably an order of magnitude lower, mainly due to a marked reduction in r_n.

At the present time it is difficult to assess the applicability of equations such as (2.24) and (2.25) because of the problems in adequately categorizing and predicting r_n from micrometeorological data. However, even if approximate values can be assigned to typical crop structure, soil water supply and growth rate situations, the expressions could be of considerable value.

CHAPTER 3

The State of Water in Soils

The water status of soil continually affects soil properties, in the long term through its influence on weathering and profile development, and in the short term through its influence on such factors as the soil's strength, friability, and permeability to water, solutes and gases.

From the viewpoint of this book, most interest in soil is centred in its function as a store of water for plant growth, but soil also provides mechanical support for plants, a reservoir of mineral nutrients and contains an active microbiological population, all of which intimately influence plant responses. This chapter deals primarily with those factors of special significance to water retention and availability and finally with methods of measuring soil water status. For additional information the reader is referred to Luthin (1957), Marshall (1959), Russell (1961) and Chow (1964).

1. Primary Soil Characteristics

Soil is a complex material consisting of four main fractions, the mineral solids and non-living organic matter which together comprise the soil matrix; and the soil solution and soil air which together occupy the voids, or pore spaces, within the matrix.

In temperate climates the primary components of the clay fraction are the clay minerals, but it can also contain appreciable quantities of hydrated sesqui-oxides of iron and aluminium; in wet tropical climates, these components can dominate its composition. The clay minerals are secondary hydrated aluminosilicates in which isomorphous substitutions of some ions may occur. The minerals are plate-like in structure and show marked basal cleavage. Their crystalline structure consists of sheets of oxygen tetrahedra. [Each tetrahedron consists of four oxygen atoms in a close packed arrangement around a silicon atom, giving sheets with the composition $(Si_2O_5)^n$]. These are bonded together, through common oxygen atoms with sheets of aluminium or magnesium octahedra as shown diagrammatically in Fig. 3.1 (see also, Russell, 1961; Thompson, 1957).

There are three major types of clay minerals, kaolinite, illite and montmorillonite. Kaolinite, which tends to predominate in mature,

weathered soils, consists of silica and alumna sheets in a 1 : 1 ratio.
This forms a non-expanding crystal lattice, so that soils containing
mainly kaolinite show little swelling or shrinking. By comparison, the
other two minerals tend to predominate in younger soils and are

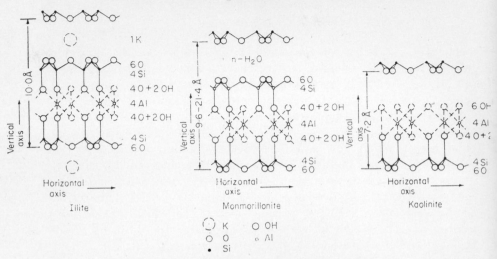

FIG. 3.1. Diagrammatic structure of the principal clay minerals. The vertical axis
dimensions are given in Angstroms. Substitution of Al and Si may occur in illite and
montmorillonite (see text) (after Thompson, 1957).

composed of silica and alumina sheets in a 2 : 1 ratio. In illite, potassium
ions occur between the adjacent silica plates and the chemical bond is
strong enough to prevent water molecules from expanding the units.
In montmorillonite, on the other hand, the chemical bond does not
exist and water molecules can expand the crystal lattice appreciably,
so that montmorillonitic soils show marked swelling and shrinkage on
hydration and dehydration.

In practice, soils usually contain all these minerals as well as others
such as vermiculite, which has partially expanded crystal lattice in a
2 : 1 ratio. Consequently most soils exhibit some swelling and shrinking
behaviour even though it is most extreme in those clay soils containing
montmorillonite as the main mineral. In such soils, cracks several
centimetres wide and a metre or more deep may develop during protracted dry periods.

The degree to which water molecules become oriented around the
clay surfaces and so influence the plastic, cohesive and swelling
properties of the clay fraction, depends on the electrical charge at the
clay surfaces, which in turn is influenced by the number of ion exchange
sites on the surfaces and the ions which are adsorbed on them.

In the $2:1$ layer minerals the negative charge appears to develop by ionic substitution, generally of Al^{+++} for Si^{++++} and of Mg^{++} or Fe^{++} for Al^{+++}. This destroys the balanced charge condition of the idealized structure and establishes cation (or base) exchange capacity. In kaolinite, isomorphous replacement of this type is uncommon and it seems probable that the negative charge which develops is due to dissociation of hydrogen ions from expanded SiOH groups at the edges of the clay crystals. Typical exchange capacities for these minerals, and also for humus, are given in Table 3.1. The high figure for humus serves as a reminder that the cation exchange capacity of a soil is influenced not only by the percentage and composition of the clay fraction but also by the organic matter content.

TABLE 3.I

Cation-exchange capacity of humus and clay minerals†

	Exchange capacity meq./100 g
Humus	200
Vermiculite	150
Montmorillonite	100
Illite	30
Kaolinite	10

† From Thompson (1957).

2. RETENTION OF WATER BY SOIL

At water contents of greatest biological significance, there are two main mechanisms by which water is retained in soil. They involve forces emanating from either the liquid-air or solid-liquid interfaces of soil-water systems and their relative importance depends largely on the degree to which the soil shrinks as water is removed from it.

The first mechanism, involving surface tension, develops when air enters the pore space of an initially saturated soil from which water is being withdrawn. In consequence an air-water interface develops and permeates the pore space, forming a series of curved interfaces between neighbouring particles of soil. If the soil does not shrink at all while the water is removed, the surface tension acting in such curved interfaces balances the force causing water extraction and constitutes the main mechanism of water retention. If, on the other hand, the soil shrinks progressively as water is removed so that no air enters the pore space, no internal air-water interface develops. Instead, the soil particles are brought into closer and closer mutual proximity. Since these particles, typically clay minerals or humus, have a similar surface charge they repel each other when suspended in water, leading to maximum hydration in the absence of water extraction forces. When they are drawn closer together by the shrinkage which accompanies water extraction, the repulsive force becomes greater until, at equilibrium, it balances the applied force. In practice, both surface tension and particle repulsion frequently act simultaneously since water extraction is generally accompanied by some shrinkage and some air entry.

Osmotically active solutes, generally salts, in the soil solution constitute a third mechanism for water retention. These solutes lower the relative vapour pressure of the soil water and hence influence its energy status, as described later. However they will not directly influence the amount of water retained against an applied pressure or suction unless the force is imposed across a solute impermeable membrane. Phenomena associated with the osmotic pressure of the soil solution will consequently not be treated at this point, even though they influence the availability of soil water to plants.

Other mechanisms for water retention are of limited significance to plant-soil relationships but may dominate the water retention characteristics of very dry soils.

When a dry soil is exposed to water vapour, for example, the exchangeable cations on the solid surfaces hydrate first, if they are capable of hydrating, and then the remainder of the surface hydrates by the formation of hydrogen bonds between the surface hydroxyl groups or oxygen atoms and the water molecules. The molecules in this initial layer are not perfectly oriented because of competition between the adsorbed ions and surface atoms for them, and their free energy is very low because of the intensity with which they are held. As other molecules come into the force field of those in the first layer they are bonded to them, generally with partially covalent hydrogen bonds, and the second layer molecules, in turn, form partially covalent bonds with their neighbours, including third layer molecules.

As additional layers accumulate, the degree of bonding and the associated structural rigidity decreases and the free energy of the water molecules increases. However, it is only when the film thickness exceeds several molecular layers that the water potential approaches a value significant to most biological processes. At this stage liquid phase continuity generally exists between many of the solid particle surfaces and the water retention characteristics of the soil can, in most cases, be adequately accounted for by the mechanisms described previously.

Further discussion of these points is given by Childs (1957) and Low (1961).

The two main water retention mechanisms will now be discussed in somewhat more detail. For the interested reader much more information is given by Baver (1956), Childs (1957) and Marshall (1959).

A. Retention by Liquid-air Interface Forces

It is first appropriate to evaluate the relationship between the suction or negative pressure applied to the water in soil and the head of water, h, in equilibrium with it. A model of this situation can be set up by inserting a porous cup, containing water (and connected to a water manometer), into the soil and permitting water to flow into or out of the soil until equilibrium is reached between the soil water potential, Ψ, and the difference in head of the manometer. In the absence of osmotic effects (the cup is assumed to be permeable to solutes), the pressure, P, (dyne cm^{-2}) at any point in the water is then given by

$$P = P^0 - \rho_w g h \qquad (3.1)$$

where ρ_w is the density of water (g cm^{-3}), g is the acceleration due to gravity (cm sec^{-2}) and h (cm) is the vertical distance of the measurement point from a reference level at which the pressure is P^0. If, as is generally the case, the reference level is pure free water at atmospheric pressure, P^0 is taken as 0 so that if P is the pressure in excess of atmospheric pressure,

$$-P = \rho_w g h = \tau \qquad (3.2)$$

where τ is termed the matric suction (Marshall, 1959) or matric pressure and is equivalent to the reduction in the soil water potential caused by the applied suction.

The matric suction can also be related to the curvature of the air-water interfaces which exist within the unsaturated soil. If it is assumed that an interface is accommodated in a capillary of circular cross-section with radius r, it would be hemispherical in form, have the same radius, and would support a soil water suction according to the relationship:

$$\tau = -P = 2\sigma/r = \rho_w g h \qquad (3.3)$$

(cf. Eqns 1.8 and 1.9) where σ is the surface tension of water (dyne cm^{-1}), r is in cm and the air in the soil is at atmospheric pressure. Owing to the irregular shape of the pore space in soils, the curvature of the interface in the various pores, channels and around points of particle contact is far from simple, and is certainly not identical from point to point. However, at equilibrium, the curvature supports the same suction at all points, since the water system is continuous. Accordingly the value of r obtained by the above expression must be regarded only as an "effective radius". Even so, the expression is of considerable value in interpreting pore size distribution from suction/water content data.

This can be appreciated by the following example. If Eq. (3.1) is re-written in the form

$$r = 2\sigma/\rho_w gh \qquad (3.4)$$

r can be regarded as the upper limiting radius of pores which can remain full when a suction equivalent to h cm of water is applied to a non-shrinking soil. The amount of water removed when the suction is increased by a finite amount represents the volume occupied by pores, the radii of which correspond to the levels of suction applied. (Since values of σ and ρ_w can be obtained from physical tables and g is 980 cm sec $^{-2}$, r can be readily calculated. It is about $0 \cdot 15/h$ at normal temperatures). The pore size distribution in a non-swelling soil can therefore be determined directly from a water characteristic curve (relating water content to applied suction). As will be discussed in Chapter 4, the size distribution of the pores has particular significance to water and air movement, since permeability is closely related to the effective cross-sectional area of pores full of water at the prevailing suction.

An important limitation to the use of suction data arises because of hysteresis phenomena, the water content of the soil differing at any given suction, depending on whether it is being wetted or dried (see Fig. 3.2). Hysteresis phenomena have been widely studied by workers concerned with the physical properties of porous media (see for example, Everett and Stone, 1958; Paulovassilis, 1962; Philip, 1964b). Generally the effects are considered to arise because the pore space consists of "caverns" interconnected by "necks" consisting of narrower pores. As a wet soil dries the water-filled "caverns" tend to remain full until the suction rises to a value which causes the "necks" to drain, according to Eq. (3.4). However, on re-wetting a dry soil, each cavern tends to remain empty until the suction has fallen to that appropriate to the widest dimension of the cavern. Consequently, the water content at any one suction tends to be less when the soil is wetting than when it is drying.

B. Retention by Solid-liquid Interface Forces

As already mentioned briefly, in soils which shrink while water is being removed, so that no air enters the system, water is retained by forces emanating from the solid surfaces rather than by surface tension.

These forces arise because the clay and humus particles dissociate loosely bound ions and behave as negatively charged particles or micelles, which tend to attract cations and repel anions. The ions in

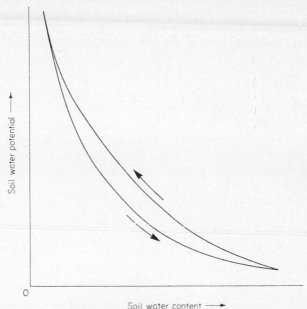

FIG. 3.2. Typical form of the water characteristic curve for a non-swelling soil showing a hysteresis loop. The arrows indicate whether the soil is being progressively wetted or dried.

the bulk solution tend to be segregated by these surface forces in a way that would result in all the cations being located on the micelles and all the anions remotely distant. However they are also influenced by kinetic diffusion activity which tends to distribute them evenly. The final result is intermediate, leading to the development of the so-called diffuse double layer, and depends on the valency and concentration of the ions already in solution and the charge density on the micelle surface (Childs, 1957; Low, 1961). In the case of non-electrolyte molecules, diffusion alone operates leading to a uniform distribution.

When two micelles are in close proximity the intervening space becomes too restricted to accommodate the layers developed by each and the total concentration of dissolved ions (and molecules) is greater at the mid-point between the micelles than in the bulk solution. In consequence the osmotic pressure in this region increases and, since the exchangeable ions are not free to migrate and so eliminate the concentration gradient, water tends to flow into the layer between the micelles along the local water potential gradient. If a confining pressure or restraining suction is not applied to balance the pressure so developed, the clay plates are pushed apart and the system swells. It is easy to see that the swelling pressure can be directly equated with suction. Much more information than provided in this necessarily simplified account is given by Childs (1957), Low (1961) and Babcock (1963).

Although it might be assumed that hysteresis would not be evident in systems of this type, similar phenomena have been observed (Holmes, 1955; Marshall, 1959). In such cases, Childs (1957) has suggested that the clay plates are not as neatly oriented in parallel crystal plates as theory proposes, so that increased suction tends not only to draw them closer together but also to improve their orientation, thus reducing porosity more than expected. He points out that reduction of suction would not be expected to cause reversible de-orientation so that the "hysteresis" observed, in such cases, may involve a degree of irreversibility.

In most real soils water retention involves both the main processes described here, even though their relative importance may change as the soil dries. Taking a typical situation of a well-structured soil of intermediate texture containing about 20% clay, for example, early stages of water removal may, to a considerable degree, be balanced by air entry, but as suction increases shrinkage will probably become more important. Finally the stage will be reached when further contraction of the space between clay plates is almost impossible and again surface tension effects may dominate the retention pattern. Regardless of the mechanism, water content remains a function of the applied pressure. Direct osmotic contributions to water retention will be measured if the pressure is applied across a solute impermeable membrane. In moist soils, the osmotic contribution is normally small, but it can become of over-riding importance in dry soils or in saline or alkaline habitats.

The relationship between water content and water retention, termed the water characteristic curve, varies widely between different soils, the water content at any given water potential generally increasing as clay content increases. Typical examples of such curves are given in Fig. 3.3.

3. CLASSIFICATION OF SOIL WATER

As the study of soil physics has progressed, the methods of classifying soil water have been based on more and more fundamental grounds which, while leading to a greater understanding of the characteristics they describe have, in many respects, provided terms which have been unsuitable for field use. In this section some of the traditional terms will

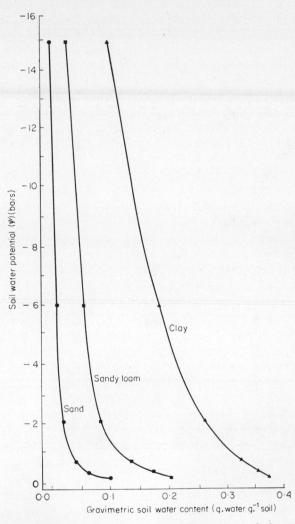

Fig. 3.3. Typical water characteristic curves for sand, sandy-loam and clay soils (after Slatyer and McIlroy, 1961).

be discussed as well as some of those recently recommended by the International Society of Soil Science (Aslyng, 1963).

A. Some Descriptive Terms and Expressions

1. Field capacity

As water enters a dry soil and drains through it after irrigation or precipitation, the horizons through which it moves are saturated until all free water disappears from the soil surface. Rate of penetration of the wetting front then commences to decline, imperceptibly at first as water from the saturated upper soil is redistributed in lower, still dry, soil. However, reduction in water content as the transmitting zone becomes a draining zone is associated with a rapid decrease in hydraulic conductivity (see Chapter 4). Finally, the continued advance of the wetting front becomes very slow. This type of situation, commonly observed, is depicted in Fig. 3.4 (after Staple and Lehane, 1954) and shows the changing form of the water content profile. When the rate of

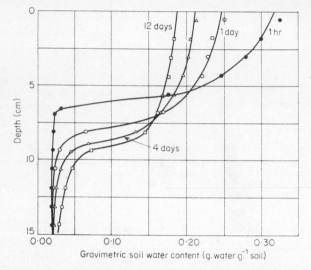

Fig. 3.4. Water content profiles of an initially dry sandy loam soil, at various times following irrigation. Evaporation from the soil surface was prevented. The decline in rate of flow and the re-distribution of water content, with increasing time, are well illustrated. The profile depicted after about 24 hr drainage is frequently referred to as "field capacity".

drainage has become much reduced, the water content is referred to as *field capacity* and has become a widely used soil "constant" indicating the maximum amount of soil water storage.

Field capacity may be determined simply by ponding water on the soil surface and permitting it to drain for 1–3 days (depending on soil type) with surface evaporation prevented. Soil samples are then collected by auger, or soil water content measured by one of the techniques discussed later in this chapter, and results expressed on a gravimetric or, preferably, volumetric basis. Field capacity frequently appears to be a quite reproduceable value when determined in this manner, as long as precautions are taken to avoid sampling in the transition zone at the end of the wetting front (Colman, 1944). As might be expected, it is a more significant and better delineated value in coarse than fine textured soils because the larger pores soon empty and the subsequent rapid decline in hydraulic conductivity tends to cause a sharp transition from wet to dry soil.

In general, however, field capacity varies too widely to be regarded as a reliable "constant". This is not only because the actual water content changes continuously with time, but is also because the upper drainage zone is drying out, while the lower zone is wetting up, so that these two parts of the profile are on different sides of the hysteresis loop. Initial soil water content can also influence the results. Even so, the measurement can be of value for field studies because it categorizes the soil water storage value following soil water recharge. For this purpose it is desirable that adequate water be applied to wet the soil to depths well below the deepest measurement point so that the part of the profile under study is on the drying curve.

Many attempts have been made to determine field capacity in the laboratory by simulating the soil water potential profiles, and values, which develop with free drainage in the field. For "undisturbed" samples, that is when the natural structure has been preserved as far as possible, the 100 cm water suction ($\tau = 0 \cdot 1$ bar) value has often been used for this purpose (Marshall, 1959). Richards (1954) has recommended the $0 \cdot 3$ bar value for use with dried, ground and sieved samples. The "moisture equivalent" has also been used for this purpose. This is the mean water content of a soil sample $1 \cdot 6$ cm thick (prepared as for the $0 \cdot 3$ bar determinations) draining to a water table at its base in a centrifuge having a centripetal acceleration of 1000 g.

It is evident that, while the suction data are of considerable value in themselves there is no simple relationship between such determinations and the determination of field capacity, including as it does complications due to textural differentiation, impeded drainage zones, etc. There is much less satisfactory agreement with the moisture equivalent data which does not even represent a single reproduceable suction value. Richards and Weaver (1944) found that the effective value of the moisture equivalent ranged from about $\tau = 0 \cdot 3$ bar in coarse textured soils to $\tau = 0 \cdot 5$ bar in clays.

2. Permanent wilting percentage

If field capacity has been widely used to refer to the upper limit of soil water storage for plant growth, the permanent wilting percentage has been just as widely used to refer to the lower limit. It is the soil water content at which plants remain permanently wilted (assuming that the leaves exhibit visible wilting) unless water is added to the soil.

Briggs and Shantz (1912a) first emphasized the importance of this soil water value and termed it the "wilting coefficient". They determined it by growing seedlings in small containers under conditions of adequate water supply until several leaves were developed. The soil surface was then sealed and the plants were permitted to deplete the soil water until wilting occurred. The containers were then placed in a humid atmosphere and, if recovery occurred overnight, were again exposed to normal conditions on the following day. This procedure was repeated until no recovery was observed, at which point the soil water content was determined.

Briggs and Shantz (1912a, 1912b) conducted a large number of measurements on a wide variety of plants and found little variation in the soil water content at which wilting occurred. Subsequently, a number of workers (see for example, Richards and Wadleigh, 1952) have determined that the soil water potential at wilting approximates $\Psi = -10$ to $\Psi = -20$ bars, with a mean value of about $\Psi = -15$ bars. In consequence the "15-bar percentage" has become identified with the permanent wilting percentage and is frequently used as a guide to it. It is usually determined with soil samples in pressure membrane equipment (see later).

Furr and Reeve (1945) modified the definition of the permanent wilting percentage to introduce a "first" permanent wilting percentage when the first (usually the lower) leaves wilted and an "ultimate" permanent wilting percentage when the entire plant was wilted. The "first" permanent wilting percentage corresponds generally with the Briggs and Shantz determination whereas the "ultimate" value may not occur until the soil water potential is much lower (Slatyer, 1957a).

Although Briggs and Shantz were careful to note that the permanent wilting percentage represented the lower limit of soil water for plant *growth*, and that water extraction may continue at lower water contents, Veihmeyer and Hendrickson (1927, 1949, 1950) and Hendrickson and Veihmeyer (1929, 1945) considered that it also marked the lower limit for water absorption. Their experiments, in the main, were conducted with deep rooted orchard crops and they considered that, at the permanent wilting percentage, measurable reductions in soil water were no longer observed in the bulk of the root zone.

These observations may have been complicated by the lack of knowledge about deeper roots and water table accessibility, and are at variance with those of most workers who have observed continued soil water depletion to much lower soil water contents (Batchelor and Reed, 1923; Taylor and Furr, 1937; Slatyer, 1957b, 1962c). There seems no physical reason why some extraction should not occur, even after the death of the plant, although it can be expected to be much reduced by more or less permanent stomatal closure (see Chapter 8). Furthermore, the appearance of visible wilting in a field crop may be due to a transient inability of the water supply system to the plant to meet the evaporative demand, rather than to conditions associated with permanent wilting. Philip (1957a) has shown how the appearance of wilting in the field depends on meteorological conditions, root density and volume of soil explored as well as the osmotic conditions in the plant.

Although this last point will be dealt with in detail in Chapter 7, it is appropriate to mention here that Slatyer (1957a) has strongly criticized the concept of the permanent wilting percentage as a soil constant. He has pointed out that, basically, wilting occurs because of loss of turgor in the leaves and that the point of zero turgor pressure associated with wilting is dependent on the osmotic characteristics of the leaf tissue sap. In consequence, wilting occurs when there is a dynamic balance between the water potential in the soil, the water potential in the plant leaves, and the osmotic pressure of the leaf cell sap (see Chapter 9). Since the latter value ranges from 5–200 bars, the (negative) values of soil water potential at which wilting occurs can be expected to vary over a similar range. Although, for most crop plants, the range will be much narrower, corresponding approximately to the values quoted earlier ($\Psi = -10$ to $\Psi = -20$ bars), it is still important to remember that the soil water potential at which wilting occurs is triggered by the osmotic characteristics of the plant leaves.

While the foregoing remarks emphasize the underlying weaknesses in the view that the permanent wilting percentage is a soil constant, two factors help to justify its use for practical purposes. The first of these is that, for most crop plants, the range of leaf osmotic pressures is only of the order of 10–20 bars, rather than the much wider range characteristics of all plant species. The second is that, because of the shape of the soil water potential/water content [$\Psi(\theta)$] curve (see Fig. 3.3), marked reductions in potential, below about $\Psi = -5$ bars, may be associated with only small changes in water content. Because of these phenomena, the permanent wilting percentage, or the "15-bar percentage" can be regarded in general terms as a soil "constant" for many practical purposes.

3. *Available soil water*

This expression, generally referring to the availability of soil water for plant growth, is taken as the amount of water retained in a soil between field capacity and the permanent wilting percentage. If field capacity can be used to represent a field index for the upper limit of soil water availability and the permanent wilting percentage a lower limit, this range has considerable practical significance in assessments of the agricultural value of soils. The water characteristic curve, however, yields much more information, since the amount of water in the soil at any one value of Ψ'_{soil} can be obtained and the pattern of water release with falling Ψ'_{soil} revealed (see Fig. 3.3).

In general, finer textured soils release water more gradually than those of coarse texture, because of a lower proportion of large pores. In sand, as an extreme case, virtually all the water drains at Ψ'_{soil} values close to zero because of the predominance of large pores. It is of interest to note, from Fig. 3.3, that the soil of intermediate texture may retain more water between values of $\Psi'_{soil} = -0.3$ bar to $\Psi'_{soil} = -15$ bars than the fine textured soil. This indicates that the "available soil water" range of such soils can be greater than in clays, due largely to the more effective pore size distribution.

A too literal interpretation of available soil water data should be avoided, for several reasons. In any one soil, increased rooting depth in the profile as a whole can compensate for a narrow available water range in each horizon. Conversely, impeded root distribution combined with a wide water range may lead to a soil with considerable hazards for plant growth. Also it should be remembered that the range of water required for transpiration and survival is substantially greater, in many soils, than that available for growth alone. In addition, within the range of available water, the degree of availability tends to decline as soil water status and Ψ'_{soil} decline (Richards and Wadleigh, 1952; Vaadia, Raney and Hagan, 1961). However this statement itself requires qualification and a detailed discussion on the general subject of soil water availability is given in Chapters 7 and 9. For the present it need only be noted that the range of soil water between the field capacity and the permanent wilting percentage constitutes an important field soil parameter when used with caution.

B. Physically Based Terms and Expressions

While the foregoing discussion of water retention by soils, in terms of equivalent pressures, has been adequate to provide a general understanding of this important aspect of soil water relations, pressures markedly different from atmospheric seldom exist in soils under natural conditions, except where gravitational heads exist in saturated soils or overburden pressures develop. When air enters the soil, the bulk pressure is generally close to atmospheric and, although the water status of the soil can be measured or changed by the application of pressure differences, it can be quite effectively described using a thermodynamic approach which permits all the forces known to be operative in the soil water system to be separately accounted for.

Soil physicists have, for many years (Buckingham, 1907), used thermodynamic treatments in basic soil water studies and these have become more and more comprehensive and applicable as time has progressed (Edlefsen, 1941; Edlefsen and Anderson, 1943; Childs, 1957; Philip, 1957a, 1964c; Bolt and Frissel, 1960; Babcock, 1963). For the plant physiologist the thermodynamic treatment is of value since it leads to a better understanding of liquid and vapour movement in soils, of the supply of water to the root surfaces, and to considerations of soil water availability for plant function.

The following account is necessarily brief and somewhat simplified. For further detail the reader is again referred to Childs (1957) and Marshall (1959) and also to Hallaire (1955), Philip (1957a, 1964c), Bolt and Frissel (1960), Gardner (1960a), Slatyer and Taylor (1960), Taylor and Slatyer (1961, 1962) and Babcock (1963).

In Chapter 1, it was shown that the capacity for water to do work, when influenced only by hydrostatic pressure, solutes and temperature, was conveniently described by the partial molal Gibbs free energy of water, $\overline{G}_w$, which can be written in differential form as (cf. Eq. 1.33):

$$\mathrm{d}\overline{G}_w = \mathrm{d}\mu_w = \left(\frac{\partial\mu_w}{\partial T}\right)_{P,\,n_s} \mathrm{d}T + \left(\frac{\partial\mu_w}{\partial P}\right)_{T,\,n_s} \mathrm{d}P + \left(\frac{\partial\mu_w}{\partial n_s}\right)_{T,\,P,\,n_j} \mathrm{d}n_s \quad (3.6)$$

where $\overline{G}_w = \mu_w$, the chemical potential of water; n_s and n_j are the numbers of moles of solute of those chemical species whole mole number changes, and does not change, respectively, during differentiation; P is the bulk pressure and T the Kelvin temperature. Since $\left(\dfrac{\partial\mu_w}{\partial T}\right) = -\overline{S}_w$, the partial molal entropy of water [cf. Eq. (1.35)]; $\left(\dfrac{\partial\mu_w}{\partial P}\right) = \overline{V}_w$, the partial molal volume of water [cf. Eq. (1.36)], and $\left(\dfrac{\partial\mu_w}{\partial n_s}\right)\mathrm{d}n_s = \left(\dfrac{\partial\mu_w}{\partial N_s}\right)\mathrm{d}N_s$, Eq. (3.6) can be written, at constant T and P

$$\mathrm{d}\mu_w = -\overline{S}_w\mathrm{d}T + \overline{V}_w\mathrm{d}P + \Sigma\left(\frac{\partial\mu_w}{\partial N_s}\right)_{n_j} \mathrm{d}N_s \quad (3.7)$$

where the summation is over those chemical species whose mole number changes.

In soil systems, however, $\overline{G}_w$ may be influenced not only by pressure, temperature, and concentration, but also by the forces arising from the soil matrix, including the surface tension of the water at the liquid-air interfaces, and, the surface forces retaining water on the charged surfaces. $\overline{G}_w$ is also influenced by external force fields such as gravity. In grouping these forces, the problem then becomes one of selecting sets of variables which can include contributing factors without omissions or duplications and, preferably, in a manner that permits their identification with such well-known terms as "suction" and with established or possible measurement techniques.

1. Non-swelling soils

A number of authors, including most of those listed above, have derived expressions of this type, (see also Day, 1947; Low, 1951, 1955; Babcock and Overstreet, 1955, 1957a, 1957b; Bolt and Miller, 1958 and Bolt and Frissel, 1960). In most cases the main problem arises in the development of a water content term which satisfactorily accounts for the surface tension and adsorption components of soil water in a way that is related to the measurement techniques used to evaluate them. Taking, in the first place, a non-swelling soil, neglecting external force fields, and following Babcock's (1963) general approach, Eq. (3.7) can be expanded to separate the specific effects of water content from those of solute concentration (associated with osmotic forces). For any change at constant volume of soil particles

$$\mathrm{d}\mu_w = -\bar{S}_w\mathrm{d}T + \overline{V}_w\mathrm{d}P + \left(\frac{\partial\mu_w}{\partial c_s}\right)_{n_j,\,P,\,T}\mathrm{d}c_s + \left(\frac{\partial\mu_w}{\partial\theta}\right)_{n_s,\,P,\,T}\mathrm{d}\theta \qquad (3.8)$$

where c_s is osmotic solute concentration (mole cm^{-3}) and θ is the volumetric water content (cm^3 water cm^{-3} soil). Again the summation of the solute molalities applies only to those solutes whose mole number changes. Also $\partial\mu_w/\partial\theta$ is defined for the change in which the mole number (n_s), but not necessarily the composition variable, of all other species is fixed (Babcock, 1963).

In isothermal situations the first term on the right hand side of Eq. (3·8) disappears. The second term describes the effect of bulk (external) pressure. This is an important term where real pressure differences exist between the point in the soil under study and the atmosphere, but, in unsaturated soils it can frequently be neglected.

The third term represents the osmotic potential component and can be related to osmotic pressure. If an equlibirium dialysate of the soil solution is considered, for example, the effect of changing solute concentration on the chemical potential of water can be balanced against applied pressure across an osmometer membrane permeable to water but not to solutes. At any one concentration there will be zero net volume flux when the applied pressure on the solution side (P^{eq}) raises the chemical potential $[\mu_w(c_s)]$ to that of the pure free water side (μ_w^0). Thus

$$\int_{\mu_w(c_s)}^{\mu_w^0} d\mu_w = \int_{P^0}^{P^{eq}} \bar{V}_w dP \qquad (3.9)$$

$$[\mu_w(c_s) - \mu_w^0] = \bar{V}_w(P^0 - P^{eq}) = -\bar{V}_w\pi \qquad (3.10)$$

[cf. Eqns (1.66) and (1.67)]. This term is generally small in non-saline soils with a high proportion of available water, and can be neglected in some cases. However, at values approximating $\Psi = -15$ bars, it is generally a significant component of the water potential. In saline soils it is frequently the principal component.

The last term on the right hand side, which describes the contribution of the sorptive forces associated with the solid-liquid and liquid-air interfaces of the soil matrix, is the matric potential component. It can be evaluated in a similar manner to the previous term, as an equivalent matric pressure.

In this case the effect of changing water content on the chemical potential of the soil water can be balanced against applied pressure across a membrane permeable to water and solutes but not to the soil matrix, as in a pressure membrane apparatus (Richards, 1954). In such a system there is zero volume flow when the applied pressure on the soil side (P^{eq}) raises the chemical potential of the soil water $[\mu_w'(\theta)]$ to that on the free solution side $(\mu_w^{0'})$. In this case, and dealing with only one side of the hysteresis loop

$$\int_{\mu_w'(\theta)}^{\mu_w^{0'}} d\mu_w = \int_{P^0}^{P^{eq}} \bar{V}_w dP \qquad (3.11)$$

$$[\mu_w'(\theta) - \mu_w^{0'}] = [\mu_w(\theta) - \mu_w^0] = \bar{V}_w(P^0 - P^{eq}) = \bar{V}_w\tau \qquad (3.12)$$

where the superscript $(')$ denotes that the chemical potential of the water on both sides of the membrane includes the osmotic component, $\bar{V}_w\pi$, but, in the estimate of the difference $(\mu'_w(\theta) - \mu_w^{0'})$ this effect disappears. The symbol τ denotes the matric pressure, synonymous with the matric suction of Marshall (1959).

Using Eqns (3.9)–(3.12), Eq. (3.8) can now be written, for practical purposes, in the form:

$$d\mu_w = \bar{V}_w dP - \bar{S}_w dT - \bar{V}_w d\tau - \bar{V}_w d\pi \qquad (3.13)$$

If the temperature is allowed to remain constant at the reference temperature and the system allowed to change from the state of pure free water at atmospheric pressure to some definite condition in the soil system where the total pressure on the soil water is different, then Eq. (3.13) can be applied to both the initial and final states and the difference written as

$$\Delta\mu_w = (\mu_w - \mu_w^0) = \overline{V}_w\Delta P - \overline{V}_w\Delta\tau - \overline{V}_w\Delta\pi \qquad (3.14)$$

Since ΔP, $\Delta\tau$ and $\Delta\pi$ are measured with respect to pure free water, Eq. (3.14) simplifies to

$$\frac{\Delta\mu_w}{\overline{V}_w} = \Psi = P - \tau - \pi \qquad (3.15)$$

where Ψ is the water potential as defined previously [Eq. (1.47)]. An alternative notation (Taylor and Slatyer, 1962) recognizing that the pressure, matric and solute terms are component potentials, is to rewrite Eq. (3.15) as

$$\Psi = \Psi_{soil} = \Psi_p + \Psi_m + \Psi_s \qquad (3.16)$$

where Ψ_p is the pressure potential, Ψ_m the matric potential and Ψ_s the solute (osmotic) potential.

$\Delta\mu_w/\overline{V}_w$, or Ψ, therefore represents the amount of work required to transfer a unit of water, reversibly and isothermally, from a pool of pure free water to the location in the soil system which is under consideration. Thus it represents the capacity of water in the system to do work as compared with pure free water. In consequence it is usually negative (except in saturated soil under positive pressure), and even though the negative sign is sometimes neglected for convenience, it will be retained throughout this book.

Both $\Delta\mu_w$ and $\Delta\mu_w/\overline{V}_w$ have been referred to as the water potential by different workers and the symbol Ψ used in both cases. In this book $\Psi = \Delta\mu_w/\overline{V}_w$ is used, with appropriate subscripts such as Ψ_{soil}, Ψ_{leaf}, so that the units of expression will be those of energy/volume, numerically equivalent to pressure units. Expression in units such as bars has the advantage that much of the existing literature is already in units of bars or atm. In the absence of gravitational effects or external pressure differences, the water potential, written in this form, is equivalent to the negative of "total suction" or "total soil moisture stress" (TSMS) (Richards and Wadleigh, 1952), which is

$$TSMS = \text{Total Suction} = \tau + \pi \qquad (3.17)$$

2. Swelling soils

In swelling soils, an additional complication arises because volume change may result from a change in the geometrical arrangement of the soil particles, as well as from water entry. Using a procedure similar to that of Bolt and Frissel (1960), Babcock (1963) introduced an unspecified term into the standard Eq. (3.8) to allow for these effects as follows:

$$d\mu_w = -\bar{S}_w dT + \bar{V}_w dP + \left(\frac{\partial \mu_w}{\partial c_s}\right)_{T, P, n_j, n_w, \chi} dc_s + \left(\frac{\partial \mu_w}{\partial \theta}\right)_{T, P, n_s, \chi} d\theta$$

$$+ \left(\frac{\partial \mu_w}{\partial \chi}\right)_{T, P, n_s, n_w} d\chi \qquad (3.18)$$

where χ is the unspecified geometric term. Since its coefficient is defined at constant T, P and the numbers of moles of water and solutes, its measurement depends upon finding a means of determining the effect of a change in particle arrangement on the chemical potential, under these conditions. Babcock (1963) suggested that if χ is expressed in volume units, the coefficient can be expressed as

$$\left(\frac{\partial \mu_w}{\partial \chi}\right)_{T, P, n_s, n_w} d\chi = \left(\frac{\partial \mu_w}{\partial V}\right)_{T, P, n_s, n_w} dV \qquad (3.19)$$

so that it could be evaluated by determining the effect of mechanical compression on the equilibrium vapour pressure. While this procedure seems satisfactory, Eq. (3.18) suggests that expression of the water content on a volume basis, as in the present treatment, may in itself adequately account for the effects of swelling, a possibility already raised by Collis-George (1961).

3. External force fields

From time to time in the foregoing discussion, reference has been made to external force fields. In soil water systems, gravity is probably the most important. It can be included as part of the water potential or regarded as an additional component potential. Common thermodynamic practice (Glasstone, 1947; Guggenheim, 1959) has been to restrict use of the $\mu_w = \bar{G}_w$ identity to systems in which pressure, temperature and composition are the only variables of concern, even though some writers have also included the effects of external force fields in μ_w. For the present, to conform with normal practice in soil physics, external force fields will be treated separately so that, in the presence of a gravitational field, at reference temperature,

$$\frac{\Delta \bar{G}_w}{\bar{V}_w} = \Phi = \Psi + \Psi_z \qquad (3.20)$$

where Φ is the total potential and Ψ_z is the gravitational potential, $(= M_w g h / \bar{V}_w)$ with M_w in g mole $^{-1}$.

In saturated soils, Ψ is generally negligible and $\Phi = \Psi_z$, for all practical purposes. However, as soon as the soil becomes unsaturated the relative importance of Ψ_z declines rapidly (see Chapter 4).

4. *Glossary of terms*

In the foregoing sections of this chapter, a number of closely related terms have been introduced. In order to draw the discussion together and to identify equivalent terms the following glossary has been prepared, based on a recent report of the International Soil Science Committee (Aslyng, 1963).

(*a*) *Water content* This term is used to express the mass (or volume) of water per unit mass (or volume) of dry soil. "Dry" normally implies that the soil has been dried to constant weight in a standard drying oven at 105°C. Water content is frequently expressed as a percentage value. Volumetric water content, expressed as cm^3 water/cm^3 soil, is given the symbol θ.

(*b*) *Total potential of soil water*, Φ Water in the soil is subject to several force fields originating from the presence of the soil solids, dissolved solutes, gas pressure and the gravitational field. These effects may be quantitatively expressed in thermodynamic terms by assigning potentials to the soil water. The sum of these potentials is called the *total potential* of the soil water and may be identified with the partial molal (or specific) Gibbs free energy of the soil water relative to pure free water at the same temperature. It therefore represents the capability of the soil water to do work compared with pure free water at the same temperature and is the amount of work that must be done per unit quantity of water in order to transport reversibly and iso-thermally an infinitesimal quantity of water from a pool of pure free water at specified elevation to the soil water at the point under consideration. It is generally expressed in units of energy per unit mass, volume, or mole.

(*c*) *Water potential*, Ψ This is identical with the total potential, as just defined, in the absence of differences in elevation, that is without inclusion of the gravitational potential term. Therefore, in comparison with the previous definition, it is the amount of work required to transport unit quantity of water from a pool of pure free water to a point in the soil water system at the same elevation.

(*d*) *Solute (osmotic) potential*, Ψ'_s The amount of work that must be done per unit quantity of pure water in order to transport reversibly and isothermally an infinitesimal quantity of water from a pool of pure water at a specified elevation at atmospheric pressure, to a pool containing a solution identical in composition with the soil water (at the point under consideration) but in all other respects identical to the reference pool. It is equal to the negative of the osmotic pressure ($\Psi'_s = -\pi$) where both are expressed in the same units.

(*e*) *Gravitational potential*, Ψ'_z The amount of work that must be done per unit quantity of pure water in order to transport reversibly and isothermally an infinitesimal quantity of water from a pool containing a solution identical in composition to the soil water at a specified elevation at atmospheric pressure, to a similar pool at the elevation of the point under consideration.

(*f*) *Matric (capillary) potential*, Ψ'_m The amount of work that must by done per unit quantity of pure water in order to transport reversible and isothermally an infinitesimal quantity of water from a pool containing a solution identical in composition to the soil water at the elevation and the external gas pressure of the point under consideration, to the soil water. It is equal to the negative of the matric pressure ($\Psi'_m = -\tau$) where both are expressed in the same units.

(*g*) *Pressure potential, potential due to external pressure*, Ψ'_p The amount of work that must be done per unit quantity of pure water in order to transport reversibly and isothermally an infinitesimal quantity of water from a pool of pure water at a specified elevation at atmospheric pressure, to a pool of water, identical in all respects except that it is under the same external pressure as the soil water. It is equal to the total pressure ($\Psi'_p = P$) where both are expressed in the same units.

(*h*) *Total suction* The negative pressure, relative to the external gas pressure on the soil water, to which a pool of pure water must be subjected in order to be in equilibrium with the soil water through a semi-permeable membrane. Total suction is thus equal to the sum of matric (or soil water) suction and osmotic suction. Total suction may also be derived from the measurement of the partial pressure of the water vapour in equilibrium with the soil water. It should be noted that this quantity may be identified with the negative of the water potential defined above when external pressure potentials can be neglected. It is expressed in pressure units such as dyne cm^{-2}, bars or atm. It is identical with *total soil moisture stress* (TSMS).

The following table, (3.II), of equivalent values is provided for conversation purposes. It should be noted that erg cm^{-3} is not identical with dyne cm^{-2} since the density of water is not exactly 1·000 g cm^{-3} at temperatures other than 4°C.

TABLE 3.II

Units for the expression of soil water potential and its components or equivalents

bars	erg g^{-1}	joule kg^{-1}	atm	cm of water
1·0	1·0 × 10^6	1·0 × 10^2	·0987	1·017 × 10^3
1·0 × 10^{-6}	1·0	1·0 × 10^{-4}	·0987 × 10^{-6}	1·017 × 10^{-3}
1·0 × 10^{-2}	1·0 × 10^4	1·0	0·987 × 10^{-2}	10·17
1·013	1·013 × 10^6	1·013 × 10^2	1·0	1·030 × 10^3
9·833 × 10^{-4}	9·833 × 10^{-2}	9·833 × 10^{-2}	9·703 × 10^{-4}	1·0

4. MEASUREMENT OF SOIL WATER

The following discussion is designed to give the reader an introduction to some of the techniques available for soil water measurement. Emphasis is placed on the principles of a few important techniques and on their more important limitations. Additional information may be found in specific references quoted in the text and in Marshall (1959), Slatyer and McIlroy (1961), and Taylor, Evans and Kemper (1961).

A. Measurements of Soil Water Content

Although the techniques referred to earlier for measuring evaporation can sometimes be used to give estimates of changes of soil water content, attention here is confined more to point measurements in both undisturbed (or relatively undisturbed) soils and on samples removed from soils. However. when such determinations are used to provide an estimate of the total change in soil water storage for water balance purposes it must be remembered that considerable spatial and vertical variability occurs in soil water distribution.

This variability is due in part to different patterns of root distribution and different rates of evapotranspiration from point to point, but also to differences in soil physical characteristics which influence the form of the soil water characteristic curve. At the same water potential, the water content in samples a few centimetres apart can vary widely. Because of this phenomenon replicated sets of samples are required for valid mean estimates of bulk soil water content. Aitchison, Butler and Gurr (1951) found that in a typical loam soil more than 10 replicate samples were required to show that differences of 1·0% soil water content (g water/100 g dry soil) between two occasions were significant at $p = 0·05$. To show that differences of 0·5% were significant at this level, more than 40 samples were required. Staple and Lehane (1962) demonstrated a similar order of variability in clay loam soils which were not regarded as being unduly variable.

In consequence, while direct determinations of the water content of extracted soil samples still provide an absolute method of soil water content determinations, field workers have tended more and more to use indirect methods which permit a number of sensing instruments or access points to be located in study areas so that repeated measurements can be made at the same point.

1. *Direct gravimetric or volumetric measurements*

These measurements are made on soil samples of known weight or volume, when the water content is expressed on the basis of g water/g dry soil or g water/cm³ soil. The volume samples are usually collected in a special container, using a tube sampler, or from the side of an exposed soil profile, see for example Coile (1936), Lutz (1944). Alternatively the bulk density of the soil can be determined separately (Vomocil, 1954).

The water is removed from the freshly sampled soil by oven drying at 105°C to constant weight or by other procedures which, if not removing as much water as oven drying, rapidly remove a more or less constant proportion so that an approximate estimate of water content can be made. Among the techniques used for this purpose are allowing the water to react with calcium carbide to form acetylene gas, or mixing the soil with a known amount of methyl alcohol, and determining the change in specific gravity of the alcohol with a hydrometer.

2. *Indirect determination by neutron dispersion*

This widely-used technique depends on hydrogen atoms having a much greater ability to slow down and scatter fast neutrons than most other atoms, so that counting slow neutrons in the vicinity of a source of fast neutrons provides a means of estimating hydrogen content. Because of this phenomenon and because the only significant source of hydrogen in most soils is the soil water, the technique offers a convenient means of estimating soil water content. In soils with high root density, or high levels of organic residues, the amount of organic hydrogen may affect the estimates, but it is generally small compared with that in the soil water.

The method has been described in detail, by among others, Belcher *et al.* (1950), Stone *et al.* (1955) and Holmes (1956). Stone *et al.* (1955) and Holmes and Turner (1958) have described details of portable instruments which include probe and counting units, the former containing a radium-beryllium fast neutron source and a boron-enriched counter tube for detecting slow neutrons. In use, the probe is lowered into a polythene lined access tube, and counting rates determined at the desired depths. Other instruments and their operational procedures are essentially similar.

The count rates, adjusted for background and referred to counts in tanks of pure water are then calibrated against direct volumetric determinations of soil water content. Theoretical calibrations have also been developed (Holmes and Jenkinson, 1959).

The method has several important advantages over most other techniques, including the absence of a lag period while the soil water equilibrates with a sensing instrument, and the fact that a large volume of soil ($\sim$20 cm radius) is monitored, thus smoothing out local variability. However it does involve the disturbing influence of an access tube and the large volume sample prevents samples being taken at a point, or near the surface unless special precautions are taken (Holmes, 1960). Also the results are influenced by other sources of hydrogen atoms, such as organic matter, and by other elements, notably chlorine, iron and boron. It is these factors and different swelling and shrinking patterns which, in the main, prevent a single universal calibration curve for all soils.

3. *Other methods*

Other methods include the use of techniques which really, directly or indirectly, measure matric potential, (such as tensiometers, and porous conductivity blocks) which can be calibrated in terms of soil water content.

Some of these methods are based on the relationship between soil penetrability and water content (Allyn, 1942), between gamma ray absorption and water content (Gurr, 1962), between the electrical capacity across two plates placed in the soil and water content (Anderson and Edlefsen, 1942; Anderson, 1943) and between thermal diffusivity and water content (Shaw and Baver, 1939; de Vries and Peck, 1958).

The gamma ray technique was initially developed for density determinations, (Vomocil, 1954; van Bavel *et al.*, 1957) but has subsequently been used for soil water content. The method is not specific for soil water (as is the neutron method) but gamma rays can be collimated, by suitable geometry and shielding, to a narrow beam which gives resolution at the point at which measurements are taken. Density changes must also be accounted for, so that field applications are still questionable, but the method may find a number of applications as a laboratory technique (Gurr, 1962).

B. Measurements of Soil Water Potential

Most methods for Ψ measurement are best suited to laboratory use and determinations are generally based on field measurements of water content or matric potential and a calibration of Ψ against the quantity measured.

However, a technique reported by Richards (1965) may lead to an effective field method. If a constant heat supply is provided to two porous surfaces, identical in all respects except that one contains water in equilibrium with water potential of the air and the other is situated in a dry atmosphere, each will rise to different equilibrium temperatures. In the case of the moist surface this equilibrium value will be maintained until the water is distilled off, when the temperature will rise to the same value as the dry surface. By plotting the temperature difference between the surfaces against time, a curve with a plateau of certain height will be obtained. A series of such curves can be constructed using a range of humidity treatments of known water potential and the calibration used to estimate the water potential of soil air, by equilibrating the moist surface at the required point in the soil.

Richards has used two glass bead thermistors as the moist and dry surfaces. These are built into a sensing probe so that one is exposed to the soil air and absorbs water in the porous glass bead of the thermistor tip, the other is located in a standard dry atmosphere, such as a dry chamber containing silica gel. As yet the technique has not been thoroughly tested but it appears to have definite possibilities.

In the laboratory the main technique of Ψ measurement is based on measuring the relative vapour pressure e/e^0 of the soil water (see Eq. 1.46).

Two main techniques for relative vapour pressure determinations have been developed, both involving the equilibration of soil samples in small containers in carefully controlled ($\pm 0.001\,^\circ$C) constant temperature baths, and the determination of the relative vapour pressure in the air space after vapour equilibration is complete. The two types of psychrometer in most common use were proposed by Spanner (1951) and by Richards and Ogata (1958).

The Spanner instrument utilizes the Peltier effect (passing a current in an appropriate direction to cause cooling) to induce dew formation on a very small thermocouple. On breaking the circuit the junction becomes a delicate wet bulb element and, while the dew evaporates, a minute e.m.f. is generated proportional to the wet bulb depression. This output, measured with a sensitive galvanometer or recording microvoltmeter, can be calibrated against solutions of known vapour pressure/concentration characteristics and values of Ψ obtained from the calibration curve. This technique has been developed by several workers including Monteith and Owen (1958) and Korven and Taylor (1959). It appears to be very suitable for determinations in the relative humidity range from about 95% to saturation corresponding to values from about $\Psi = -75$ to $\Psi = 0$ bars at normal temperatures. Determinations at values lower than $\Psi = -75$ bars are limited by the efficiency of Peltier cooling.

The Richards and Ogata instrument is based on the temperature difference between dry and wet junctions and so is more orthodox. The thermocouples are made from chrome-p/constantan wire, soft soldered to a small silver ring containing a tiny water droplet which forms the wet bulb.

As distinct from the Peltier effect instrument, a true vapour pressure equilibrium is not normally reached. Instead, readings are made when the rate of evaporation from the water droplet (and hence the vapour flux to the soil) reaches a steady value and the temperature depression on the droplet is constant. This stage is usually reached in less than two hours and appears to be maintained for some time, even though it can be appreciated that changes in the surface area and size of the drop will affect droplet temperature.

As with the other instrument, calibration is achieved by the use of aqueous solutions of known concentration and vapour pressure. This procedure appears to be satisfactory, although Rawlins (1964) has found problems in its application, when plant material rather than soil is being examined. As explained in more detail in Chapter 6, this is attributed to the influence, on total vapour transfer, of the diffusive resistance in that part of the vapour pathway within the leaf. This resistance is high and variable, and its characteristics can dominate the net rate of evaporation from the water droplet. The greater mass of water in soil samples, and the absence of a boundary such as the leaf cuticle, probably reduce this effect in soils, but it is important that the applicability of the calibration procedure shall be carefully checked before the instrument is used for measurement purposes.

Apart from these problems, the instrument has the relative advantage over the Peltier effect instrument in that it is sensitive over a wider range of e/e^0 values and hence of soil water content.

C. Measurements of Matric Potential (or Matric Pressure)

1. Tensiometers

Direct field measurements of Ψ_m or τ can be made only with tensiometers. They are direct reading instruments which consist of a water-filled porous ceramic cup which is buried in the soil at the desired depth of measurement and connected by water-filled tubing to a manometer or pressure gauge on which the pressure deficiency or suction is indicated. It is the best measuring instrument in wet soils when suctions are smaller than about 0·8 bar. Air tends to enter the cup at higher suctions (lower potentials) and the instrument breaks down.

Although this range of suction is of considerable interest to the soil physicist and is of value in irrigation studies, in most ecological and agricultural situations lower potentials are of equal interest and the instruments, in consequence, cover only a limited part of the appropriate soil water range. This limitation and the need to re-charge the instruments following air entry are their biggest drawbacks.

Details of tensiometer design and operation have been given by Richards (1949, 1954). Because they are direct reading no calibration is necessary to obtain pressure measurements, but they must be calibrated against soil water content if required for that purpose. In regular use they appear to be subject to two main sources of error, both of which can be eliminated without undue difficulty. The first of these is the diurnal fluctuation in readings which is sometimes observed, even when matric potential is maintained constant. This appears to be caused by heat conduction along the water-filled tubes and can be minimized if plastic components, rather than metal, are used (Haise and Kelley, 1950). The second influence results from preferential root growth at the surface of the cups where water is more readily available than in the bulk soil. Care should be taken to check on this point and, if necessary, periodically change the location of the instruments.

2. Porous conductivity units

The most common indirect field method for matric potential measurement is the porous conductivity block. This method, first developed by Bouyoucos and Mick (1940) involves burying in the soil a small block containing a pair of electrodes surrounded by a porous matrix, and running the lead wires to a resistance bridge. The water in the block reaches matric potential equilibrium with that in the soil and the resistance measured at the bridge gives the electrical conductivity of the solution within the electrical influence of the electrodes.

The porous matrix usually consists of gypsum, or regular plaster of paris, sometimes impregnated with water-proofing resins to reduce solubility (Bouyoucos, 1954) or a fabric sandwich of nylon (Bouyoucos, 1952, 1954) or fibreglass (Colman and Hendrix 1949).

Gypsum blocks have a sensitive range from about $\tau = 0.5$–15.0 bars, although Bouyoucos (1954) claims to have improved the wet-end sensitivity to some degree. By comparison, the fabric units are usually sensitive down to less than 0.1 bar. However the solubility of the gypsum exerts a buffering effect which avoids the otherwise considerable sensitivity to solute effects and consequent variability experienced with the fabric units.

Hysteresis effects are common to all porous block techniques but are probably more pronounced with the finer pore structure of the gypsum units, and restrict the method to drying cycle studies, except that the progress of the wetting front can be easily and accurately followed by observing the sudden reductions in block resistance as the wetting front reaches the depth of block placement.

3. *Pressure membrane and pressure plate techniques*

Laboratory measurements of matric potential are almost invariably made with the pressure membrane and pressure (or suction) plate equipment developed primarily by Richards (1949, 1954). In these references considerable detail as to procedure is given and only a brief account is given here. With both the pressure and suction equipment a sample of soil, previously wetted, is placed upon a membrane which is permeable to water and solutes but impermeable to the soil matrix. A pressure difference is then imposed across the membrane by imposing a suction beneath it, or enclosing the membrane system in a cell and increasing the internal gas pressure. When water outflow from the sample ceases and equilibrium is reached between the matric potential and the imposed pressure difference, the sample is removed and the water content determined.

Determinations are usually made with small soil samples which can be contained in plastic or metal cylinders about 5 cm diameter and 1.5 cm high. The samples may be natural aggregates, but, if not, are usually ground and passed through a 2 mm sieve to remove rock fragments. Elrick and Tanner (1955) have shown that the sieving procedure can overestimate the water retention at any one suction by up to 30% in the $\tau < 1$ bar range. At pressure in the $\tau > 1$ bar range the reverse tendency was noted but the underestimation did not exceed

10%. Although this latter error can be tolerated for some purposes, the influence of sample pre-treatment should always be borne in mind.

In an effort to avoid contact changes between sample and membrane as the sample dries, a phenomenon sometimes noticed with swelling soils, Richards and Richards (1962) have introduced a new radial cell in which the sample takes a doughnut shape around a cylindrical membrane. As shrinkage proceeds this arrangement tends to maintain, or even improve, membrane-sample contact. As yet this instrument has not been widely used but it appears to have considerable potential.

D. Measurements of Solute (Osmotic) Potential

Measurements of solute potential, Ψ_s, or osmotic pressure, π, of the soil solution can only be made after the solution is extracted from soil samples. Extraction usually involves making up the water content to a specified super-saturated level and then dialysing a "saturation extract" (Richards, 1954). Using the vapour pressure psychrometers just described for Ψ measurements, π can then be determined if the extract is used in place of soil samples. Alternatively it can be determined by the freezing point technique as used for determinations of π of tissue cell sap extracted from plants (see Chapter 5).

Electrical conductivity measurements have also been used for π measurements because of the close agreement which has been found between electrical conductivity and osmotic pressure of the saturation extract (Campbell et al., 1949, Richards, 1954).

Adjustment of the osmotic pressure of the saturation extract to that of field soil water content is normally accomplished by simple proportional correction. It must be recognized that this procedure, while satisfactory for some purposes, is crude since the degree of dissociation and the osmotic coefficients of the individual salts vary with their own concentration, and the concentration of the other salts represented, as well as with temperature. The actual osmotic pressure of the soil solution is dependent on all of these factors.

CHAPTER 4

Movement of Water in Soils

From the time water enters the soil as precipitation or irrigation until it is removed from the root zone by drainage, direct evaporation or extraction by plant roots, it is involved in a number of distinct processes, all of which are determined, to a considerable extent, by the rate of water movement. In consequence, it is of value to identify and discuss the more important of these processes, initially in terms of moisture re-distribution in the soil after the soil surface has been wetted, and then in terms of the subsequent movement of water out of the system. A number of accounts of soil water movement have been prepared in recent years. These include comprehensive works by Luthin (1957), Marshall (1959), and more specific contributions by Hallaire (1955), Philip (1957a, 1964c), and Gardner (1960a, 1960b, 1965).

1. WATER ENTRY INTO SOIL

A. Terminology

Many studies have been made of the downward entry of water into soil, a process commonly referred to as infiltration. Because various terms have been introduced from time to time to describe aspects of infiltration, it is of value briefly to summarize them.

The term *infiltration capacity* has been used for a number of years by hydrologists to describe "the maximum rate at which a given soil, in a given condition, can absorb rain as it falls" (Horton, 1940). The definition refers to rate rather than capacity and *infiltration rate* has been brought into common use among soil physicists in a sense synonymous with infiltration capacity. Another term, *infiltration velocity*, has been introduced by a Soil Science Society of America Terminology Committee (Richards, 1952) to designate the instantaneous local rate of infiltration. Thus the infiltration rate can be used in the limited technical sense to refer to the infiltration velocity at any instant when a thin layer of water is maintained on the soil surface over a wide area. This difference can be illustrated by considering the case where infiltration rates are being measured with small cylindrical infiltrometers. In this situation it is sometimes not possible, even with the use of guard rings, to control divergent flow in the soil below the infiltrometer cylinder and in such cases infiltration velocities will exceed infiltration rates.

Other words in common use in connection with infiltration phenomena are percolation and permeability. Percolation is qualitatively used in the same context as infiltration, but with emphasis towards the downward movement of water in saturated (or very nearly saturated) soils. *Percolation rate* is therefore synonymous with infiltration rate with the qualitative proviso of saturated or near saturated conditions. *Permeability*, on the other hand, refers to the soil rather than to the soil water. Qualitatively, it is the characteristic of the soil relating to the readiness with which it transmits water. Quantitatively, Richards' (1952) committee expressed it as the specific property denoting the rate at which the soil transmits water under standard conditions. According to this definition, equations used for expressing flow, which take into account the properties of the fluid, should give the same soil permeability value for all fluids which do not alter the medium.

In most cases however (see below), a permeability coefficient is introduced as a proportionality factor in a flow equation relating flow velocity to the driving force. For this coefficient, the term *hydraulic conductivity* has been introduced for water flow under saturated conditions, and *unsaturated conductivity* or *capillary conductivity* for unsaturated conditions (Richards, 1952). However, in recent years, *hydraulic conductivity* has also been used to describe unsaturated flow (Childs, 1957; Philip, 1964c). Ordinarily hydraulic conductivity refers to film and liquid flow but not vapour movement.

The term *intrinsic permeability* is used as a permeability factor independent of the fluid, and for soils of stable structure, but it must be recognized that any factors which tend to change the permeability of the soil matrix to water will influence this value and prevent its use unless they can be measured or evaluated separately. Finally, *drainage* describes the discharge of water from an area of soil by sheet or stream flow (surface run-off) or the removal of excess water from within the soil by downward flow through it (internal drainage). The word is also used to describe the means for effecting the removal of water, and it is applied to the water which is removed.

Further details covering these and other common terms can be found in the report by Richards (1952), which still provides a standard basis for most terms concerned with permeability. One notable exception is the more general use of hydraulic conductivity to include saturated and unsaturated water movement and this custom is followed herein.

Other terms, used to describe more specific relationships, will be introduced during the discussion.

B. The Process of Infiltration

The path of downward water movement in the soil following surface application has been described in detail by Bodman and Colman (1944) for a uniform profile and by Colman and Bodman (1945) for a non-uniform profile. They found that the wetted portion of a column of uniform soil, into which water was entering at the top and advancing downwards, appeared to comprise a transmitting zone through which water was transmitted, with little change in water content, from a saturated zone at the top to a wetting zone at the bottom, with a

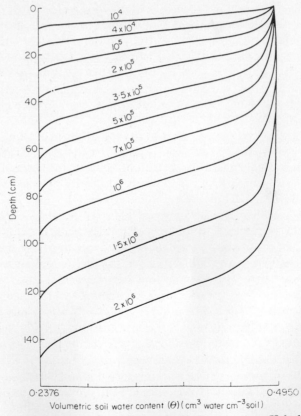

FIG. 4.1. Computed soil water profiles during infiltration into Yolo light clay soil The numbers on each profile give the time in seconds at which the profile is realized (after Philip, 1957d).

sharply defined wetting front at the lower end. Five zones in series were categorized as (1) a saturation zone, a zone presumed saturated which reached a maximum depth of 1·5 cm; (2) a transition zone, a region of rapid decrease of water content extending to a depth of about 5 cm from the surface; (3) the main transmission zone, a region in which only small changes in water content occurred; (4) a wetting zone, a region of fairly rapid change in water content; and (5) the wetting front, a region of very steep gradient in water content which represents the visible limit of water penetration.

A diagrammatic representation of this situation is given in Fig. 4.1 from a theoretical analysis by Philip (1957d) which satisfactorily accounts for all of these zones except the transition zone. This discrepancy results, presumably, from differential air entrapment which means that the water potential, Ψ, is not a unique function of water content, θ, in the small surface layer. It can be seen that the transmission zone is a continually lengthening unsaturated zone of fairly uniform water content and potential. Marshall (1959) considers that in this zone the matric potential Ψ_m is probably higher (closer to zero) than $-0\cdot025$ bars and that the degree of pore-space saturation is of the order of 80%. The gradient of Ψ is usually very small, once the water has penetrated to a reasonable depth (Taylor and Heuser, 1953) so that movement within the transmitting zone appears to be due primarily to gravity, which is the gravitational potential term, Ψ_z, of the total potential, Φ. However, if the water is ponded above a zone of low permeability, positive pressures may develop in the overlying water, as shown by Colman and Bodman (1945).

The infiltration rate decreases with time following application of surface water and the rate of decrease also decreases with time so that a fairly stable minimum value is frequently observed in infiltration experiments. This is well illustrated in the three curves of Fig. 4.2. For practical purposes, infiltration data are often presented in units such as cm hr $^{-1}$ (cm^3 cm $^{-2}$ hr $^{-1}$) but really represent the total amount of water entering in a given time, so the type of presentation in Fig. 4.2 provides much more information than a simple statement of the minimum rate.

Several equations relating infiltration to the parameters most commonly measured have been developed, for example, by Kostiakov (1932), Lewis (1937), and Horton (1940). Philip (1954; 1964c) compared the Lewis and Horton equations and devised a simplified equation which has the form

$$I = st^{1/2} + at \qquad (4.1)$$

where I is the cumulative infiltration (cm^3 cm $^{-2}$) at time t (sec) and the constants s and a have a physical meaning related to the diffusion analysis of infiltration, developed by Philip (1957d). The first term on the right hand side of Eq. (4.1) describes the contribution arising from capillarity, s being termed the "sorptivity" (a measure of the uptake of water by the soil, related to initial water content), while the second term represents the contribution arising from gravity.

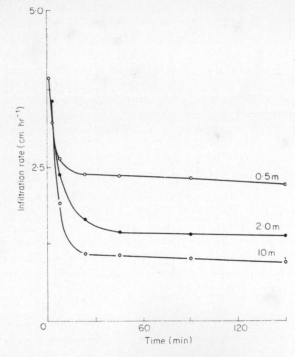

FIG. 4.2. Infiltration characteristics of a red earth soil at points 0·5 m, 2·0 m and 10 m from the trunk of an isolated tree (after Slatyer, 1962a).

The two constants can readily be evaluated experimentally. If Eq. (4.1) is written in differential form, it becomes

$$v = \frac{d\mathbf{I}}{dt} = \tfrac{1}{2}st^{-1/2} + a \tag{4.2}$$

where v is the velocity in units such as cm sec^{-1}. The values of $\mathbf{I}$ and t can then be plotted and the slope of the line and its displacement from the origin used to obtain s and a.

C. Factors Affecting Infiltration

Those soil and water conditions which affect permeability generally, and which are discussed in more detail in the next section, also affect infiltration. Two factors worth separate mention are initial water content and zones of low permeability.

Initial water content effects have been investigated by Tisdall (1951) and subsequently by Reinhart and Taylor (1954) and Ayres and Wikramanayake (1958), all observing marked reductions in infiltration rate as initial soil water content increased, and the latter authors finding a linear relationship between cumulative infiltration and the water storage capacity of the soil. These effects are to be expected because the potential gradient across the wetting front is reduced and swelling reduces the cross-sectional area available for flow.

Philip (1957c) has analyzed the situation further and shown that, as initial water content increases, the initial infiltration rate is reduced

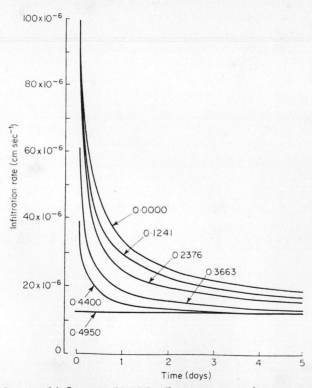

FIG. 4.3. Computed influences of initial soil water content, θ_0, on infiltration rate, I, for the soil used in Fig. 4.1. The numbers on each curve refer to values of θ_0. The curve for $\theta_0 = 0.4950$ is that for saturated soil (after Philip, 1957c).

considerably. These results are illustrated in Fig. 4.3 from which it can also be seen that as time increases, the effect becomes less pronounced and all curves approach the value for initially saturated soil. However during the period normally involved in infiltration experiments ($< 10^5$ sec), the computed rates for initially dry soil were more than twice as great as those for initially saturated soil, with the loam soil examined by Philip.

Zones of low permeability always exert profound effects on infiltration rate. These zones can develop as surface crusts (McIntyre, 1958), as clay pan, cultivator pan and hardpan zones (Fischbach and Duley, 1950), and can arise from dispersed clay structure under alkali conditions (Richards, 1954), compaction due to farm implements (Diebold, 1954) and other conditions. Frequently permeability characteristics change during infiltration, not only because of change in water content, but also because of re-orientation of surface particles, in-washing of finer material to soil cracks, etc. (McIntyre, 1958).

Partly because unimpeded soil water recharge is so necessary to successful agriculture, infiltration characteristics and associated soil structural features are of considerable practical importance. Protective mulches are frequently used to improve surface permeability (Fischbach and Duley, 1950) and organic matter is frequently incorporated into the surface soil (Pillsbury and Richards, 1954; Schiff, 1953). The data of Fig. 4.2, taken at different radial distances from the trunk of an isolated tree, give an impressive demonstration of the effect of organic matter and root activity on initial and final infiltration rates.

In some areas the primary purpose of all cultivation is to improve infiltration characteristics and some procedures appear to do so to considerable depths and for lengthy periods (Diebold, 1954; van Duin, 1955). Because infiltration characteristics can change with time and management practice, it is difficult to establish any simple relations between infiltration rate and soil conditions. However Musgrave (1955) proposed that four main groups of soils could be distinguished in terms of their expected minimum infiltration rates. These groups are presented in Table 4.I and indicate that a 10-fold range can be expected between, for example, deep sands on the one hand and sodic soils on the other.

D. Measurement of Infiltration Rate

Infiltration is commonly measured under field conditions by ponding water on the soil surface and measuring the rate of decline in the depth of standing water as a function of time. Suitable methods for this procedure, which require fairly large areas and spraying or other water-

TABLE 4.I

Minimum infiltration rate of various groups of soils†

Soils	Minimum infiltration rate (in hr^{-1})
Deep sands, deep loess, aggregated silts	0·45–0·3
Shallow loess, sandy loams	0·3 –0·15
Many clay loams, shallow sandy loams, soils low in organic matter, soils of high clay content	0·15–0·05
Soils of high swelling capacity, sodic soils	< 0·05

† from Musgrave (1955).

applying equipment are given by Richards (1954). Because the "edge" effect from lateral water movement increases as the area decreases and the area/perimeter ratio declines, large ponding measurements are frequently the most accurate.

However, the area or water supply may be limited and a variety of alternative devices have been developed (Parr and Bertrand, 1960). Of these, probably the most commonly used is the ring infiltrometer, which consists of a metal ring usually about 30 cm diameter which is inserted 2–3 cm into the soil. Inside the ring a pond of water is maintained at constant head by a self-syphoning burette leading from a calibrated storage tank.

Infiltration below this type of instrument generally wets a much larger volume of soil than that directly below the ring so that the apparent infiltration rates can be much higher than those measured without edge effects, unless low surface permeability dominates the infiltration characteristics (Slater, 1957). Since this is unusual, most measurements require a considerable degree of correction. A number of procedures for error correction are described by Marshall and Stirk (1950); the most common is that due to Kachinsky (1936) who attempted to calculate the fraction of the water which was found to penetrate directly below the ring.

Marshall and Stirk showed that this simple technique reduced the variability to an order commonly found for much larger plots. However, the correction is very crude since it does not permit a check on the patterns of internal water distribution at intermediate times during a set of measurements. It is of interest to mention here that the effect of depth of ponded water on the infiltration rate has been examined by Philip (1958a; 1958b) and found to be of minor influence. His results indicated that the initial infiltration rate could be increased by about

2% per cm of water depth (giving an error of only 10% with water as deep at 5 cm, an extreme figure). In addition he found that there was no effect on the final rate.

2. MOVEMENT OF WATER WITHIN SOILS

After the surface water disappears during infiltration, the water within the soil continues to move downward under the influence of the potential gradient which, as has been shown, is strongly influenced by the component potential due to gravity when the soil is near saturation. If there is a water table fairly close to the surface, the water will continue to drain through the soil until the potential at each point corresponds to the distance above the water table so that $\tau = \rho_w g h$

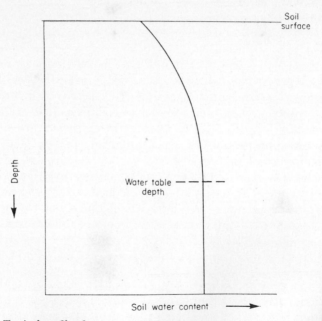

FIG. 4.4. Typical profile of water content in a homogeneous soil after drainage to water table.

as shown in Chapter 3. [If the water table is at considerable depth, the overburden pressure of the soil can cause positive pressures to develop in the water (Rim, 1954; Babcock and Overstreet, 1957a), but this is seldom evident in the root zone.] In this situation, and in the absence of water removal by roots or evaporation, the final profile of soil water in a homogeneous soil has a form resembling that of Fig. 4.4.

When the water table is at too great a distance to influence the form of the profile, the rate of downward movement decreases considerably and the situation regarded as field capacity develops, as described in Chapter 3. In this situation, depicted in Fig. 3.4 (p. 75), the water content above the wetting front may reach a value which is fairly constant with depth although less than at saturation. As also shown in Fig. 3.4, subsequent drainage in the absence of water removal from the soil will lead to further downward movement and hence to further change in the shape of the soil water profile. However, these changes are slow, and this is why field capacity has practical significance.

It is now proposed to discuss the factors which influence the subsequent movement of soil water, firstly with respect to the basic water flow relations of soils, and then in terms of thermal gradients, vapour flow, the movement of salts, capillary rise phenomena, and the movement of water to plant roots.

A. Saturated Flow

Soil water movement tends to occur along gradients of decreasing total water potential. In saturated soil the voids are filled with water so that vertical gradients of $d\Psi'/dz$ are effectively zero and since $\Phi = \Psi' + \Psi_z$ (see Eq. 3.20), the gradients of $d\Phi/dz$ are determined largely by gradients of gravitational potential, $d\Psi_z/dz$. The theory of water movement is based on Darcy's law, or a generalized form of it, which states that the quantity of water passing unit cross-section of soil in unit time is proportional to the gradient of hydraulic head. Written in appropriate terms for present purposes we replace the hydraulic head with the difference in total potential $\Delta\Phi$ and introduce a proportionality coefficient, K, to give

$$v = -K\Delta\Phi/\Delta z \qquad (4.3)$$

where v is the velocity of flow ($v = V/At$ where V is the volume, A the cross-sectional area and t the time). When v is in cm sec^{-1} and Φ is in units of equivalent cm of water (obtained from Φ expressed in units of erg cm^{-3} or dyne cm^{-2} by dividing by $\rho_w g$, see Chapter 3), the coefficient K is the *hydraulic conductivity*, defined previously, in cm sec^{-1}. However, because the value of K will vary not only with the properties of the soil but also with the properties of the water, an *intrinsic permeability* K', also defined previously, may be recognized which is a property of the soil alone. This is given by

$$v = -\frac{K'}{\eta} d\Phi/dz \qquad (4.4)$$

where η is the viscosity of the water (dyne sec cm^{-2}). K' is normally expressed in cm^2 and for this purpose Φ must be in units of erg cm^{-3} or dyne cm^{-2}.

There is a very wide range of permeability in different soils, K values ranging, for example, over several orders of magnitude as can be seen from Table 4.II taken from Smith and Browning (1946). As might be expected, permeability is primarily influenced by pore-size distribution, but its calculation from the geometry of the pore space alone is very difficult. However several alternative approaches have been developed using the Darcy and Poisseuille equations combined with experimental Ψ (θ) data (Wyllie and Rose, 1950; Childs and Collis-George, 1950; Marshall, 1958). These have been reasonably successful, particularly in fairly coarse textured soils, but direct measurements of permeability either in the field or on soil cores in the laboratory (Richards, 1954; Luthin, 1957; Marshall, 1959) are still the chief means of permeability determination.

TABLE 4.II

Permeability classes for saturated soils†

Class	Hydraulic conductivity K (in hr^{-1})	Comments
Extremely slow	$< 0\cdot001$	So nearly impervious that leaching process is insignificant.
Very slow	$0\cdot001–0\cdot01$	Poor drainage results in staining; too slow for artificial drainage.
Slow	$0\cdot01.0\cdot1$	Too slow for favourable air-water relations and for deep root development.
Moderate	$0\cdot1–1$	Adequate permeability.
Rapid	$1–10$	Excellent water holding relations as well as excellent permeability.
Very rapid	> 10	Associated with poor water holding conditions.

† from Smith and Browning (1946).

In most soils, K is not a constant but changes with time and management history, leading to changes in infiltration rates as well as in specific permeability characteristics.

Because the pore space distribution determines the permeability characteristics, any factor which affects pore space, and particularly the degree of swelling of the soil colloids, influences hydraulic conductivity. Such swelling is strongly influenced by the nature of the cations absorbed on the exchange surfaces, sodium in particular causing pronounced effects. Also, for the reasons given in Chapter 3 concerning the nature of the repulsive and attractive forces between surfaces, the extent of such swelling depends on the concentration of the electrolytes in the soil solution. Quirk and Schofield (1955) and Gardner *et al.* (1959) have shown that, as long as a relatively high concentration is maintained, permeability remains high but as concentration drops from, say, $1 \cdot 0N$ to $0 \cdot 01N$, the individual particles tend to become dispersed rather than aggregated and the hydraulic conductivity may decrease by 2–3 orders of magnitude.

Entrapped gases also affect the hydraulic conductivity, primarily by reducing the effective cross-sectional area available for flow (Christiansen, 1944; Orlob and Radhakrishna, 1958). The effect of temperature is usually by way of influence on the viscosity of water, but Pillsbury (1950) and Peck (1960) have shown that it can also influence the pressure-volume relations of entrapped air.

B. Unsaturated Flow

As air enters the soil pores the contribution of the gravitational component of the total potential rapidly becomes less important and the gradient of matric potential begins to dominate flow. The effect of pressure is generally negligible because of the continuous nature of the air space and, although the solute concentration can reach appreciable values, this does not affect liquid flow since the both solutes and water are moving.

Under unsaturated conditions Darcy's law is still applied but with some modifications and qualifications. Its applicability was assumed by Richards (1931) a number of years ago, and has since been supported by several workers including Low and Deming (1953) and Philip (1957e). It has been experimentally confirmed by Childs and Collis-George (1950), Youngs (1957), and Gardner and Mayhugh (1958); Childs and Collis-George showing that it held satisfactorily as long as K was not regarded as a constant but was assumed to be a function of the volumetric water content, θ. Theoretical validity of this assumption depends, in turn, on the reasonable assumption that the drag at the air-water interfaces in the soil is negligibly small (Philip, 1957e).

The form of the $K(\theta)$ function is now well established (Moore, 1939; Childs and Collis-George, 1950a) and a typical relationship is given in Fig. 4.5. K is found to decrease rapidly as θ decreases from saturation, primarily for the following reasons (Philip, 1957a, 1964c): (1) the total cross sectional area available for flow decreases with θ; (2) the largest pores are emptied first as θ decreases; since the contribution to K per unit area varies roughly with the apparent pore radius, K can be expected to decrease much more rapidly than θ; (3) as θ decreases, the

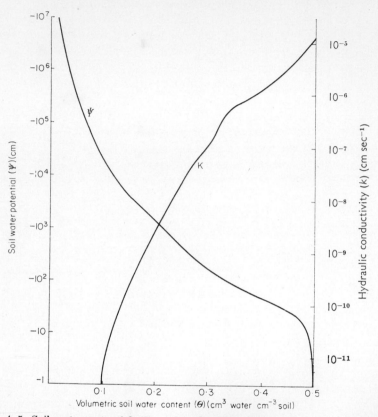

FIG. 4.5. Soil water potential, Ψ, and hydraulic conductivity, K, plotted against volumetric water content, θ, for Yolo light clay (based on experimental observations and theoretical estimates) (after Philip, 1957b).

possibility increases of water occurring in pores or wedges isolated from the general three-dimensional network of water films and channels. Once continuity fails, of course, there can be no flow in the liquid phase other than through liquid islands or necks in series-parallel with the vapour transfer, as shown by Philip and de Vries (1957). Flow of this type is more appropriately considered as part of vapour transfer and is usually negligible in the absence of significant temperature gradients.

The relationship between hydraulic conductivity and water potential for three soils of different texture is given in Fig. 4.6. The plot is on a log/log scale and it is apparent that as Ψ falls, K declines even more rapidly, so that, at $\Psi = -15$ bars it is about 10^{-3} of its value at saturation. Gardner (1960a) has pointed out that the conductivity of

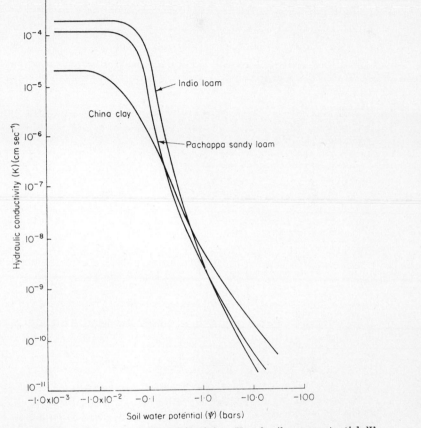

Fig. 4.6. Relationship between hydraulic conductivity, K and soil water potential, Ψ, for Indio loam, Pachappa sandy loam and Chino clay (after Gardner, 1960a).

coarse textured soils, initially higher, decreases more rapidly than that of finer textured soils so that there is a tendency for the curves to cross over in the region between $\Psi = -0.1$ and $\Psi = -1.0$ bar. Gardner (1958) has related K to Ψ for several soils by the empirical expression

$$K = a/(\Psi^n + b) \qquad (4.5)$$

where a, b and n are constants. The ratio a/b gives the hydraulic conductivity in saturated soil when Ψ is zero. For most soils he considers that n varies from 2 for fine textured soils to 4 or more for coarse textured soils and up to 10 or 15 for sands and gravels. A somewhat similar expression has been derived by Hallaire (1949, 1950).

Apart from development and use of the Darcy equation itself, a considerable amount of progress has been made in developing a general equation describing soil water transfer in the liquid phase. In this context, combination of the Darcy law with the continuity equation provides a means of relating changes in water content (or potential) with time to known initial and boundary conditions and soil characteristics (Childs and Collis-George, 1950; Klute, 1952; Philip, 1957b, 1964c).

The continuity expression equates the change in water content with time at a given position to the difference between the rate of flow into and out of the soil at that position. For flow in the vertical (z) direction,

$$\frac{\partial \theta}{\partial t} = -\frac{\partial v}{\partial z} \tag{4.6}$$

Since, from Darcy's law

$$v = -K\frac{\partial \Phi}{\partial z} \tag{4.7}$$

one can obtain

$$\frac{\partial \theta}{\partial t} = \frac{\partial}{\partial z}(K\,\partial\Phi/\partial z) \tag{4.8}$$

Substituting the relationship $\Phi = \Psi + z$ in Eq. (4.8), when Φ and Ψ are expressed in cm of equivalent head of water, gives, for flow positive downwards

$$\frac{\partial \theta}{\partial t} = \frac{\partial}{\partial z}(K\,\partial\Psi/\partial z) + \partial K/\partial z \tag{4.9}$$

If the flow is in a horizontal direction, the last term, which accounts for the effect of gravity, can be eliminated. In applying Eq. (4.9) only one side of the hysteresis loop can be examined at a time because it is necessary to assume a unique relationship between Ψ and θ and between θ and K.

Equation (4.9) has also been written in the more familiar form of a diffusion equation by Childs and Collis-George (1950) and subsequently by Klute (1952) and others. They distinguish the product of K and $\partial\Psi/\partial\theta$ as a diffusion coefficient or diffusivity D_{liq} and re-write Eq. (4.9) in the form

$$\frac{\partial \theta}{\partial t} = \frac{\partial}{\partial z}(D_{liq}\partial\theta/\partial z) + \partial K/\partial z \tag{4.10}$$

D_{liq} is analogous, in general terms, to diffusivity as used elsewhere but is not a constant. Also, although the equation has the same form as a diffusion equation it should be remembered that "diffusion" of water in soils involves bulk water movement as well as molecular motion. This expression, and in fact Eqs. (4.6) to (4.9) and the initial Darcy Eq. (4.3) can be used for three dimensional flow considerations by use of appropriate vector notation (Childs, 1957; Gardner, 1960a; Philip, 1964c) and, although it is not an overriding factor, it is worth stating that the vector presentation, once understood, leads to more straight-forward and realistic applications to water flow problems. The present form is used here to retain consistency with the treatment of other aspects of water movement.

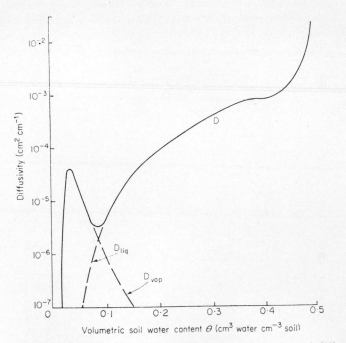

Fig. 4.7. Liquid diffusivity D_{liq}, vapour diffusivity D_{vap}, and total diffusivity D, plotted against volumetric water content θ, at 20° for Yolo light clay (after Philip, 1957b).

C. Simultaneous Flow of Liquid Water, Water Vapour and Heat

Movement of soil water in unsaturated soils involves both liquid and vapour phases. Although vapour transfer is insignificant at high soil water contents, it increases as void space increases and, at a value approximating the − 15 bar percentage, becomes the dominant mechanism in many soils. When liquid continuity fails it becomes, for all practical purposes, the sole means of water transfer even though the total flow may be less, by more than two orders of magnitude, than in saturated soil. This is well illustrated in Fig. 4.7 after Philip (1957b) which shows the diffusivity for both liquid, D_{liq}, and vapour, D_{vap}, for the same soil as used for Fig. 4.5.

This diagram refers to isothermal conditions and normal atmospheric pressure, and shows clearly that, for values of θ in excess of about 0·12, total diffusivity in Yolo clay is almost entirely due to liquid diffusivity D_{liq}, whereas at values less than about 0·07 vapour diffusivity accounted for virtually all D. The reason for negligible vapour flow under isothermal conditions, at moderate to high water contents, is primarily due to the fact that even quite steep gradients of Ψ are associated with relatively small gradients of vapour pressure, because of the shape of the Ψ (e/e^0) relationship. Thus it is only when D_{liq} is very small and

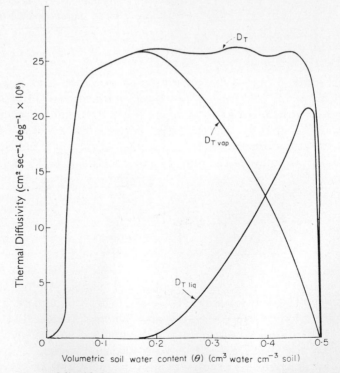

Fig. 4.8. Thermal liquid diffusivity, D_{Tliq}, thermal vapour diffusivity D_{Tvap}, and thermal moisture diffusivity, D_T, ($= D_{Tliq} + D_{Tvap}$) plotted against volumetric water content θ, for the same soil of Fig. 4.7 (after Philip, 1957b).

extremely steep gradients of Ψ develop that significant vapour pressure gradients also are established.

Steep vapour pressure gradients can be established, however, even in moist soils, if adequate gradients of temperature, pressure or solute concentration exist. Of these phenomena, temperature is the most important since very steep temperature gradients can develop, particularly in the surface soil and it has already been shown how rapidly increasing temperature leads to increasing saturation vapour pressure (see Eq. 1.1). Solute concentration and atmospheric pressure gradients are seldom steep enough to cause much change in vapour pressure, but short term changes in atmospheric pressure may lead to bulk flow of water vapour in and out of the soil (Schroeder *et al.*, 1965).

The effect of temperature on the liquid and vapour diffusivities in Yolo clay is shown in Fig. 4.8, after Philip (1957b). The figure shows clearly the dominance of the thermal vapour diffusivity coefficient, $D_{T\text{vap}}$, at water contents as high as $\theta = 0.4$, and that, at $\theta = 0.17$, total thermal moisture diffusivity, D_T, is indistinguishable from $D_{T\text{vap}}$.

1. *Mechanism of vapour flow*

Early studies of vapour movement in soils assumed that it would be described by a simple diffusion equation, modified to apply to porous media, in which the rate of flow, q, (g cm^{-2} sec^{-1}) would be proportional to the water vapour concentration gradient, $d\rho_y/dx$, (g cm^{-4}), and the diffusion coefficient of water vapour in air, D_w, (cm^2 sec^{-1}) multiplied by factors to take account of the reduced cross sectional area available for flow in soil, and by the extra path length caused by the tortuosity of the pathway. Thus

$$q_{\text{vap}} = -af D_w (d\rho_y/dx) \qquad (4.11)$$

where a is the tortuosity factor and f the fractional cross sectional area available for diffusion ($=$ total porosity $- \theta$) (Penman, 1940; van Bavel, 1952). Penman suggests a value of 0.66 for af and van Bavel, 0.58.

Experimental measurements of observed vapour diffusion have, however, usually given transfer rates several times as high as predicted by this equation (Gurr, Marshall and Hutton, 1952; Taylor and Cavazza, 1954; Rollins, Spangler and Kirkham, 1954). Furthermore, it is sometimes observed that water transfer under temperature gradients is negligible in very dry or very wet soils and attains a fairly well-defined maximum at an intermediate water content, which appears to depend on the water potential, Ψ, and on the air filled porosity

(Gurr, *et al.*, 1952; Jones and Kohnke, 1952; Smith, 1943; Hadley and Eisenstadt, 1955).

Explanations of this discrepancy have been advanced by Philip and de Vries (1957) who showed that two phenomena were of importance. When a soil dries and liquid phase continuity is disrupted, it has already been noted that liquid water remaining in the system tends to be located almost wholly in small pores or wedges which form necks or islands between the solid particles ("liquid" water is used here to denote that capable of viscous flow, as distinct from stationary absorbed water on the surfaces themselves). Philip and de Vries first proposed that water vapour condenses on one meniscus of each neck and evaporates from the other so that, instead of the necks blocking vapour flow, it proceeds without interruption in a series-parallel fashion through regions of liquid and vapour. The flux is determined by the vapour pressure gradient across the air-filled pores and the flux through the liquid necks adjusts accordingly.

Secondly, when they examined in detail heat transfer through the solid, liquid and vapour phases of the soil by the methods of de Vries (1952a, 1952b), it was found that the temperature gradients across air-filled pores could be approximately twice as steep as the mean temperature gradient in the medium as a whole. The effect of both of these factors is to increase the rates predicted from theory by three to eight times the previous estimates, enabling much better agreement with experimental observations.

2. *Combined expressions for liquid and vapour flow*

The general flow equations for one-dimensional liquid water transfer (Eqns 4.7–4.10) have been modified by Philip (1957b) to take account of temperature effects by including a term $D_{T\text{liq}}$ which equals $K/\lambda\Psi$ (and has the dimensions of $cm^2\ sec^{-1}\ deg^{-1}$) where λ (deg^{-1}) is the temperature coefficient of surface tension. For liquid flow this gives

$$\frac{q_{\text{liq}}}{\rho_w} = -D_{\text{liq}}\frac{\partial\theta}{\partial z} - D_{T\text{liq}}\frac{\partial T}{\partial z} - K \qquad (4.12)$$

where q_{liq} is the vertical one dimensional flow in $g\ cm^{-2}\ sec^{-1}$ and ρ_w is the density of liquid water ($g\ cm^{-3}$). D_{liq} is $K(\partial\Psi/\partial\theta)$ as before. Similarly, for vapour flow, the appropriate expression is

$$\frac{q_{\text{vap}}}{\rho_w} = -D_{\text{vap}}\frac{\partial\theta}{\partial z} - D_{T\text{vap}}\frac{\partial T}{\partial z} \qquad (4.13)$$

where D_{vap} ($cm^2\ sec^{-1}$) is an isothermal vapour diffusity factor (Philip,

1955) and D_{Tvap} (cm² sec⁻¹ deg⁻¹) is the thermal vapour diffusity (Philip and de Vries, 1957). Both vapour diffusities are obtained from Eq. 4.11 and are single valued functions of θ arising from the molecular diffusion of water vapour down gradients of specific humidity induced by water content and temperature differences (see Figs. 4.7 and 4.8).

Since the total flow, $q = q_{liq} + q_{vap}$; total diffusivity, $D = D_{liq} + D_{vap}$ and thermal moisture diffusivity $D_T = D_{Tliq} + D_{Tvap}$, Eq. 4.10 can be re-written as a general equation describing total liquid and vapour movement under combined water content and temperature gradients

$$\frac{\partial\theta}{\partial t} = \frac{\partial}{\partial z}(D\partial\theta/\partial z) + \frac{\partial}{\partial z}(D_T\partial T/\partial z) + \frac{\partial K}{\partial z} \qquad (4.14)$$

Where soil temperature gradients are unimportant, this reduces to the original form except that D_{liq} replaces D.

3. *The importance of vapour movement in nature*

A wide variety of problems in soil physics can be investigated using equations similar to Eq. (4.14) (see for example, Klute, 1952; Philip, 1954, 1955, 1957b, 1964c; Staple and Lehane, 1954; Gardner, 1960a). For one dimensional steady state flow Eq. (4.14) becomes (Philip, 1957b)

$$\frac{q}{\rho w} = -D\frac{d\theta}{dz} - D_T\frac{dT}{dz} - K \qquad (4.15)$$

The amount of vapour transfer reported by different investigators varies considerably but, in some cases, particularly those concerned with vapour movement to the surface of dry soils during periods of heat flow to the surface, substantial values have been observed (Lebedeff, 1927; Edlefsen and Bodman, 1941). Lebedeff concluded that over 70 mm of water could condense in the surface soil at Odessa annually, although his experimental technique, involving interruption of downward heat flow, is suspect as a measurement procedure for net flow. Edlefsen and Bodman showed substantial upward and downward movements in early winter and early summer in response to temperature gradients.

Although considerable amounts of water movement can undoubtedly occur in the vapour phase, it is important to remember that whenever liquid phase continuity exists, both liquid and vapour transfer processes operate simultaneously and, in some cases, may do so in opposite directions. Figures 4.7 and 4.8 for isothermal diffusivity and thermal diffusivity indicate that this can readily occur.

Several very interesting experiments have given useful demonstrations of this phenonomen (Gurr *et al.*, 1952; Smith, 1943; Kuzmak and Serader, 1957; Richards, Gardner and Ogata, 1956). Those of Gurr *et al.* will be described briefly. These workers used horizontal columns of soil, each of initially uniform water and salt content, the water content in 10 columns ranging from practically dry to practically saturated levels. A uniform temperature gradient was then applied across all columns and the net transfer of water and salt after a period of five days used as an indication of net water movement and total

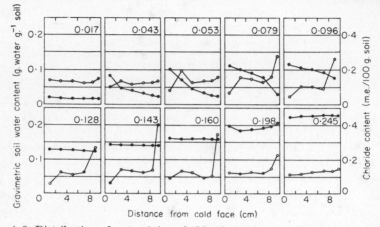

FIG. 4.9. Distribution of water (●), and chloride (o) in columns of a loam soil after being subject to a temperature gradient for 5 days. Figures in each diagram indicate the initial (gravimetric) water content of each column (after Gurr, Marshall and Hutton, 1952).

liquid movement, respectively. The results showed that, in general, there was a net transfer of water towards the cold side but that the salt moved in the reverse direction. Moreover the greatest net movement of water was in the low to medium water content treatments, and the greatest reverse movement of salt in the medium to high water contents. There was little change of salt or water distribution in the very dry and very wet treatments (see Fig. 4.9).

These apparently contradictory results are readily explained on the basis of the flow equations just given. At the stage when there was no liquid phase continuity, the net transfer of water was the total transfer and was exclusively due to vapour flow. Hence there was no salt movement and the water movement increased rapidly with water content (as does the $D_{T\text{vap}}$ curve of Fig. 4.8). Once liquid continuity started to develop, however, the net flow consisted of both vapour and liquid components and any temperature-induced flow associated with $D_{T\text{vap}}$ and $D_{T\text{liq}}$ towards the cold face tended to be opposed by a return flow of liquid associated with D_{liq} (as in Fig. 4.7). The net transfer of water towards the cold face could therefore be expected to decrease rapidly as θ increases, even though for a time there would still be a susbtantial flow of vapour in one direction and liquid in the other, leading to salt accumulation at the warm face, but a final apparent absence of gradient in θ. Once complete liquid continuity developed and the soil approached saturation, vapour flow would become negligible and any movement of liquid water (containing salt) associated with $D_{T\text{liq}}$ would be compensated for by a return flow associated with D_{liq}. Hence neither salt nor water content would appear to change.

4. *Alternative approaches to simultaneous flow*

Before considering other phenomena associated with soil water movement, it should be mentioned that an alternative approach to water flow problems may be made using the techniques and theory of the thermodynamics of irreversible processes. Although not used to any great extent in soil physics at the present time, this approach provides a means of describing linked transport phenomena, such as occur in the situations just described, where a water content gradient not only causes a flow of water but also a flow of heat and, similarly, a temperature gradient not only causes a flow of heat but also a flow of water. This approach has been used recently to describe heat and water flows in soils (Taylor, 1962; Cary and Taylor, 1962; Taylor and Cary, 1964) and is capable of being extended to include solute and other recognizable flows.

The general theory of irreversible thermodynamics concerns the relationships between flows (of matter, energy, electrical charge, etc.) in a system and the "forces" responsible for them. It assumes that the flows are linear functions of all the forces operative in the system, provided that they are correctly defined and that the system is not too far from equilibrium. A set of phenomenological equations can then be written in the following form (de Groot, 1952)

$$J_1 = L_{11}X_1 + L_{12}X_2 + \ldots + L_{1n}X_n$$

$$J_2 = L_{21}X_1 + L_{22}X_2 + \ldots + L_{2n}X_n$$

$$J_n = L_{n1}X_1 + L_{n2}X_2 + \ldots + L_{nn}X_n$$

where each flow J_i is dependent on its conjugate force X_i through a straight coefficient L_{11}, which is always positive, and may also be linked to non-conjugate forces X_j through the cross-coefficients L_{ij} ($i \neq j$) which may be positive or negative as long as the determinant formed by the coefficients is positive, that is

$$|L_{ij}| \geqslant 0$$

Also the Onsager reciprocal relation states that the "matrix" of the cross-coefficients is symmetrical so that

$$L_{ij} = L_{ji}$$

This final relationship is of particular value because it reduces the number of coefficients to be determined.

The advantage of expressing the transport equations in this form is that all the parameters known to influence the flow of any component can be included and possible interactions between the flows of different components can be described and brought out.

The choice of forces and fluxes is to some extent open, and can be modified to suit the particular system under consideration, as long as the products of any pair of forces and fluxes have the dimensions of entropy production and the sum of all the products should leave the total entropy production invariant.

In applying the theory to considerations of heat and water flow in soils, the chief problem is one of non-linearity, since the hydraulic conductivity varies with water content, as has already been shown. (The equilibrium requirement can usually be met since, in most situations $\Delta\mu_w \ll RT$ and $\Delta T \ll T$). Despite this problem Taylor (1962) and Taylor and Cary (1964) have developed phenomenological equations to describe the observed flows of water and heat, and the following brief account gives some aspects of their derivation and use.

In the first place it can be recognized that flows of matter $J_m = \mathrm{d}n/\mathrm{d}t$ (mole sec^{-1}) and energy $J_u = \mathrm{d}U/\mathrm{d}t$ (erg sec^{-1}) are involved where n is the number of moles and U is total energy. An expression for rate of entropy production ($\mathrm{d}S/\mathrm{d}t$) can then be written in the form

$$\frac{\mathrm{d}S}{\mathrm{d}t} = J_m \Delta(\mu/T) + J_u \Delta(-1/T) \qquad (4.16)$$

where the driving forces for flow of matter and energy are (μ/T) and ($-1/T$), respectively. In practice, a heat flow, $J_q = J_u - HJ_m$ [where H is the molal enthalpy of the substance under study (erg mole^{-1})], is measured. Inserting this relationship into Eq. 4.16 gives

$$\frac{\mathrm{d}S}{\mathrm{d}t} = J_m[\Delta(\mu/T) + H\Delta(-1/T)] + J_q \Delta(-1/T) \qquad (4.17)$$

Since the first term on the right hand side simplifies to $J_m(\Delta\mu)_T/T$ (de Groot, 1952; Taylor and Cary, 1964; Spanner, 1964), and $\Delta(-1/T) = T/T^2$, Eq. (4.17) becomes

$$\frac{dS}{dt} = J_m(\Delta\mu)_T/T + J_q(\Delta T/T^2) \tag{4.18}$$

Appropriate phenomenological equations for the flow of a component, i, and heat can now be written in the form

$$J_i = L_{ii}(\Delta\mu_i)_T/T + L_{iq}(\Delta T/T^2) \tag{4.19}$$

$$J_q = L_{qi}(\Delta\mu_i)_T/T + L_{qq}(\Delta T/T^2) \tag{4.20}$$

and

$$L_{iq} = L_{qi} \tag{4.21}$$

For the case of water and heat flow in soil, it is more convenient to use as driving forces the gradient of soil water potential $-(d\Psi/dx)_T$ and the reciprocal of the temperature gradient

$$-(1/T)dT/dx = -d \ln T/dx.$$

This gives, instead of Eqns. (4.19) to (4.21)

$$J_w = -L_{ww}(d\Psi/dx)_T - L_{wq}(d \ln T/dx) \tag{4.22}$$

$$J_q = -L_{qw}(d\Psi/dx)_T - L_{qq}(d \ln T/dx) \tag{4.23}$$

and

$$L_{wq} = L_{qw} \tag{4.24}$$

with the water flux, J_w, in units of mole sec^{-1}, and the heat flux J_q in erg sec^{-1}. It can now be seen that three coefficients are required to describe water and heat flows in soils. The first, L_{ww}, can be related to the hydraulic conductivity, K, by taking an isothermal situation where Eq. (4.22) becomes

$$J_w = -L_{ww}(d\Psi/dx)_T \tag{4.25}$$

This expression is in the same form as the Darcy's law equation (4.3) except that in saturated soils, and for one dimensional vertical flow, the water potential gradient should be replaced with the gradient of total potential, $d\Phi/dz$. In a similar manner, by taking a situation where $d\Psi/dx$ is zero, L_{qq} can be obtained from Eq. (4.23) which simplifies to

$$J_q = -L_{qq}(d \ln T/dx) \tag{4.26}$$

in which L_{qq} is seen to be related to a normal heat transfer coefficient. The third coefficient can be evaluated using Eq. (4.23) by assuming that d $\ln T/dx$ is zero. Under these conditions of isothermal water flow there will still be a flow of heat given by

$$J_q = -L_{qw}(d\Psi/dx)_T \tag{4.27}$$

This interdependence of the two flows is most clearly brought out by envisaging a situation in which a uniform temperature gradient and a uniform water potential gradient are imposed across a layer of soil in such a way that $J_w = 0$. Then from Eq. (4.22)

$$\frac{(\mathrm{d}\Psi)_T}{(\mathrm{d}\,ln\,T)} = -\frac{L_{wq}}{L_{ww}} = Q*$$

where $Q*$ is the heat of transfer.

Thus the isothermal flow of unit quantity of water is associated with a flow of $Q*$ units of heat. Put another way, this quantity of heat would have to be supplied to one side of the system in order to keep the temperature gradient constant during water flow. $Q*$ can be introduced to Eq. (4.22) to give

$$J_w = L_{ww}\,[\,(\mathrm{d}\Psi/\mathrm{d}x)_T + Q*(\mathrm{d}\,ln\,T/\mathrm{d}x)\,] \qquad (4.28)$$

If desired, soil water content may be used as the independent variable in these equations, instead of water potential. In this case the results are analogous to those obtained previously using diffusion equations. To do this, one writes (Taylor and Cary, 1964)

$$J_w = -D'(\mathrm{d}\theta/\mathrm{d}x + Q*'(\mathrm{d}\,ln\,T/\mathrm{d}x)\,) \qquad (4.29)$$

where $D' = L_{ww}(\mathrm{d}\mu_w/\mathrm{d}\theta)$ is a coefficient of diffusion and $Q*'$ is the appropriate heat of transfer. The similarity of this expression to Eq. (4.14) is immediately apparent.

Taylor and Cary (1964) have shown that $Q*$ or $Q*'$ can be obtained experimentally and consider that the foregoing equations have wide applicability to soil studies. It must be pointed out, however, that the problem of non-linearity in the J_w ($\mathrm{d}\Psi/\mathrm{d}x$) relationship seriously limits the straightforward application of the equations at the present time, and that most workers prefer to use the more mechanistic approach described previously. Apart from the question of non linearity, a satisfactory general analysis of water and heat flow should separately include the liquid and vapour components of the water flux. As now written, the transfer coefficients in Eqns (4.22)–(4.29) represent the sum of the two components, but it should be possible to separately account for them without undue difficulty.

D. Capillary Rise and Related Phenomena

Equations established in the foregoing sections of this chapter may now be applied to some other water flow phenomena of general interest. The upward movement of water through the soil from a water table, commonly referred to as "capillary rise", will be considered first. For zero flow, or vertical flow upwards under steady state conditions, and isothermal conditions, $d\theta/dt = 0$, and the general flow Eq. (4.9) becomes (positive upwards)

$$\frac{d}{dz}\left(K\left[\frac{d\Psi}{dz}-1\right]\right) = 0 \qquad (4.30)$$

so that for zero flow, as in the case of a water table, $d\Psi/dz = 1$. Thus the water potential at a point (in cm) is equal in magnitude (but opposite in sign) to the height of the point above the water table (also in cm). For steady state flow, by combining Eq. (4.30) with Eq. (4.5) the relation between Ψ, (in cm) the height above the water table z, and the rate of capillary rise, v_z, (in units such as cm sec^{-1}) is seen to be:

$$v_z = a[(d\Psi/dz)-1]/(\Psi^n + b) \qquad (4.31)$$

On integration, and solving for z, this equation becomes (Wind, 1960):

$$z = \frac{a}{bv_z + a}\int \frac{d\Psi}{[v_z\Psi^n/(bv_z+a)]+1} \qquad (4.32)$$

A similar expression can be derived from Eq. (4.15), which under isothermal conditions becomes:

$$\frac{d\theta}{dz} = -(K+q/\rho_w)/D \qquad (4.33)$$

and, on integration, solving for z, gives

$$z = \int_{\theta_{tab}}^{\theta_{surf}} \frac{D}{(K+q/\rho_w)}\,d\theta \qquad (4.34)$$

where θ_{surf} and θ_{tab} are the water contents at the surface and at the water table, respectively.

If $n = 1$ the solution of Eq. (4.32) will have a logarithmic form, and if $n = 2$ an arctan form. For higher values of n, a combination of these forms is found. Wind (1955a) has given solutions for $n = 1\cdot5$ and $b = 0$. Gardner (1958) published solutions for values of n up to 4. For typical coarse and fine textured soils with $n = 4$ and $n = 1\cdot5$, respectively, Wind (1960) has computed the values of Fig. 4.10. In the coarse textured soil $K = 400$ mm day^{-1} [$= a/b$ in saturated soils, see Eq. (4.5)]; a/b was not given for the clay soil but it was about 100 times less.

This figure gives the relation between Ψ and z when a constant flow of water takes place from a water table, at constant depth z, to the surface where it is removed by evaporation. The form of the curves shows that at values of Ψ lower than about -1 bar, in the fine textured soil, the maximum flow appears to be determined to a much greater extent by water table depth than by Ψ. In the coarse textured soil this is true at Ψ values lower than about -0.1 bar. As the figure shows, a maximum rate of 5 mm day $^{-1}$ can be achieved in the coarse

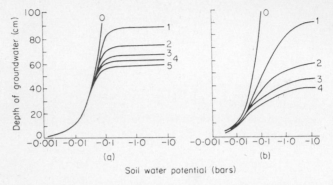

Soil water potential (bars)

FIG. 4.10. The relationship between soil water potential, Ψ (bars), and height of capillary rise, z (cm) at different rates of upward flow, v_z (mm day $^{-1}$) for (a) coarse and (b) fine textured soils (after Wind, 1960).

soil from a water table at about 60 cm but comparable rate in the clay is only about 2 mm day $^{-1}$. At 90 cm however, the rate is 1 mm day $^{-1}$ in both soils. It is important to note that, although Ψ becomes progressively less important as depth increases, some flow will occur as long as the value of Ψ, expressed in cm, exceeds the depth to the water table expressed in the same units. The limiting rate of upward movement, as a function of depth, is given by (Gardner, 1960a):

$$\lim.v_z = \frac{ma}{z^n} \tag{4.35}$$

where a and n are the constants from Eq. (4.5), (b is negligible here) and m is a constant depending on n.

It can be seen, therefore, that liquid water can move from even deep water tables at measurable rates and data such as those of Lebedeff (1927), attributed to vapour movement, may need to be reconsidered in this context. [The effect of including vapour flow in (4.35) is to increase flow by about 10 per cent, according to Gardner (1960a)]. Gardner and Fireman (1958) have shown that water can reach the surface, at significant rates, from water tables as deep as 7 metres.

In many cases the rate of capillary rise could probably be adequate to sub-irrigate crops, as long as the water table is less than 1 metre below the surface (Wesseling and van Wijk, 1955; Wind, 1955a). In such cases, and even when the water table is too low for this purpose, the movement of soluble salts to the surface can present serious problems in agriculture unless sufficient leaching occurs, by rain or irrigation, and appropriate drainage procedures are employed to prevent an undesirable degree of accumulation. On a more microscopic scale, removal by roots of water containing less salts than exist in the bulk solution should also result in salt accumulation, in this case directly adjacent to the adsorbing areas of the roots. The degree of accumulation will depend on the relative rates of uptake and back diffusion. Whether or not the accumulation has detrimental effects depends on the rate of root extension into fresh soil and on the leaching effect of rain and irrigation. Comprehensive accounts of water table and salinity phenomena have been prepared by Fireman (1957) and van't Woudt and Hagan (1957).

3. Movement of Water to Plant Roots

Water tends to move from the soil mass to the root surfaces, then through the plant and into the atmosphere along a gradient of decreasing water potential. In some segments of this pathway the effective driving force causing flow is not necessarily a simple gradient of total potential, and different component potentials may be responsible for disproportionate amounts of flow, particularly when the flow of water is accompanied by flows of heat, electricity or solutes (see Chapter 6 and earlier sections of this chapter). However, the general tendency remains and, because the water potential in the plant has a direct effect on plant growth and metabolic processes, its magnitude at different parts in the plant is of particular significance.

In Chapters 6, 7 and 9 a detailed discussion of the relationship between Ψ and plant growth is given, as is an account of the magnitude of the gradients of Ψ through the plant and the factors which influence them. In this chapter, the view that the soil water potential at the root soil interface, Ψ_{root}, appears to be the main soil characteristic controlling the availability of soil water to plant growth will be accepted (Kramer, 1956a; Slatyer, 1957a; Gardner, 1960b). In so doing it is recognized that this value is, in turn, influenced not only by the water potential in the soil mass, Ψ_{soil}, but also by the potential gradient from the bulk soil to the root, which develops as a result of water removal by the root. Its slope is controlled by the relative rates of flow to, and removal by, the roots, and with steady rates of water uptake reaches a constant value inversely proportional to the hydraulic conductivity of the soil in any segment of the water pathway.

Because the root system of physiologically active plants is continually exploring new regions of soil, and because zones of most rapid water uptake are located close to the elongating root tips (see Chapter 7) soil water content and water potential may differ widely in different parts of the root zone. It follows that much of the uptake by any one root system may take place from a relatively small proportion of the total root zone. This is accentuated if soil water recharge by rain or irrigation only re-wets part of the soil mass explored by roots. Associated with this variability in uptake will be a range of values of Ψ_{root} and the effective value for the root system as a whole is probably impossible to obtain by direct methods.

The root system itself integrates the range of external values and the final value of Ψ_{root}, measured at the base of the stem, generally reflects a value close to that of Ψ_{root} in the wettest areas of the root zone from which most uptake has taken place (Slatyer, 1960c). Although the problem of soil water flow to the roots appears complicated, it is of interest that simple models recently proposed (Philip, 1957a; Gardner, 1960b; Slatyer and Gardner, 1965; Cowan, 1965) appear to give good agreement with experimental data. The following gives a brief description of the approach used by Gardner (1960b) and some of the applications which arise from it.

In the first place, it is recognized that in order for water to move from soil to root there must be a gradient of decreasing potential. Further, as the soil water content falls, Ψ_{soil} is lowered so that Ψ_{root} must be lowered still further in order to maintain the gradient (since hydraulic conductivity also declines, a relatively greater reduction in Ψ_{root} is required to maintain the same flow rate). Finally, a critical value of Ψ_{root} is reached, and physiologically significant uptake will cease when Ψ_{soil} also reaches this value. Water movement to the root should conform to the previously given flow equations and their solution, subject to the appropriate boundary conditions, should provide answers of practical significance.

The model developed by Gardner assumes cylindrical geometry, with the root envisaged as an infinitely long cylinder of uniform radius and water absorbing proportions, and soil water is assumed to move only in the radial direction. Although these are approximations, they do not appear to affect the conclusions to any great extent. Gardner (1960b) considered first a single root of radius r, in an infinite two dimensional medium. The flow equation (4.10 or 4.14) then becomes

$$\frac{\partial \theta}{\partial t} = \frac{1}{r} \frac{\partial}{\partial r} \left(r D \frac{\partial \theta}{\partial r} \right) \tag{4.36}$$

where r is the radial distance to the root axis, and the initial soil water content is assumed uniform throughout the soil. Solution of this equation yields the information given in Fig. 4.11–4.13. Much more detail is given by Gardner (1960b).

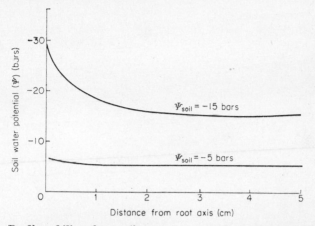

Distance from root axis (cm)

FIG. 4.11. Profiles of Ψ_{soil} from soil mass to root surface, associated with soil mass values of $\Psi_{soil} = -5$ and -15 bars. Estimates refer to Pachappa sandy loam and an assumed water uptake rate of $0 \cdot 1$ cm³ cm⁻¹ root length day⁻¹ (after Gardner, 1960b).

In Fig. 4.11 the change of Ψ_{soil} as a function of distance from the root is shown (using the same Pachappa sandy loam shown in Fig. 4.7) for initial values of Ψ_{soil} of -5 and -15 bars and assuming a typical flow rate of $0 \cdot 1$ cm³ cm⁻¹ root length day⁻¹ (Ogata, Richards and Gardner, 1960). Gardner has shown that the distance at which any given value of Ψ_{soil} occurs increases as the square root of time. Consequently the distance from which water can be expected to move can be estimated from the total time allowed for movement. For typical times expected under natural conditions, water can be expected to move about 4 cm to a root surface. However the water transmitting properties of the soil, and the evaporative demand, jointly determine the optimum root density for any one situation. Figure 4.11 shows that very little gradient is required to move water at the desired rate when bulk values of $\Psi_{soil} = -5$ bars; but at -15 bars a much steeper gradient is required because of the reduced hydraulic conductivity. In the two cases depicted $\Delta \Psi = (\Psi_{soil} - \Psi_{root}) = 2$ and 13 bars, respectively, the intercepts of the curves indicating values of $\Psi_{root} = -7$ and -28 bars.

The influence of different flow rates on $\Delta\Psi$ is shown in Fig. 4.12, for the same soil. The value of $\Delta\Psi$ depends on K and the actual flow rates required. Because K decreases with Ψ_{soil}, the values of $\Delta\Psi$ required to increase the flow from 0·1 to 0·5 cm³ cm⁻¹ root day⁻¹, initially small, increase rapidly as Ψ_{soil} drops below $-0·5$ bars. In finer textured soils the increase in $\Delta\Psi$ could be expected to be more gradual due to the different K (Ψ) relationship.

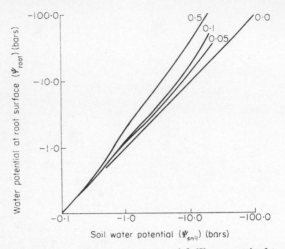

FIG. 4.12. Estimated values of root water potential, Ψ_{root}, required to cause soil water flow at different nominated rates of uptake (ranging from 0 to 0·5 cm³ cm⁻¹ root length day⁻¹) as a function of soil mass water potential, Ψ_{soil}. Estimates refer to Pachappa sandy loam (after Gardner, 1960b).

Where a range of flow rates and values of ($\Psi_{soil} - \Psi_{root}$) are considered in relation to a critical level of Ψ_{root}, it can be seen that the critical value is reached at markedly different values of Ψ_{soil}. In Fig. 4.13 this situation is depicted for an arbitrary critical level of $\Psi_{root} = -20$ bars. At zero flow rates the value is clearly not reached until $\Psi_{soil} = \Psi_{root} = -20$ bars, but as flow rate increases to 0·1 cm³ cm⁻¹ root day⁻¹ the value rises to $\Psi_{soil} = -12$ bars and at 0·4 cm³ cm⁻¹ root day⁻¹ it is at $\Psi_{soil} = -8$ bars.

The question has been raised from time to time as to whether or not a vapour gap develops at the root-soil interface as the soil and root shrink with decreasing water content while absorption proceeds from an initially wet soil. Philip (1958c) proposed that such a gap might develop and its presence could explain some of the phenomena associated with saline soils. Although it may develop in some situations, Bernstein,

Gardner and Richards (1959) considered that rates of vapour flow were too slow to equal observed water uptake rates, except in soils much drier than the permanent wilting percentage; and Slatyer (1961) showed that the particular salinity effects referred to by Philip (see Chapter 9) could be found in water culture where no possibility of a vapour gap exists. In consequence, it seems unlikely that vapour gaps, even if they do develop, are of much significance in most situations.

The continued extension of roots in intimate contact with soil particles and the presence and elongation of root hairs would also minimize the effects.

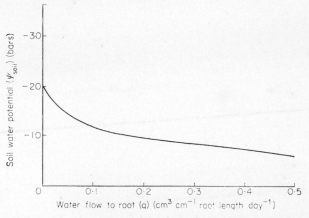

FIG. 4.13. Influence of water potential in the soil mass, Ψ_{soil}, on the rate of water flow to the root, when Ψ_{root} is -20 bars (after Gardner, 1960b).

To summarize, it may be stated that even simplified models of the type developed by Gardner (1960b) appear to give useful illustrations of water flow from soil to root. Under most conditions it seems that when bulk values of Ψ_{soil} are higher than -1 bar, only small gradients can be expected to develop between Ψ_{soil} and Ψ_{root}: even for high rates of flow but, as Ψ_{soil} and hydraulic conductivity fall, the value of $\Delta\Psi$ needed to maintain flow increases rapidly and may cause critical values of Ψ_{root} to develop, even when the soil mass is moist.

It appears unlikely that water will move more than about 4 cm through the soil to a root surface although, when a water table is close to the surface, the distance may approach 1 metre. In this connection Wind (1955a, 1955b) concluded, on the basis of his own and Emerson's (1954) data, that the reason for shallow rooting of grasses on heavy clay soils, in the presence of shallow water tables was not aeration difficulty but the fact that flow of water through the soil encountered less resistance than through the fine metaxylem elements of the grass roots. This conclusion is open to question and is certainly not valid in the absence of a shallow water table, as the above discussion shows, but it does emphasize that relatively high resistances can develop in plant roots. This question will be considered again in the general discussion of the absorption process, the factors influencing absorption and movement of water and solutes in the plant, which is given in later chapters, particularly Chapter 7.

CHAPTER 5

Water as a Plant Component

Water constitutes more than 80% of most plant cells and tissues in which there is active metabolism, rising in some cases to over 90%. It forms a continuous liquid phase through the plant from the root hairs to the leaf epidermis, and liquid phase continuity generally extends into the soil or substrate in which the plant is rooted.

By comparison with soil, the proportion of water in plant material is high; furthermore, for continued active metabolism it can usually only fluctuate within relatively narrow limits; a change of water content of 20–25% of the value at maximum hydration ($\Psi_{plant} = 0$) frequently resulting in a cessation of most growth processes.

1. DISTRIBUTION OF WATER IN CELLS AND TISSUES

A. Water Content of Tissues and Organs

The range of water contents found in different cells, tissues and organs is illustrated in the data of Table 5.I, collected by Kramer (1956b). Dry grains can be seen to contain as little as 5% water, as a percentage of the total weight, whereas actively growing tissue approaches 95%. In the latter category, the meristematic regions of tissue such as root apices generally have slightly lower water contents than vacuolated cells further away from the growing point, even though the general water content level is still very high. This is presumably due to the absence of a large vacuole in such cells.

B. Distribution of Water in Cells

When a young cell is first formed it is small, thin-walled and packed with cytoplasm, with a centrally placed nucleus, and no vacuoles are visible with the light microscope. Mitochondria are conspicuous in the cytoplasm but chloroplasts are absent. As the cell grows, the most apparent change is the development of vacuoles, which gradually coalesce into one large vacuole surrounded by cytoplasm. Plastids appear and develop and there is also continued synthesis of cytoplasm so that the ratio of cytoplasm to surface area remains more or less the

TABLE 5.I

Water content of various plant tissues expressed as percentage of total weight†

	Plant parts	Water content	Authority
Roots	Barley, apical portion	93·0	Kramer and Wiebe (1952)
	Pinus taeda, apical portion	90·2	Hodgson (1953)
	Pinus taeda, mycorrhizal roots	74·8	Hodgson (1953)
	Carrot, edible portion	88·2	Chatfield and Adams (1940)
	Sunflower, av. of entire root system	71·0	Wilson *et al.* (1953)
Stems	Asparagus stem tips	88·3	Daughters and Glenn (1946)
	Sunflower, 7 wks. old av. of entire stems	87·5	Wilson *et al.* (1953)
	Pinus banksiana	48–61	Raber (1937)
	Pinus echinata, phloem	66·0	Huckenpahler (1936)
	Pinus echinata, wood	50–60	Huckenpahler (1936)
	Pinus taeda, twigs	55–57	McDermott (1941)
Leaves	Lettuce, inner leaves	94·8	Chatfield and Adams (1940)
	Sunflower, av. of all leaves on 7 wks. old plant	81·0	Wilson *et al.* (1953)
	Cabbage, mature	86·0	Miller (1938)
	Corn, mature	77·0	Miller (1938)
Fruits	Tomato	94·1	Chatfield and Adams (1940)
	Watermelon	92·1	Chatfield and Adams (1940)
	Strawberry	89·1	Daughters and Glenn (1946)
	Apple	84·0	Daughters and Glenn (1946)
Seeds	Sweet corn, edible	84·8	Daughters and Glenn (1946)
	Field corn, dry	11·0	Chatfield and Adams (1940)
	Barley, hull-less	10·2	Chatfield and Adams (1940)
	Peanut, raw	5·1	Chatfield and Adams (1940)

† Collected by Kramer (1956b).

same (Robertson and Turner, 1951). The cell wall increases in thickness, with the original wall forming the middle lamella. On the inside of the original wall, layers of cellulose are added. Impregnation with lignin or cutin frequently occurs. Diagrammatic sketches of young and adult cells of typical higher plants are given in Fig. 5.1.

In assemblages of cells, the cytoplasm of adjoining cells is interconnected by firm cytoplasmic strands, termed plasmadesmata, which penetrate the walls. Thus cells in tissues cannot be regarded as being independent of one another. It is also of significance that practically all cells have part of their surface abutting on an intercellular air space.

Because of this, gas exchange is facilitated, and gas pressures are close to atmospheric.

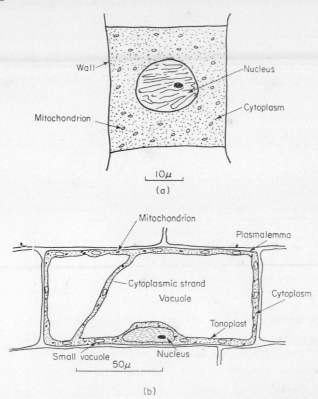

(a)

(b)

FIG. 5.1. Diagrammatic sketches of (a) young and (b) adult cells from typical higher plants (redrawn from Briggs, Hope and Robertson, 1961).

The structure of the cell wall consists of a mesh of cellulose microfibrils. The length of the fibrils and arrangement of the mesh may differ widely in different tissues and different species, but the fibrils are generally of the order of 10–30 mμ in diameter and the spaces between of the order of 1–100 mμ (Preston et al., 1948; Meyer and Anderson, 1952). In addition to cellulose, the wall may contain pectin, lignin and, in the outer wall layers exposed to air, cutin. Cutin generally develops on the outer wall surfaces but may penetrate the interfibrillar spaces to some degree (Esau, 1960). On the outer surface of epidermal cells, it forms a dense layer, with very small pores of low permeability to water.

Briggs, Hope and Robertson (1961) have described how the cytoplasm develops during enlargement, with plastids and mitochondria increasing in size and number and cytoplasmic streaming commencing. The nucleus itself may move to one side of the cell. It may alternatively be somewhat mobile, or remain suspended in the cytoplasmic strands which traverse the vacuole at different points. Bounding the cytoplasm, which is generally not more than a few microns thick, are two extremely important membranes. At the wall–cytoplasm interface, the membrane is termed the plasmalemma; at the cytoplasm–vacuole boundary, it is termed the tonoplast. Although evidence for the existence of the plasmalemma has not been found in all cells, the tonoplast is well defined. Both membranes appear to be of the order of 10 mμ thick.

The cytoplasm is, however, more than an aqueous phase bounded by membranes with mitochondria and other organelles suspended in between. The nucleus, mitochondria and chloroplasts have their own characteristic membranes and internal properties, and the surrounding medium is better regarded as a gel or sol than as a simple solution. According to Frey-Wyssling (1953), the solid phase consists of a network of their strands of polypeptide chains, sometimes referred to as a "brush heap" of micelles. This mesh is occupied by insoluble materials such as oil droplets, as well as by water and solutes.

C. Forces Holding Water in Cells

1. Water in cell walls

The final structure of the wall in adult cells is generally considered to be composed of three phases: the middle lamella, consisting almost entirely of calcium pectate; the primary wall, consisting of a matrix of cellulose fibrils impregnated by pectic materials; and the secondary wall, which is deposited in the primary wall, and may contain, as well as cellulose and pectins, lignin and cutin. The cellulose of both the primary and secondary walls appears to be a continuous matrix of cellulose fibrils around which lie the deposited materials. The pectic compounds, like the cellulose, have long chain molecules made up of galacturonic acid, methylated to various degrees in different types of cells and in cells of different ages (Briggs, Hope and Robertson, 1961). The ionizible carboxyl groups probably play an important part in ion exchange since they are balanced by divalent cations, which may be replaced in exchange processes. These surfaces, and the hydroxyl groups on the cellulose molecules, are strongly hydrophilic substances, and adsorb water, primarily by hydrogen bonding. However, in turgid cells, most cell wall water is probably held by surface tension in the voids created by the interfibrillar spaces (see Chapter 1).

There are very few data dealing with water retention in the walls but, in Fig. 5.2, sorption isotherms are given for pine wood flakes, which can be expected to show sorption characteristics generally similar to those for the cellulose wall components (Christensen and Kelsey, 1958). The hysteresis which is exhibited by these sorption and desorption curves can certainly be expected in living cell walls. However, it is important to recall that hysteresis is less pronounced in the range of relative vapour pressures above 0.98%, which represents the range of equivalent water contents of most interest physiologically.

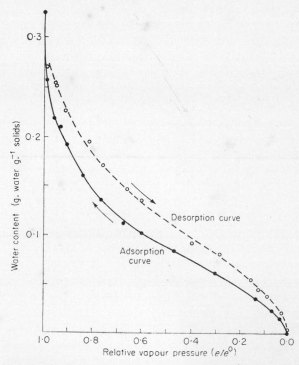

FIG. 5.2. Sorption isotherms for Klinki pine wood at 25 °C (after Kelsey, 1957).

In dehydrated tissue at very low relative vapour pressure, it can be expected that most of the water in the wall would be adsorbed on the solid surfaces. As the relative vapour pressure increases and extra molecular layers of water are added to those already adsorbed at the solid-liquid interfaces, the water becomes progressively less strongly oriented and by the time the relative vapour pressure reaches levels of physiological interest most of the water is probably retained by the surface tension acting arcoss the liquid-air menisci of the interfibrillar spaces. Stamm (1944) considers that these strongly oriented water molecules represent less than 10% of the total cell wall water in turgid cells.

Table 1.III gives an indication of the size of pores which can be expected to drain at different water potentials or relative vapour pressures. The table shows that, at $\Psi_{wall} = -15$ bars, pores larger than 100 mμ radius would be expected to drain, and at $\Psi_{wall} = -150$ bars, the appropriate limiting radius would be 10 mμ. Pores of 1mμ radius, would be expected to retain water at $\Psi_{wall} = -1500$ bars. Corresponding relative vapour pressures (at 20°C) would be 0.99, 0.89 and 0.33, respectively. Since most interfibrillar spaces appear to be of the order of 10–100 mμ diameter, these figures suggest that there would be little drainage of water from the walls at relative vapour pressures above about 0.98. In heavily cutinized walls, with interfibrillar spaces of the order of 1 mμ or less, drainage would not be expected to occur until after the death of the plant.

Although data of Fig. 5.2 provide a useful guide to the sorption characteristics of the cellulosic components of cell walls, living cell walls may retain significantly more water at any one relative vapour pressure, particularly those near saturation, because of the hydrophilic character of other, non-cellulosic, wall components. However, it is difficult to prepare cell wall material without disturbing its structure and hence its pore size distribution, so that data such as those of Gaff and Carr (1961), which show water contents up to 150% (g water g^{-1} solids) at saturation, must be interpreted with caution.

The volumetric water content of cell walls (cm^3 water cm^{-3} wall volume) of turgid cells exceeds 50% in *Avena* coleoptiles (Preston and Wardrop, 1949) and is probably higher in cells with loosely constructed walls and relatively large interfibrillar spaces. In walls where substantial deposition of non-cellulosic materials has occurred, however, the water content may fall significantly below 50%.

The water content of the cell wall is also influenced by the osmotic concentration of the cell wall sap. In turn, this is influenced by the nature and concentration of the immobile ions adsorbed on the exchange surfaces of the solid wall components, since the wall phase constitutes a Donnan system in which classical Donnan equilibrium exists between the free aqueous phase of the vascular system and the Donnan phase in the other cell components. Under any set of conditions, the Donnan equilibrium will be reflected in a specific vascular sap composition, a specific distribution of immobile ions adsorbed on the cell wall surfaces and a specific concentration and ion distribution of mobile ions in the interfibrillar solution. The sum of the molalities of the mobile ions in the Donnan phase exceeds that in the open phase, because the sum of the accumulation ratios in the former phase exceeds the simple sum of the ionic molalities in the latter. Hence, the osmotic pressure in the Donnan phase exceeds that in the external solution. Furthermore, since the accumulation ratio decreases as the external solution increases, this discrepancy is greatest when the external solution is dilute and decreases as the concentration of the open phase solution increases.

It follows that, as the system approaches final equilibrium, there is a tendency for the Donnan phase to swell and water content to increase, the degree to which this occurs depends on the concentration and composition of mobile and immobile ions in both phases and the geometrical restrictions imposed on the wall structure itself.

2. *Water in the protoplasm*

By comparison with the cell wall, the water content of active protoplasm is as high as 95% even though it may drop to much lower levels when there is inactivation due to temperature, water stress or other factors. The properties of protoplasm are dependent, to a considerable degree, on its water content, which determines whether it exists in fluid state or as a gel, and hence influences its elastic, cohesive and structural properties. Useful reviews of protoplasmic structure include those of Virgin (1953), Frey-Wyssling (1953), and Seifriz (1956).

Protoplasm consists predominantly of proteins and water, but also contains lipids, sugars, salts and other compounds. Proteins are hydrophilic colloids and the strong affinity of proteins for water is due largely to the hydrophilic groups of the amino acid side chains which are located along the polypeptide backbone of each protein molecule. Some of these side chains contain non-polar hydrocarbon units which lack affinity for water, but the oxygen and nitrogen atoms are hydrophilic and water is strongly oriented around them, mainly by hydrogen bonding.

The protein chains themselves appear to be folded and coiled into a loose molecular network in which van der Waal forces, hydrogen bonds, and other bonds, such as salt and ester bridges, provide linkages between the polypeptide chains. During protoplasmic streaming, there must be a continual breaking and reforming of such linkages, with the activation energy presumably arising in oxidative reactions (Stocking, 1956a). Within the mesh so established, a considerable amount of water is held. Up to distances of 1 mμ from the protein surfaces, this water is strongly bound in an ice-like structure. Beyond that, up to about 10 mμ, Bernal (1956) considers that the water contains some ions, but still has partially restricted movement.

Although the principal hydrophilic colloids in cytoplasm are proteins, Stocking (1956a) points out that there are numerous other compounds with polar properties, the presence of which may greatly alter protoplasmic hydration. The cytoplasm, bounded by the plasmalemma and tonoplast, and containing extensive ion binding sites, constitutes a Donnan system with characteristics similar to, but, in most respects, more strongly developed than, the cell wall. Thus the volume and water content of the cytoplasm can also be influenced by the nature and concentration of the solutes in the external solution.

Apart from the· examples already given, cytoplasmic volume will always increase at the expense of vacuolar volume if the external solution contains ions which enter the cytoplasm but not the vacuole. Kappen-plasmolysis is a terminal stage of such a phenomenon. It also follows from what was said about the Donnan phase in the wall that, if the valency of the mobile cations present in the cytoplasm is greater than that of the cations in the external solution, replacement by the latter will increase the osmotic pressure in the cytoplasm since the exchange is on an equivalent and not a molar basis (Briggs, 1957).

The plastids, mitochondria and nuclei are separated from the remainder of the cytoplasm by distinct membranes (Granick, 1956; Briggs, Hope and Robertson, 1961), and exhibit osmotic volume changes, the whole remaining in equilibrium with the surrounding cytoplasm. The water content of chloroplasts is generally less than that of the cytoplasm and approximates to 50% (Rabinowitch, 1945) because of the concentration of lipids and lipophilic materials. These can amount to 20–40% of the dry weight. Few data are available on mitochondria but, in view of the similar lipid content, comparable figures can be expected.

3. *Water in the vacuole*

The water content of the vacuole is the highest of the three cell phases, frequently reaching levels of about 98% (Stocking, 1956a). The solids consist of a mixture of sugars, organic acids and salts, inorganic salts and colloidal material, and the vacuolar sap can, under these circumstances, be regarded as a true solution. In some species, however, or in tissues inactivated by senescence or stress, the proportion of colloidal material becomes significant and much lower water contents occur. The vacuolar fluid then has many of the properties of a gel (Roberts and Styles, 1939; Guilliermond, 1941).

In vacuoles of high water content, the forces retaining water in the vacuole are primarily osmotic, associated with the concentration of the vacuolar solutes, the differential permeability of the tonoplast to

ion and water transport and the confining pressure imposed on the
water by the cell wall. In those with a high proportion of colloidal
material, however, much of the water will be adsorbed on the colloidal
surfaces and the sap cannot be regarded as an aqueous solution.

4. *Water in the vascular system*

The composition of xylem and phloem sap is generally quite different,
as follows from the different functions of these tissues; the xylem
containing mainly water and mineral salts absorbed from the soil, and
the phloem containing mainly metabolites produced in the leaves and
redistribution products.

The xylem is of particular interest in cell and tissue water relationships since it provides the distributory system by which all tissues
receive their water supplies. In angiosperms, the xylem tissues are
composed of vessels, tracheids, fibres, wood parenchyma and xylem
ray cells. The proportion of each tissue type in different species and
different organs of the one plant varies widely, but the vessels
characterize the angiosperm xylem more than any other component.

These tubular structures contain no protoplasm, and may extend
for several metres of the xylem, often with perforated cross-walls at
intervals along their length. Their radius in woody plants generally
ranges from < 10 to > 100 μ and in vines and lianas may exceed 1 mm.
The walls are pitted and, although the function of the pits is not fully
understood, they may exceed 1 μ radius. In some cases, they are covered
by membranes, but the membranes themselves appear to be perforated
by pores of > 0.1 μ radius (Eicke, 1954; Liese and Johann, 1954). These
structures undoubtedly facilitate water transport from vessel to vessel.
In grasses and more ephemeral plants, vessel radii may be as small as
1–2 μ (Emerson, 1954), but Wind (1955b) found with several grasses
that, as long as root radius exceeded 0.1 mm, vessels of 10 μ radius
existed.

Tracheids, found in the wood of many angiosperms, also contain
no protoplasm. They are typically spindle-shaped cells with pitted
lignified walls. They occur in longitudinal stacks neatly fitting into
one another to form effective water conducting tissue. Individual
tracheids reach 5 mm in length and 10–20 μ in radius. In gymnosperms,
vessels do not occur and highly developed tracheid systems constitute
the water conducting system.

In the walls of the vessels and tracheids, water is retained in much
the same way as in normal cell walls, except that a secondary thickening
mainly involving lignification, reduces the volume and extent of the
voids. In consequence, such water as occurs in the walls is tightly

retained by surface forces emanating from the wall matrix. Under normal conditions it is probable that there is little water flow longitudinally or laterally in the thickened areas of the walls, although free movement occurs through perforations in the walls.

Most of the water in the vessels and tracheids themselves is free in aqueous solution. During transpiration, the water can be subjected to tensions in excess of 100 bars, the columns remaining intact due to cohesive bonding between adjoining molecules, and due to the low air entry values of those outer walls exposed to intercellular air. Perforations on such walls must therefore be much smaller than those connecting individual xylem elements.

Occasionally, however, the water columns become ruptured as a result of mechanical damage or air entry. In such cases, the affected vessel or tracheid tends to drain if the air entry value appropriate to the vessel radius is less than the tension prevailing on the water. Drainage under such circumstances does not usually involve the entire vessel and frequently terminates at the first cross-wall. In xylem elements primarily composed of tracheids, drainage is often restricted to a single tracheid, depending on the size of the perforations between adjoining cells.

5. *Entire cells and tissues*

In whole cells or tissue segments the various forces retaining water, in each of the phases already discussed, interact continuously to produce a general relationship between relative water content and water

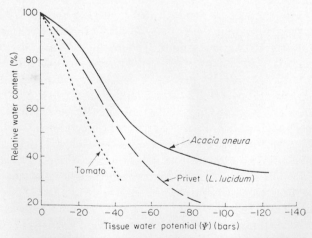

Fig. 5.3. Relative water content/water potential relationships for three types of leaf tissue (after Slatyer, 1962c).

potential of the form shown in Fig. 5.3. These curves are effectively sorption isotherms for the tissue segments concerned and hence are similar in form to the Ψ (θ) curves for soils given in Chapters 3 and 4 and to the pine wood data of Fig. 5.2.

Curves of this type clearly demonstrate the inadequacy of the "bound water" concept of water retention. The amount of water retained after any particular bound water determination reflects nothing more than a point on the Ψ (θ) curve. Even so, the foregoing discussion will have brought out clearly that a significant proportion of the water in a cell is not free in the sense of an aqueous solution. In the same way as with animal cells (Perutz, 1946; Savitz, Sidel and Solomon, 1964) this water is likely to be unavailable for participation in osmotic concentration changes. Almost all of it may be involved in isotopic exchanges (Raney and Vaadia, 1965a) although there is evidence that the water in some cell fractions exchanges less readily than in others (Vartapetyan and Kursanov, 1959; Vartapetyan, 1960a, 1960b; Vasil'eva and Burkina, 1960; Slatyer, 1962b; Slavik, 1963c).

2. TERMINOLOGY IN PLANT WATER RELATIONS

An effective and comprehensive terminology of plant water should take into account the major groups of forces which operate in the plant water system. Further, this should be done in a manner which permits the evaluation of the various forces by experimental measurements, and which is closely linked to water transport and metabolic function. However, the structure of plant cells and their combination into different types of tissue, each with different amounts of vascular elements, cell wall, cytoplasm and vacuole, makes this extremely difficult and a number of simplifying assumptions have been introduced in past years. The most important of these involve the concept of the "typical" plant cell as an ideal osmometer, comprising an elastic cell wall of negligible dimensions, a single differentially permeable "membrane" also of negligible dimensions (comprising the cytoplasm and its bounding membranes) and a vacuole containing an aqueous solution which conforms to the ideal gas equations.

Although these assumptions involve considerable simplification, they do not necessarily involve oversimplification for many cells, and even for assemblages of cells. In fact, for large coenocytic algal cells and for storage tissue such as carrot or potato tuber, they can provide close approximations to actual situations. Applied to leaf and root tissue, the main general principles of water equilibrium and exchange usually can be evaluated on the basis of the typical cell, but a number of important exceptions occur. As will be discussed later, these result partly from the cell membranes being somewhat permeable to solutes, so that flows of water are generally associated with flows of other sources of matter or energy (such as solutes, electricity and heat) and partly to the finite volume and specific water distribution and retention characteristics of the non-vacuolar components. The following discussion will deal first with the characteristics of simplified cell systems and then with those of other cells and tissues.

A. Terminology for the Ideal Osmotic Cell

In an idealized cell, such as has just been described, the difference between the chemical potential of the water inside (μ_w) and outside (μ_w^0) the cell is assumed to be determined only by the difference between the hydrostatic pressure (ΔP) and the activity of water ($RT\Delta \ln a_w$) inside and outside the cell, so that, from Chapter 1, an appropriate expression for this situation is,

$$\Delta\mu_w = \mu_w^i - \mu_w^0 = \overline{V}_w\Delta P + RT\Delta \ln a_w \tag{5.1}$$

where μ_w is in units such as erg mole^{-1}, $\overline{V}_w$ is the partial molal volume of water in cm^3 mole^{-1}, P is in dynes cm^{-2}, R is the ideal gas constant (erg mole^{-1} deg^{-1}) and T is the Kelvin temperature. If the cell is immersed in pure free water so that μ_w^0 is zero, ΔP becomes P, the pressure inside the cell in excess of atmospheric pressure and $RT\Delta \ln a_w$ becomes $RT\ln a_w$ the activity of the water inside the cell. At equilibrium ($\mu_w^i = \mu_w^0$) the equation gives the classical definition of the osmotic pressure, π, developed in an osmometer in which, for an ideal solution, $a_w = N_w$ (where N_w is the mole fraction of water):

$$P = \pi = \frac{-RT}{\overline{V}_w} \ln N_w \cong \frac{RT}{\overline{V}_w} \ln N_s \cong RTc_s \tag{5.2}$$

where the approximations give the commonly used expression for osmotic pressure ($\cong RTc_s$), and N_s and c_s are the mole fraction and molal concentration of non-electrolyte solute, respectively.

When there is pure free water outside the cell, or if pure free water is used as a reference state, the term $(\mu_w^i - \mu_w^0)/\overline{V}_w$ is the water potential, Ψ, which can then be expressed as

$$\Psi = P - \pi \tag{5.3}$$

where Ψ is in units of energy/volume such as erg cm^{-3}, equivalent to pressure units of dynes cm^{-2} and usually expressed in bars or atm.

Since the chemical potential of pure free water is arbitrarily assigned the value of zero energy, Ψ, in plant systems, is generally less than zero. It is probably only positive during the process of guttation or similar phenomena.

As in the case of soils, the component potentials need not be written as P and π but can be expressed, if desired, in a form similar to Eq. 3.16, used for soil water

$$\Psi = \Psi_p + \Psi_s \qquad (5.4)$$

where Ψ_p is a pressure potential identified with turgor pressure and Ψ_s is a solute or osmotic potential, which is the negative of π, so that

$$\Psi_s = \frac{RT}{\overline{V}_w} \ln N_w = -\pi$$

As shown in Chapter 3, a matric potential, Ψ_m, gravitational potential, Ψ_z, and other component potentials may also be included if necessary.

The relationships between hydrostatic pressure (usually referred to as turgor pressure) P, osmotic pressure, π, and water potential, Ψ, in the ideal vacuolate plant cell have frequently been portrayed in a diagram of the type shown in Fig. 5.4 after Höfler (1920).

With decreasing volume the gradual increase in π due to solute concentration can be seen, as can the more rapid decline in P. Ψ is

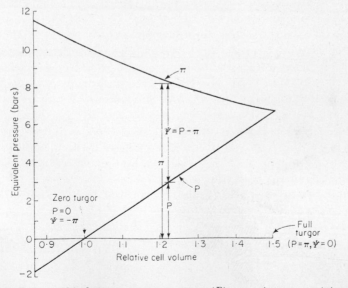

FIG. 5.4. Relationship between turgor pressure (P), osmotic pressure (π), and water potential (Ψ) in an idealized osmotic cell.

zero at full turgor, when $P = \pi$, and falls progressively with decreasing cell volume. The P/V relationship is shown as being more or less linear in this model. While this is reasonable as a first approximation, in some cells and tissues non-linear behaviour undoubtedly occurs, particularly when the cell approaches full turgor and P approaches π ($\Psi \to 0$; $P \to \pi$) (Crafts, Currier and Stocking, 1949). At $P = 0$, $\Psi = -\pi$, and the cell is at incipient plasmolysis.

If the observed volume change occurs when the cell is immersed in a plamolysing solution, the concentration of which is being progressively increased, continued water efflux causes plasmolysis, cytoplasmic detachment from the cell wall being associated with continued shrinkage of the protoplast and entry of the bulk external solution into the space between the outer protoplast surface and the wall. This phenomenon is considered in more detail in Chapter 6. For the present, however, it should be noted that plasmolysis is, in many respects, a laboratory phenomenon, which seldom occurs during dehydration under natural conditions.

During plasmolysis, as more water is removed from the protoplast, adhesion between the protoplast and the wall seldom seems to cause the development of significant negative pressures on the water within the protoplast (Crafts, Currier and Stocking, 1949) although this point warrants further study. Under natural conditions, however, it is frequently suggested (see, for example, Bennet-Clark, 1959) that substantial negative pressures may develop.

Further discussion of cell water relations on the basis of the completely vacuolated plant cell is not warranted at this stage, but it is important to note that many valid general principles have been established with it. Moreover, for many considerations of water movement in and out of cells, the water potential, and the relative values of P and π, inside the vacuole and outside the cell, are the important experimental parameters (see Chapter 6) since the vacuole is a sink (or source) for movement of both water and solutes. Consequently, while it is desirable to understand, and be able to classify more comprehensively, the characteristics of the water located elsewhere, the vacuolar characteristics remain of special importance. Because of this it is surprising to find that very few direct measurements of P have ever been made, particularly since Eq. 5.3 has beer so widely accepted.

B. Other Terms Used to Describe the Chemical Potential of Water

As plant physiology has developed, many terms have been used to describe the chemical potential of water in plants. Although it was not, at first, appreciated that differences in the chemical potential of water were being evaluated, it was recognized that any effective term had to include not only the effect of osmotic solutes, but also the effects of hydrostatic pressure, and terms such as "net osmotic pressure" (Shull, 1939; Lyon, 1941) and "water absorptive power" (Thoday, 1918) were introduced for this purpose. "Saugkraft" was introduced by Ursprung and Blum (1916) and "suction force", "suction tension" and "suction pressure" by other workers (Stiles, 1922; Beck, 1928). The most commonly used term, up until the present time has been *"diffusion pressure deficit"* (DPD) introduced by Meyer (1938) and subsequently slightly modified by Meyer (1945) and Meyer and Anderson (1952). Meyer (1945) defined diffusion pressure as "that physical property of a substance which is responsible for its diffusion whenever other prevailing conditions permit the occurrence of this process", and the diffusion pressure deficit of water in a solution or biological system as "the amount by which its diffusion pressure is less than that of water at the same temperature and under atmospheric pressure".

As Crafts, Currier and Stocking (1949) point out, diffusion pressure can be regarded as a function of the partial Gibbs free energy of water and, to this extent, can be related to the chemical potential of water, μ_w, as used here. However, to avoid the use of negative quantities, Meyer and Anderson defined DPD in a manner opposite to conventional thermodynamic usage so that

$$DPD = \frac{(\mu_w^0 - \mu_w)}{\bar{V}_w} \qquad (5.6)$$

where μ_w^0 and μ_w are the chemical potentials of pure free water and water in the system, respectively.
Hence

$$\Psi = \frac{(\mu_w - \mu_w^0)}{\bar{V}_w} = -DPD \qquad (5.7)$$

and

$$DPD = \pi - P \qquad (5.8)$$

This last expression has generally been written

$$DPD = OP - TP \qquad (5.9)$$

where OP and TP are osmotic pressure and turgor pressure, respectively.

There are several reasons why it is thought the term DPD should be abandoned. Among them is the point that diffusion pressure is no longer used as a term in physical chemistry or thermodynamics, and that diffusion is no longer considered as the only mechanism by which water will move under the influence of free energy or chemical potential gradients. The fact that, for most biological work, DPD is regarded as being equal to the difference between the chemical potential of water in the system and in pure free water makes it desirable to use a more basic term since from Eq. (5.7) it is seen that the sign used for DPD is opposite to conventional thermodynamic usage. Also, the use of a three-symbol abbreviation is unnecessary and somewhat confusing. For such reasons the term is not used in this book.

A number of thermodynamic expressions, other than chemical potential, have been used from time to time, including Edlefsen's (1941) "specific free energy" and Broyer's (1947) "net influx specific free energy". However, the widespread use of chemical potential in physical chemistry and thermodynamics, and the adoption of potential terminology in soil physics, and by a number of workers in plant water relations (Owen, 1953; Slatyer and Taylor, 1960; Taylor and Slatyer, 1961, 1962; Dainty, 1963a; Ray, 1960; Slatyer, 1966c), appear to provide more than sufficient justification for its widespread use in this field. For this reason the term water potential, $\Psi[= (\mu_w - \mu_w^0)/\overline{V}_w]$ is used in this book with the dimensions of energy per unit volume so that the existing units of bars and atmospheres can be retained, but use of μ_w, in units of erg mole^{-1} or of μ_w/M_w in units of erg gm^{-1} (where M_w is the molecular weight of water) presents no difficulty.

One additional term which warrants mention is "hydrature" first introduced by Walter (1931). Subsequently, Walter has modified its derivation and use to some degree and now (Walter, 1955, 1963; Kreeb, 1963) considers the "hydrature" to be identical with relative humidity, defined as "identical with that specific water state which is given by the relative vapour pressure (expressed as a percentage) of hydrophilic colloids or solutions with a free surface". Thus:

$$hy = 100 \ e/e^0 \qquad (5.10)$$

This definition means that the pressure component of Ψ is never included in the hydrature, so that the hydrature will vary from point to point in the cell, depending on the magnitude of the pressure component, even though Ψ is constant throughout.

This leads to ambiguities since, in the case of an entire cell or tissue segment, hydrature can be uniquely related to Ψ by the expression

$$\Psi = \frac{RT}{\overline{V}_w} \ ln \ (hy/100) \qquad (5.11)$$

whereas in the case of the vacuole, hydrature can be uniquely related to π by a similar expression

$$\pi = \frac{-RT}{\bar{V}_w} \ln (hy/100) \qquad (5.12)$$

This is because the outer surface of the entire cell or tissue is only subject to atmospheric pressure, hence there is no pressure component of Ψ, but vacuolar sap is under positive turgor pressure. In the first case therefore, hydrature can be measured with normal measuring technique for Ψ. In the second case, it is measured with normal measuring techniques for π. Thus, confusion can easily result from use of the term hydrature unless the location of the point where it is being considered is indicated.

C. Expanded Terminology for Other Cells and Tissues

In arranging the various forces concerned with water retention in normal plant cells and in tissues into groups which can be recognized and, preferably, measured, it seems that three main groups can be set up. These will be associated with the pressure on the water, the influence of osmotically active solutes (as have already been discussed) and with the constraining influence of the cell and tissue matrix. The latter will include primarily those surface forces which are associated with protoplasmic hydration, with the water in the cell walls and intercellular spaces and with the water in the vascular elements. The effect of electrical force fields and effects due to forces such as gravity can also be included when appropriate but the three main groups of forces will generally dominate equilibrium water relations.

With these groupings, an expression for the chemical potential of water can be written along the same lines as used for swelling soils. However, the extreme heterogeneity of plant tissue, with its three intra-cellular phases of wall, cytoplasm and vacuole, and the extra-cellular vascular system phase, means that a general expression for a whole cell, or assemblage of cells, does not account for differential changes in the different phases. The situation, therefore, is different to that of swelling soils, which are, by comparison, quite homogeneous. In consequence, any description which aims to be quantitative applies really to only one phase, since the component potentials contributing to the water potential of a phase may be of different magnitude, and may change to different degrees with changing water potential, to those in other phases.

With this proviso, however, the assumption of homogeneity within phases appears reasonable, at least at first approximation level, and a more microscopic approach is probably unwarranted at the present time, when measurement techniques to evaluate component potentials in different phases are, in many cases, non-existent.

In developing an expression for the water potential for a phase, two points to be borne in mind are the nature of the pressure term, and the adequacy of the expression to account for volume changes in the tissue at constant external pressure. The pressure term, regarded as an external gas pressure in the case of non-swelling systems, is now identified with a swelling pressure. It is therefore the pressure that needs to be applied to the system under consideration to prevent entry of water, when all other conditions are held constant. Inside the cell it is therefore equal to turgor pressure.

The phenomenon of volume change at constant external pressure may arise following a simple entry of water, in which case it can be adequately accounted for by including a term for the dependence of water potential on volumetric water content. It may also arise, however, due to an altered geometrical arrangement of the solid matrix. If this occurs, it may not be adequately accounted for by the above procedure, in which case an additional term must be included in the expression for the water potential. This leads to an expression similar to that of Eq. (3.18):

$$d\mu_w = -\bar{S}_w dT + \bar{V}_w dP + \left(\frac{\partial \mu_w}{\partial c_s}\right)_{T, P, n_j, n_w, \chi} dc_s + \left(\frac{\partial \mu_w}{\partial \theta}\right)_{T, P, n_s, \chi} d\theta$$

$$+ \left(\frac{\partial \mu_w}{\partial \chi}\right)_{T, P, n_s, n_w} d\chi \qquad (5.13)$$

where the symbols have the same meaning as in Chapter 3, except that P, the external pressure, is now regarded as a swelling pressure, and θ is the volumetric water content (cm^3 water cm^{-3} tissue) and can be related to the relative water content (see page 150).

For many purposes it may be unnecessary to include the geometric term, particularly if unique $\Psi'(\theta)$ relationships can be established, under specified but realistic conditions, for the various cell phases. In such cases, Eq. (5.13) simplifies to

$$d\mu_w = -\bar{S}_w dT + \bar{V}_w dP + \left(\frac{\partial \mu_w}{\partial c_s}\right)_{T, P, n_j, n_w} dc_s + \left(\frac{\partial \mu_w}{\partial \theta}\right)_{T, P, n_s} d\theta \qquad (5.14)$$

where the first term on the right hand side is the entropy function, not present when isothermal situations are described; the second term describes the dependence of the water potential on pressure; the third term describes the osmotic contribution of the dissolved solute species, and the final term is the matric function.

When the temperature is held at the reference temperature and the system is allowed to change from the state of pure free water to some definite condition in the plant water system, Eq. 5.14 can be applied to both the initial and final states and rewritten in a form similar to Eq. 3.11.

$$\Delta\mu_w = \mu_w - \mu_w^o = \overline{V}_w\Delta P - \overline{V}_w\Delta\tau - \overline{V}_w\Delta\pi \tag{5.15}$$

where P, τ and π are actual pressure, matric pressure ($=$ matric suction) and osmotic pressure ($=$ solute suction), respectively. Since the water potential, Ψ, $= \Delta\mu_w/\overline{V}_w$, Eq. 5.15 can be written, with respect to free water

$$\Psi = P - \tau - \pi \tag{5.16}$$

or substituting from Eq. 3.16

$$\Psi = \Psi_p + \Psi_m + \Psi_s \tag{5.17}$$

where Ψ_p, Ψ_m and Ψ_s are the pressure potential, matric potential and solute potential, respectively. The units, as before, are of energy/volume, equivalent to dyne cm^{-2} and usually expressed in bars or atm. Ψ_p may be positive, but Ψ_m and Ψ_s are always negative since the matrix and solutes reduce the capacity of the water to do work. The total water potential is generally negative or zero, only reaching positive values during exudation, guttation and similar phenomena.

D. Relative Importance of the Various Component Potentials in Different Regions of Cells and Tissues

It is now of value to examine the relative importance of these three groups of forces in different regions of cells and tissues. It can be assumed for this purpose that the system is at equilibrium and that the water potential is everywhere the same.

1. Component potentials in the protoplast

Where the vacuolar sap can be regarded as a true solution, the matric term, τ, can be disregarded so that Eq. 5.3 provides an adequate description of the forces contributing to the vacuolar value of Ψ. Where the vacuolar sap contains a significant proportion of colloidal material, and in the cytoplasm, all three groups of forces may be represented. It is unlikely that the pressure on the water in the cytoplasm is much, if at all, different from that in the vacuole, although some differential can conceivably occur if the tonoplast has definite elastic properties. Any significant pressure difference would be avoided by water transfer between the cytoplasm and vacuole with a corresponding change in their relative volumes and an equilibration of the water potential in each phase.

Because a component of the potential is due to the retention of water by the surface forces associated with the cytoplasmic matrix, however, the osmotic pressure of the relatively free water in the cytoplasm can be expected to be less than that of the normal (true solution) vacuolar sap since $(\tau_{cyto} + \pi_{cyto}) = \pi_{vac}$. It is appropriate to mention again, at this point, that it is the orientation and concentration of ions between the charged surfaces of the cytoplasmic matrix which, by lowering the water potential in the immediate vicinity, induces an inward water flow and results in the swelling of the cytoplasm. The increased local concentration of ions maintains a higher osmotic pressure close to the surfaces than at points further away because swelling cannot proceed freely due to the restraining influence of the cell wall.

The Donnan equilibrium, established between the cytoplasm and the external solution, is associated with these surface forces. The resultant higher ionic concentration inside the cytoplasm, compared with outside the cell, does not invalidate the $\pi_{vac} > \pi_{cyto}$ relationship proposed above since the diffusivity of the vacuolar solutes is even more restricted and, in general, the accessibility of the different cell regions to ions declines from wall to cytoplasm to vacuole (MacRobbie and Dainty, 1958; Pitman, 1963).

In non-vacuolated cells, typically meristematic, or in cells in which the vacuole is itself a gel, much of the water will be retained by matric forces and relatively little free "solution" may exist. In extreme cases it is probable that there will be no osmotic pressure term and the water potential within the cell will be influenced only by the matric potential and the pressure exerted on the cell contents by the cell wall.

As cell water content decreases, the osmotic and matric forces increase in magnitude since both are dependent on water concentration. The positive pressure component drops rapidly and may become negative if the water in the protoplast passes into a state of tension

The progressive changes in Ψ and π, with decreasing water content, are shown in Fig. 5.5 for leaf tissue of a mesophytic species, tomato, and a xerophyte *Acacia aneura*.

In the case of tomato, an initial divergence of the Ψ and π curves for each species, due to positive turgor pressure, P, is apparent and, as water is removed and P falls, the two curves approach, presumably intersecting the stage of $P = 0$; $\Psi = -\pi$. Subsequently, the two curves

follow one another very closely, suggesting that negative values of
P do not develop. In the *Acacia*, however, the Ψ and $-\pi$ curves
diverge as more water is removed. It is this divergence which is regarded
as evidence of negative turgor pressure and it is proposed that it arises
because of the resistance of the wall to inward contraction. The
examples of the fern sporangium (Ursprung, 1915) and the liverwort
elator (Renner, 1915) also are quoted as evidence of this phenomenon.
However, it is not established that negative wall pressures can develop
of the magnitude necessary to explain observed differences between
Ψ and π or that such pressures, if they are of importance, are the only
forces involved.

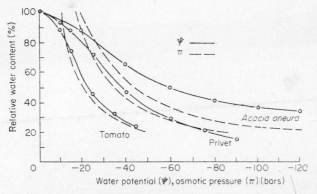

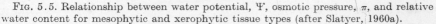

FIG. 5.5. Relationship between water potential, Ψ, osmotic pressure, π, and relative
water content for mesophytic and xerophytic tissue types (after Slatyer, 1960a).

It is doubtful, for instance, whether cell walls, in general, resist
compression and infolding since they are constructed to resist expansion.
Furthermore, if the water is under tension, there is the possibility that,
as the pressure in the cell drops below atmospheric pressure, dissolved
gases in the vacuolar and cytoplasmic fluids may come out of solution
and effectively prevent the pressure from becoming negative. There
is also the point that, since not all the cell water contributes to the
effective osmotic concentration, data of the type in Fig. 5.5, based on
assumed changes in π caused by relative water content changes,
should be recalculated on the basis of "exchangeable" water content
alone, leading to an enhanced value of π at any relative water content,
and hence reducing, or eliminating, the discrepancy between the two
sets of curves (Slavik, 1963c).

Even so, if there is no gas evolution and if contraction of the cell is
prevented, the water in the protoplast must be under tension. Although

this can be regarded as a negative value of the pressure component, it can also be treated as a matric component of the total potential, as in the case of water in the vascular system.

2. Component potentials in cell walls

In examining the components of Ψ in the wall, it is preferable to consider not just the wall material but all the volume external to the protoplasts, with the exception of the vascular elements. This, therefore, includes specific wall material abutting on intercellular spaces, inter-woven common walls adjoining cells and the intercellular spaces themselves.

In this region the water may also be influenced by the three groups of forces of Eq. 5.16. Now, however, the pressure component can probably be neglected, since the presence of air spaces means that there is continuity with the outside air, and, regardless of the tortuosity of the pathway, gas pressure will be almost the same as atmospheric pressure, as is the case with unsaturated soil.

If the extra-protoplast volume is saturated with water, positive pressures may possibly develop, but it is doubtful if they can be much above atmospheric pressure. Positive pressures could arise from the tendency of the surface forces on the cell wall matrix to draw in water and hence develop a swelling pressure. Donnan phenomena, which cause higher osmotic pressures in the Donnan phase than in the intercellular solution, can also tend to increase the local pressure if water influx is restricted by compression between cells. Guttation and root pressure phenomena may also cause positive pressures, but it is doubtful (Kramer, 1956a, 1959) if they exceed 1–2 atm because excess pressure can be avoided by exudation of water from the system. Any positive pressures will only develop in effectively turgid tissue. As soon as the water potential falls below $\Psi = -2$ bars there is virtually no liklihood of positive pressures developing.

The osmotic pressure, π, of the fluid in the interfibrillar spaces, will be greater than that in the effectively free solution in the xylem because of Donnan effects, but less than that in the cytoplasm or vacuole, where solute accumulation is much more pronounced. In consequence, much of the water in the wall phase is retained by matric forces emanating from the solid surfaces of the phase and arising from the curvature of the liquid-air menisci in the interstices between the wall fibrils.

As was the case with the protoplast water, both matric and osmotic forces will increase as the cell wall water content is reduced. There will initially be some shrinkage of the wall, but the matric forces will increase rapidly as the solid surfaces are drawn into closer proximity and the rate of loss of water will be progressively reduced. If the interfibrillar spaces are smaller than 100 mμ radius, drainage will not commence until the water potential falls below $\Psi = -15$ bars. If spaces are smaller than 10 mμ, the limiting value will be $\Psi = -150$ bars, a value which would result in the death of many types of tissue.

3. *Component potentials in the vascular elements*

The osmotic pressure of the xylem sap is usually of the order of $\pi < 2$ bars (van Overbeek, 1942; Broyer, 1951; Arisz, Helder and van Nie, 1951), but in the phloem, substantially higher osmotic pressures may develop (Weatherley, 1962). The relative importance of pressure and matric component potentials therefore differs in these two types of tissue.

In the phloem positive pressures generally occur, as evidenced by exudation experiments, and values of the order of $P = 30$ bars have been recorded (Weatherley, 1962). It is probable that negative pressures only develop transiently, if at all, in the phloem under normal conditions. By comparison, in the xylem, positive pressures only reach values of $P = 1$–2 bars, as indicated from root pressure, guttation and exudation experiments (Kramer, 1956a), and the normal situation is for substantial negative pressures to exist.

These negative pressures may be assigned to either a pressure component or a matric component of the total water potential, and, since the pressure term has been reserved for external gas pressure, or a "swelling pressure" in the case where positive pressures develop, it is convenient to assign the negative pressures which develop in the xylem, when the water comes under tension, to the matric component.

These forces arise from cohesion between individual water molecules and adhesion between the water molecules and the vessel or tracheid walls, as discussed previously. Because of the size of most of the conducting elements (radius $> 10 \mu$), they would drain even before the water potential was reduced to $\Psi = -1$ bar, if the liquid-air menisci at the peripheries were of similar radius. Clearly the menisci are of much smaller dimensions both in the outer walls of the vessels and tracheids, and at their terminals in the leaf or root, and are probably of a similar order to the interfibrillar spaces of cell walls. Thus it is possible for the water to pass into a state of tension as water loss exceeds water entry and the water potential falls. There appears to be relatively little resistance to liquid flow through the xylem (see Chapter 6) so the tension in the water is presumably approximately equivalent to the water potential in the leaves. This suggests that values of the order of $\Psi = -100$ bars may occur in desiccated woody plants, and that values equivalent to $\Psi = -10$ bars may commonly occur in crop plants. Values in this range have been estimated by MacDougal, Overton and Smith (1929) and Stocking (1945).

3. Measurement of Plant Water Status

As with soil water measurements, the following treatment is not designed to give a detailed account of various techniques and instruments for measuring plant water but rather to discuss the principles of the more important methods and the more important limitations which they contain. For additional information the reader is referred to Crafts, Currier and Stocking (1949) and Slatyer and McIlroy (1961).

A. Measurements of Tissue Water Content

Tissue water content measurements are generally made on leaves but have also been reported for most tissues and organs including, particularly, stem sections and fruit. For non-photosynthetic tissue, expression of the water content on a dry weight basis, analagous to that used for soils, is frequently quite satisfactory. The tissue is generally sampled and weighed to give a fresh weight (W_f), oven dried to constant weight at a temperature of 85–90°C and re-weighed to give a dry weight (W_d), and the water content $(W_f - W_d)$ expressed as a percentage of the dry weight, thus:— $100 \ (W_f - W_d)/W_d$. Expression of the results on a fresh weight basis [i.e. $100 \ (W_f - W_d)/W_f$] is unsatisfactory for almost all purposes because the denominator is not a constant. For leaf tissue, expression on a dry weight basis may also be unsatisfactory for the same reason, since dry weight changes diurnally due to photosynthesis and over longer periods due to growth.

For leaf tissue a much better denominator appears to be turgid water content, generally obtained by floating the tissue in water until the water deficit existing at the time of sampling is eliminated. The turgid weight (W_t) is then introduced into the "relative turgidity" expression developed by Weatherley (1950, 1951), and now often called "relative water content", thus

Relative water content (or relative turgidity)

$$= 100 \left(\frac{W_f - W_d}{W_t - W_d} \right) \qquad (5.18)$$

It can also be introduced into the "wasser defizit" or water deficit expression developed by Stocker (1929)

$$\text{water deficit} = 100 \left(\frac{W_t - W_f}{W_t - W_d} \right) \qquad (5.19)$$

Water deficit has also been called water saturation deficit (Oppenheimer and Mendel, 1939). Both of these terms are usually expressed as percentages (as shown), and each complements the other (i.e. relative water content $= 100 -$ water deficit).

The validity of both methods depends on obtaining reproducible and reliable estimates of turgid water content, W_t. In this regard Stocker (1929) suggested placing intact leaves, after sampling and weighing, in small humid chambers with their petioles in water until they became turgid. They were then removed, weighed to give turgid

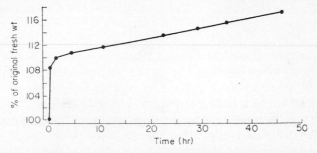

FIG. 5.6. Water uptake by floating leaf discs of *Ricinus* (after Barrs and Weatherley, 1962).

weight, oven dried and reweighed. Although this procedure has been used with success by some investigators and appears to be especially suitable for tissue such as pine needles, the elimination of the water deficit was observed to take 2–3 days in some cases and, during this time, important metabolic changes may occur in the tissue. In order to improve on this procedure, Weatherley (1950, 1951) proposed using leaf tissue segments and discs instead of entire leaves, and floating them on water in covered petri dishes to hasten water uptake. Catsky (1960) has, instead, placed discs in contact with water filled polyurethane foam, with satisfactory results.

Barrs and Weatherley (1962) subsequently showed that water uptake could be divided into two fairly distinct phases, the first associated with elimination of the passive water deficit and the second with renewed uptake associated with continued tissue growth (see Fig. 5.6). They considered these two phases to be independent (i.e. phase two did not commence until phase one was completed); phase one was generally completed in 4 hr, and phase two could be minimized by metabolic inhibitors. Although the first point is open to argument, the rapid elimination of the water deficit and the relatively slow uptake in phase two means that for most purposes it is unnecessary to use inhibitors as long as the water uptake characteristics of the tissue to be used in an experiment can be evaluated beforehand and an appropriate floating time adopted. Since water temperature affects both the rate (Werner, 1954) and amount (Barrs and Weatherley, 1962) of phase one uptake in some tissues, its use as a metabolic inhibitor is precluded even though it is most effective in suppressing phase two uptake. However in some mature tissue, e.g. *Acacia aneura*, Slatyer and Barrs (1965) found that the amount of phase one uptake was the same at 5° and 20° and in such cases low temperature floating can be a desirable procedure.

Other sources of error in turgid water content determinations are due to injection of water into the intercellular spaces, to changes in dry weight during the floating period and to surface drying after floating. Injection is not detectable in some tissues but is important in others. It is always reduced if sharp cutting tools are used and in some cases can be eliminated if the tissue is illuminated, due perhaps to continued photosynthesis and oxygen evolution.

Although dry weight errors are often unimportant with a floating time of 4 hr, illumination of the tissue at a light intensity approaching the compensation point also minimizes this source of error. The importance of surface drying procedure varies with the compressibilty of the leaf tissue and with its surface conformation. Slatyer and Barrs (1965) showed that with smooth rigid leaf tissue such as privet or *Eucalyptus*, little error was introduced by a range of drying procedures, but with tomato leaves it was most important to adopt a standard procedure which, as far as possible, removed the surface water without pressing out any sap. Apparent differences of up to 5% relative water content could be caused by different observers or by the same observers in the absence of a standard procedure.

In summary, relative water content appears to be the best basis for water content expression, and the disc (or segment) technique can be considered as a sound and quantitative procedure, provided values which frequently enable estimates of water potential to be made. Much additional information can be found in papers by Weatherley (1950, 1951), Werner (1954), Weatherley and Slatyer (1957), Slatyer (1955, 1962c), Catsky (1962, 1963), Barrs and Weatherley (1962), Hewlett and Kramer (1963) and Slatyer and Barrs (1965).

The only effective non-destructive method so far developed for water content measurements is a thickness gauge based on beta radiation absorption. It effectively measures leaf density thickness when a beta source such(as C^{14}, Pm^{147}, Tc^{99} or Tl^{204}) is placed on one side of the leaf and a detector on the other side. This technique was first used for leaves by Yamada et al. (1958) and Mederski (1961) and has since been further developed (Whiteman and Wilson, 1963; Nakayama and Ehrler, 1964; Jarvis and Slatyer, 1966b) and employed by a number of workers. Although leaf water content is not always proportional to leaf density thickness, because of differential contraction during removal of water and the fact that leaf density changes when air enters the system, the gauge readings give close agreement with relative water content observations, and the technique is an extremely important tool in water relations investigations.

Limitations to its use arise from changes in leaf thickness (and density) from point to point on the leaf, and with time, as leaf age increases, so that measurements need to be made at the one point, or on a sampling basis. Since the instrument must, in many applications, be removed from the proximity of the leaf between readings, care must be taken to set it up in the same position each time. Also, because of ontogenetic changes in the leaf and accumulation of photosynthate, periodic re-calibrations against relative turgidity are required. The calibration procedure has, however, been much simplified by Jarvis and Slatyer (1966b).

B. Measurements of Water Potential, Ψ

When a cell or tissue segment is at equilibrium with the vapour or liquid which surrounds it, the water potential, Ψ, is assumed to be constant throughout the entire system. The most convenient procedure for the measurement of Ψ is, therefore, to determine its value in the liquid or vapour with which the tissue is in equilibrium. This usually is achieved by one of two main techniques, involving either the direct determination of the water vapour pressure of the air with a thermocouple psychrometer or similar device, or determination of the concentration of the vapour or solution in which the tissue neither gains nor loses length, weight or volume.

The former technique generally utilizes one of the thermocouple psychrometers previously described for determinations of soil water potential (see Chapter 3). No further details are required here except to mention that problems may arise because leaf tissue sometimes appears to behave differently to the wet filter paper used for calibration.

It is thought that this is due in part to the fact that there is much less water in the psychrometer when leaf tissue is being used than when soil or wet filter paper is being measured, and the amount of water which appears to be adsorbed on the psychrometer walls during equilibration may significantly reduce the apparent water content and Ψ of the tissue at the time when the measurement is made. This problem appears to be minimized if the test quantity of leaf tissue is increased.

The leaf is also continuously generating small quantities of heat through respiration, although this is probably negligible in some cases and can be minimized, by the use of metabolic inhibitors, in others (Barrs, 1964). There is also the possibility with the Richards and Ogata (1958) instrument that, since thermocouple output is related to the steady state evaporation of water from the psychrometer wet bulb to sinks in the leaf, the internal diffusive resistance of the leaf to vapour flow can influence the results (Rawlins, 1964). A similar criticism may apply to the technique suggested by Macklon and Weatherley (1965), although Barrs (1965a) has not found evidence of this phenomenon in a comparative study of measurement techniques.

Despite the precautions necessary, the technique appears to be well suited for measurements of Ψ_{leaf} and has also been applied to measurements on intact plants (Lang and Barrs, 1965; Lambert and van Schilfgaarde, 1965). In general, equilibrium methods, rather than steady state flux methods, appear the most satisfactory. Further details can be found in papers by Spanner (1951), Ehlig (1962) and Barrs (1965b).

The liquid or vapour exchange techniques have been used for many years. In practice, several tissue samples of segments, discs, strips, or individual cells are immersed in a range of graded aqueous solutions (or in vapour over a range of solutions), the osmotic pressure, vapour pressure and hence water potential characteristics of which are known (see Eq. 1.46). Water exchange occurs between the tissue and the surrounding liquid, or vapour, and after sufficient time has elapsed to allow measurable changes in weight, length or volume (or in the density or refractive index of the solutions) the samples are removed, re-weighed or re-measured, and the results expressed as shown in Fig. 5.7 after Slatyer (1958). The tissue water potential, Ψ, is read off from the intercept on the diagram.

Of these methods, those involving liquid exchange may incur errors whenever Ψ is lower than the plasmolytic value of some of the cells of the tissue since further shrinkage of the protoplast of such cells in the more concentrated solutions leads to plasmolysis and the entry of external solution without any further change of volume or weight. When all the cells are at Ψ values lower than the plasmolytic value when the measurements are commenced, the tissue may actually increase in volume and weight regardless of the external solution concentration, so that no determination can be made. Thus, as soon as plasmolysis commences in some of the cells, there is a tendency to underestimate Ψ, (the apparent intercept in Fig. 5.7 moving further along the abcissa) and when all cells are plasmolysed no measurements can be made (Slatyer, 1958). Other sources of error can arise in the same way as in relative water content determinations, and are associated with removal of surface water and injection of external solution into intercellular spaces. However, these can usually be corrected by appropriate experimental technique.

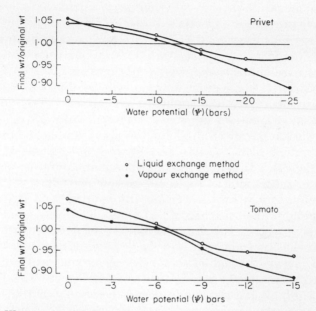

FIG. 5.7. Water potential determined by gain or loss of weight of tissue samples floated in, or suspended over, a series of graded osmotic solutions (after Slatyer, 1958).

By comparison, those liquid exchange methods involving determination of a solution characteristic, such as refractive index or density, are not invalidated by the above factors since there is still a net water influx or efflux, depending on the initial values of Ψ_{tissue} compared with Ψ_{soln}, and injection and surface wetting are of no consequence. Exudation from cut surfaces remains a problem, however (Gaff and Carr, 1964).

The other main source of error, common to all liquid exchange methods, occurs when the cell membranes are sufficiently permeable to the solute used in the external solution that, instead of a simple water exchange phenomenon operating, there is a net volume exchange. As a result a volume efflux, consisting of water alone, is associated with a volume influx consisting of solute and water. At the (quasi-) equilibrium situation of zero volume flow, tissue volume (and weight) is therefore greater than would be the case if there had been no solute uptake. In a similar way to the previous example this shifts the apparent point of intercept in Fig. 5.7 along the abscissa, leading to an estimate of Ψ_{tissue} lower than the true value.

The effect can be reduced to some extent by plotting change of water content, rather than change of fresh weight, against external solution concentration, particularly if the solute has a high molecular weight, as occurs with sucrose (Weatherley, 1955). However, this does not eliminate the problem and it is ineffective when the tissue is plasmolysed. This phenomenon is accentuated if the solute and water interact in their passage through the cell membranes so that the velocity of solute influx tends to give the water a velocity in the same direction. The effect of solute entry in reducing the apparent osmotic pressure of the external solution below its expected value (on the basis of concentration) is considered in Chapter 6. A term, the reflection (or selectivity) coefficient, is introduced to account for these effects of permeable solute when they are apparent. They are only detectable when solute permeability is rapid and appear to be relatively insignificant in some storage tissues but quite important in some leaf tissues (Slatyer, 1966b).

As an alternative to the direct methods of water potential measurement, indirect measurements can be of considerable value in some situations. The most commonly used indirect measurement involves establishing a water content/Ψ curve for the tissue under study (see for example Fig. 5.3) and using this to calibrate Ψ in terms of relative water content. This procedure, first suggested by Weatherley and Slatyer (1957) has been used by Slatyer (1960b) and is implicit in the general use by most researchers of relative water content measurements.

The accuracy of this prediction depends on the stability of the calibration and the sensitivity of Ψ to changes in water content. The calibration can be expected to vary with changes of internal osmotic pressure as well as of matric characteristics (Jarvis and Jarvis, 1963a, 1963c). All these factors change with age, but for crop plants, or any situations where a plant is grown in a fairly uniform environment and leaves of the same physiological age can be sampled during the experiment, quite reliable results can be obtained, particularly if the calibration curve is rechecked. Slatyer (1960b, 1962c) noted that even in a desert plant, *Acacia aneura*, the calibration did not shift substantially but different results have been found with other species (Jarvis and Jarvis, 1963c).

The sensitivity of Ψ to changes in water content is generally low at Ψ values from zero to about -5 bars (see Fig. 5.3) but subsequently increases. For this reason indirect methods may be unsatisfactory under conditions of high water status.

Other indirect methods involve calibrating Ψ against tissue volume. This is done with the beta gauge technique and also with measurements of stem circumference. The resistance of the stem to compression has also been measured (MacDougal, Overton and Smith, 1929). With these measurements, it is assumed that repeated wetting and drying involves the same degree of water removal and shrinkage. The stem circumference measurements can be calibrated against the compression resistance or with sorption isotherm data (Kelsey, 1957). Although neither necessarily gives a true representation of tensive cohesion forces, they can be very useful as indicators of Ψ_{stem}. Measurements involving liquid exchange in the stem, such as those of Currier (1945), in which liquid exchange in hollow petioles of squash was measured, are subject to similar errors to the liquid exchange methods mentioned previously.

C. Measurements of Water Potential in Different Tissue Regions

Measurements of the water potential of entire organs or tissue segments can be achieved by the methods just described, by equilibrating the material in vapour or osmotic solutions. Measurements in different regions of cells and tissues are much more difficult to make because simple equilibration is no longer adequate and measurements of the different component potentials are required. These measurements are considerably more complicated than the corresponding measurements in soils because of the complexity of the water distribution in intra- and extra-cellular compartments. The following discussion deals briefly with the measurements more commonly made.

1. *Measurement of vacuolar potentials*

Vacuolar measurements have been made far more frequently than in any other tissue region. Probably this is partly because the vacuole occupies a large proportion of many cells, partly because its water potential is frequently considered to be influenced only by pressure and solute concentration and so is a relatively simple system, and partly because of analogies with classical cell osmometers. Also, such measurements are key ones in understanding the dynamics of water and solute exchanges because the vacuole is both a sink and source for water and solute movement between cells and their surroundings.

Considering first measurements of vacuolar osmotic pressure, two main procedures have been used; one involving extraction of the vacuolar sap and the other measurements on intact cells in more or less undisturbed assemblages or tissue segments.

The first approach is subject to errors of variable magnitude in sap extraction. The normal procedure is to rupture the cell structure by freezing (or some other procedure) and then to express the sap under pressure, but the material so obtained contains cytoplasmic and extra-cellular sap too, generally of lower osmotic pressure. In consequence, the first aliquot may be dilute compared with the final one (Gortner, Lawrence and Harris, 1916; Mason and Phillis, 1939) but this is not always observed.

In addition, rapid chemical changes, such as inversion of sugars, may occur after the sap has been expressed and it is important to reduce these as much as possible by low temperature storage. The osmotic pressure of the sap is then determined by measuring one of the colligative properties of the solution, generally its vapour pressure or freezing point depression. A number of techniques for this purpose have been described in recent years (Weatherley, 1961; Vaadia and Marr, 1961; van Andel, 1953) and the thermocouple psychrometers previously described for water potential measurements can also be used if sufficient sap is obtained. Commercial equipment also is available. A useful general account is given by Crafts, Currier and Stocking (1949) and, in summary, it may be stated that the method has been of considerable value for many studies and appears to be particularly useful when tissue of high vacuolar content is under study. For tissue of low vacuolar content, however, significant errors may be introduced by contamination of the vacuolar sap with extra-vacuolar material.

The alternative procedure involves measurements of the osmotic pressure at incipient plasmolysis, and its adjustment to the volume of the normal cell. Two methods are available. The plasmolytic method involves microscopic observations on the cells, or thin strips of tissue immersed in a series of graded osmotic solutions. An external solution which reduces the turgor of the cell to zero is considered to be isotonic with the vacuolar osmotic pressure and the cell is assumed to be in a condition of *limiting* plasmolysis. In practice a slightly more advanced stage is detected, known as incipient plasmolysis, in which the proto-plast visibly commences to shrink away from the cell wall. This too is taken as isotonic, even though a small discrepancy exists of a magnitude equal to the adhesion between the protoplast and the wall. Measurements of cell and protoplast volume are made in the original and plasmolysed states and a proportional adjustment made to give the osmotic pressure at the original condition. The plasmometric method involves similar general procedures except that severe plasmolysis is induced by strongly hypertonic solutions, followed by measurement of the change in volume and proportional correction.

Although these techniques permit observations on single cells and can also be conducted in conjunction with studies of cell permeability, they are subject to the errors reported earlier for the liquid exchange measurements of Ψ. These are often insignificant when solute penetration is very slow but should always be checked. Also, errors are introduced by inaccurate volume estimation, adhesion between the cell wall and the protoplast, and other sources (Craft, Currier and Stocking, 1949). Volume measurement can be a particularly difficult proposition since vacuolar volume has to be measured and its shape cannot necessarily be assumed to be regular. Frequently, protoplast volume is measured instead, this being satisfactory when the cytoplasm is minimal in volume but obviously most unsatisfactory in some cells with significant amounts of cytoplasm, or when "cap" plasmolysis occurs, the cytoplasm swelling to occupy the space between the vacuole and the wall. Adhesion effects, mentioned earlier in this chapter, can also introduce errors. Good accounts of these limitations are given by Crafts, Currier and Stocking (1949) and Mercer (1955).

Few direct measurements of turgor pressure have been made because of the small size of most cells in normal tissue. However, Ahrens (1939) measured turgor pressure of *Nitella* with a manometer and as more micro-surgical equipment becomes available this approach may be used with many smaller cells. Bennet-Clark and Bexon (1940) estimated turgor pressure of leaf tissue segments from the hydrostatic pressure necessary to cause sap exudation. Fully turgid cells should exude as soon as pressure is applied, less turgid ones when the extra pressure raises the water potential inside the cells to zero. This technique has possibilities in densely packed tissue segments but it is very difficult to avoid some exudation from extra-cellular regions and the initial vacuolar exudation cannot be detected until it first fills the extra-cellular spaces. Direct measurements of turgor pressure are needed to check many basic assumptions in plant water relations, and new techniques are urgently needed.

2. *Other measurements*

Measurements of component potentials outside the vacuole are so meagre that it is of little value to classify them into specific cell and tissue regions. The main measurements have been determinations of negative pressures in the vascular elements made in conjunction with cohesion theory investigations (MacDougal, 1933; Preston, 1958; Scholander *et al.*, 1965). Such measurements combined with measurements of the osmotic pressure of the xylem sap (usually on exudation collections) should give an adequate picture of the water potential in continuous liquid-filled columns which are under tension. It should be remembered, though, that much of the water in vascular tissue, as the water potential falls, may be retained by matric forces rather than by cohesion.

The pressure measurements can be made by studying the time course of dye (or isotope) flux after injection into the xylem elements of the stem. If there is no cohesive tension in the system, entry is due to gas under reduced pressure in the vessels, and the only liquid moving after injection will be the test solution, so the flow should conform to Poiseuille's law. On the other hand, if there is tension in the system, a different relation will hold. In studies with ring-porous trees, Preston (1958) has found that in most vessels entry appears to be due to sub-atmospheric, but positive, gas pressure, of the order of 0·3 to 0·9 bar, but in some vessels liquid tensions of the order of about −3 bars developed. In these studies, the water potential in the tree was probably close to zero, and greater tensions can almost certainly be sustained.

The most profitable way to investigate the component potentials in non-vacuolar regions may be the extension of sorption isotherm studies of the type used for whole tissue relationships (Weatherley and Slatyer, 1957; Slatyer, 1958, 1960a; Jarvis and Jarvis, 1963a, 1963c). In this regard extraction procedures for the different cell fractions are being improved continuously (Plaut and Ordin, 1961). Separate free space and osmotic volume determinations with different solutes may also be of value. Gaff and Carr (1961) have recently examined the water retention characteristics of extracted dried and ground cell wall material of *Eucalyptus globulis*. If techniques of this type can be validly used, a considerable amount of information can be gained on the component potentials of water in various regions of cells and tissues.

CHAPTER 6

Water Exchanges in Plant Cells and Tissues

In recent years, a great deal of progress has been made in better understanding and describing the flows of water and solutes across biological membranes. In this chapter, general principles of water and solute transport are discussed in relation to water movement in and out of plant cells. Water movement through specific tissues and organs, and through the plant, is considered in Chapter 7. The influence of flow rates and resistances to water flow in the plant on the magnitude of water deficits developing in the plant at any one flow rate is deferred until Chapter 9.

Background information for the material covered in the present chapter can be found in Crafts, Currier and Stocking (1949), Ruhland (1956a, 1956b, 1956c), Steward (1959), Briggs, Hope and Robertson (1961), Kleinzeller and Kotyk (1961), and Kozlowski (1964).

1. General Aspects of Water and Solute Movement Across Membranes

Traditionally, plant physiologists, in common with biologists generally, have used two independent expressions to describe solute and water flow across membranes, and in and out of cells. Considering an isothermal system consisting of an outer (superscript[o]) and inner (superscript[i]) compartment separated by a membrane, the equation for net inward flux of a simple non-electrolyte solute is commonly written in a form analogous to Fick's equation:

$$\frac{dn_s^i}{dt} = k_s A (c_s^o - c_s^i) \qquad (6.1)$$

where n_s^i is the number of moles of solute in the inner compartment, t is time (sec), k_s is the conventional solute permeability coefficient (cm sec^{-1}), A is the membrane area (cm^2) and $(c_s^o - c_s^i)$ is the solute concentration difference (mole cm^{-3}) which provides the driving force.

For water flow or, more generally, net volume flow, the corresponding equation is written in a similar fashion, so that

$$\frac{dV^i}{dt} = k_w' A (\mu_w^o - \mu_w^i) \qquad (6.2)$$

where V^i is the volume of the inner compartment (cm^3), μ_w is the chemical potential of water (erg mole^{-1}) and k'_w is a proportionality factor related to the water permeability coefficient. Since the water potential difference $(\Psi^o - \Psi^i) = (\mu_w^o - \mu_w^i)/\overline{V}_w$ where $\overline{V}_w$ is the partial molal volume of water (cm^3 mole^{-1}), this is more commonly written

$$\frac{\mathrm{d}V^i}{\mathrm{d}t} = k_w A (\Psi^o - \Psi^i) \tag{6.3}$$

where Ψ is in bars and k_w is the conventional water permeability coefficient (cm sec^{-1} bar^{-1}). When there is no hydrostatic pressure difference across the membrane it is assumed that $\Psi = -\pi = -RTc$, where π is the osmotic pressure (see Eq. 1.69) and c is the sum of the molal concentrations of all the solutes regardless of whether the membrane is permeable to them or not. This situation $(P = 0)$ exists in permeability studies with cells at incipient plasmolysis when Eq. (6.3) becomes

$$\frac{\mathrm{d}V^i}{\mathrm{d}t} = -k_w A (\pi^o - \pi^i) \tag{6.4}$$

When a hydrostatic pressure difference $(P^o - P^i)$ exists across the membrane, as in plant cells when there is positive turgor pressure, Eq. (6.3) becomes

$$\frac{\mathrm{d}V^i}{\mathrm{d}t} = k_w A [(P^o - P^i) - (\pi^o - \pi^i)] \tag{6.5}$$

Although these equations have been widely used for studies of water and solute exchanges [Eq. (6.1) is generally used for studies of self-diffusion of water using D_2O or THO which effectively act as solutes], their inadequacy to account for observed phenomena when a solute flow occurs simultaneously with a solvent flow was pointed out a number of years ago (Frey-Wyssling, 1946; Laidler and Shuler, 1949; Pappenheimer, 1953; Ussing, 1952). This is because the flow of water, for example, is a unique function of the water potential difference, $\Delta\Psi$, only when water alone is moving in the system. When other flows occur, a net volume flux is measured rather than a net water flux, and the water flow itself is no longer influenced by $\Delta\Psi$ alone.

A. Combined Expressions for Water and Solute Flow

1. Development of equations

The incompleteness of the above equations for water and solute movement results from the fact that they involve only two of the three coefficients required to categorize flow in such a system. Kedem and Katchalsky (1958) have described the need for three coefficients by comparing free diffusion with passage through a membrane. In the case of free diffusion, the hydrodynamic resistance to flow is caused by the friction between the solute and solvent alone, so that diffusion of a single solute is determined by a single diffusion coefficient. In the case of a membrane, however, two additional influences develop due to the friction between solute and membrane, and between solvent and membrane. While the relative importance of these different processes may vary considerably between, say, coarse membranes with large pores and fine membranes without apparent pore development, all need to be considered.

Although Onsager (1931) had established a sound and convenient means of coupling linked flows with the forces responsible for them, it was not until the work of Staverman (1951, 1952) that this approach was applied to solute and solvent flows across membranes and to osmotic pressure phenomena. Subsequently, a considerable amount of research has been conducted with special reference to transport across biological membranes. Important contributions in this area have been made by, amongst others, Kedem and Katchalsky (1958, 1961, 1963a, 1963b, 1963c), Katchalsky (1961), Katchalsky and Kedem (1962), Dainty (1963a, 1965), Dainty, Croghan and Fensom (1963) and Dainty and Ginzburg (1964a, 1964b).

The following account, based on these papers, presents a brief and necessarily simplified treatment of this approach. For the interested reader general texts on the thermodynamics of irreversible processes, such as those by Denbigh (1951), de Groot (1952), Prigogine (1961), de Groot and Mazur (1962) and Fitts (1962) are recommended.

As stated in Chapter 4, the general theory of the thermodynamics of irreversible processes concerns the relationships between flows (of matter, energy, electrical charge, etc.) in a system and the "forces" responsible for them. It assumes that the flows are linear functions of all the forces operative in the system, provided that they are correctly defined and the system is not too far from equilibrium. The choice of forces and fluxes is arbitrary to a certain extent but the product of any pair of conjugated forces and flows should have the dimensions of entropy production and the sum of all the products should leave the entropy production invariant.

For present purposes, the entropy production equation is conveniently written in terms of the dissipation function, $T\mathrm{d}S/\mathrm{d}t$, where T is absolute temperature and S is the internal entropy (erg deg^{-1}) of the system. For the transport of a single solute and water, such as is considered below, the dissipation function per unit area can be written (Kedem and Katchalsky, 1958):

$$\frac{1}{A}\frac{TdS}{dt} = \frac{1}{A}\frac{dn_w}{dt}\Delta\mu_w + \frac{1}{A}\frac{dn_s}{dt}\Delta\mu_s \qquad (6.6)$$

where the flows of water $J_w = \frac{1}{A}\frac{dn_w}{dt}$ and solute $J_s = \frac{1}{A}\frac{dn_s}{dt}$ (expressed

in mole $cm^{-2}\,sec^{-1}$) are conjugated to the forces $\Delta\mu_w$ and $\Delta\mu_s$ (expressed in erg $mole^{-1}$), which are the differences between the corresponding potentials across the membrane. The relation between the flows and forces can then be written as a set of phenomenological equations (cf. Chapter 4, p. 115), so that all the forces known to influence the flow of any component can be included and possible interactions between the flows of different components can be described. In Eqns. (6.1–6.5), only the straight coefficients were used and the interactions were neglected. The importance of including the cross coefficients will now be demonstrated.

In the present system, it is generally assumed that the correct driving forces are the differences in hydrostatic pressure $\Delta P = P^o - P^i$ and solute concentration $RT\Delta c_s = RT(c_s^o - c_s^i)$ which exist across the membrane. These are obtained from the differences in chemical potentials, $\Delta\mu$, with the expressions for solute and water (c.f. Eq. 5.1)

$$(\mu_s^o - \mu_s^i) = \bar{V}_s\Delta P + RT\Delta \ln N_s \qquad (6.7)$$

$$(\mu_w^o - \mu_w^i) = \bar{V}_w\Delta P + RT\Delta \ln N_w \qquad (6.8)$$

In the case of dilute solutions

$$\Delta \ln N_s = \Delta N_s/N_s \cong \Delta c_s/\bar{c}_s \qquad (6.9)$$

where $\bar{c}_s$ is a mean of the concentrations of solute in the two compartments, $(\cong [c_s^o + s_s^i]/2)$, so Eq. (6.7) becomes:

$$(\mu_s^o - \mu_s^i) = \bar{V}_s\Delta P + \frac{RT\Delta c_s}{\bar{c}_s} \qquad (6.10)$$

and since

$$\Delta \ln N_w = \Delta \ln(1 - N_s) = \Delta(1 - N_s)/(1 - N_s) \cong \qquad (6.11)$$
$$- \Delta N_s/(1 - N_s) \cong - \Delta c_s/\bar{c}_w$$

where $\bar{c}_w$ is the mean concentration of water, the corresponding equation for water is:

$$(\mu_w^o - \mu_w^i) = \bar{V}_w\Delta P - \frac{RT\Delta c_s}{\bar{c}_w} \qquad (6.12)$$

In incorporating these driving forces in an expression for the transport of a single solute and water, different flows can now be used. These are (Kedem and Katchalsky, 1958) a volume flow, J_v, (in units of cm^3 cm^{-2} sec^{-1}):

$$J_v = J_w\overline{V}_w + J_s\overline{V}_s \qquad (6.13)$$

and an exchange flow, J_D, (also in cm^3 cm^{-2} sec^{-1}), equivalent to cm sec^{-1}, which describes the relative velocity of solute and solvent, given by

$$J_D = \frac{J_s}{\bar{c}_s} - \frac{J_w}{\bar{c}_w} \qquad (6.14)$$

The appropriate dissipation function now becomes (cf. Eq. 6.6):

$$\frac{1}{A}\frac{TdS}{dt} = J_v\Delta P + J_D\Delta\pi_s \qquad (6.15)$$

and the appropriate phenomenological equations are (Kedem and Katchalsky, 1958):

$$J_v = L_p\Delta P + L_{pD}\Delta\pi_s \qquad (6.16)$$

$$J_D = L_{Dp}\Delta P + L_D\Delta\pi_s \qquad (6.17)$$

where L_p and L_D are the "straight" coefficients linking volume and solute flow to ΔP and $\Delta\pi_s$, respectively; and L_{pD} and L_{Dp} are the "cross-coefficients" which describe the volume flow associated with $\Delta\pi$ and the exchange flow associated with ΔP, respectively. It should be remembered that $L_{pD} = L_{Dp}$. The meaning of the coefficients can be seen from the following.

When there is no difference in solute concentration across the membrane, $\Delta\pi$ is zero, so that $J_v = L_p\Delta P$, and the observed flow is a result of ΔP alone. Therefore, L_p is the filtration coefficient of the membrane and can be identified with k_w as used in Eqns. (6.3)–(6.5). At the same time, a difference in the flows of water and solute will be observed according to the relationship $J_D = L_{Dp}\Delta P$ in which L_{Dp} is seen to be the ultrafiltration coefficient of the membrane. By comparison, when ΔP is zero, an osmotic volume flow $J_v = L_{pD}\Delta\pi_s$ will result and a different exchange flow $J_D = L_D\Delta\pi_s$ will occur.

It is apparent that an inadequate description of the observed flow will be given if only one coefficient is used for volume flow and one for exchange flow unless the membrane is completely permeable to the solute, in which case $\Delta P = 0$ under all conditions; or unless the membrane is completely impermeable to the solute, in which case $\Delta\pi_s = \Delta P$ so that $L_p = -L_{pD}$.

These extreme situations, and the range of intermediate ones which are generally encountered in plant physiology, can be handled by introducing a selectivity or reflection coefficient of the membrane for a particular solute (Staverman, 1951). Consider a typical experiment in which a cell, originally immersed in pure water, is immersed in an aqueous solution, to the solute of which its effective membranes are somewhat leaky. As shown schematically in Fig. 6.1, there is initially a loss of volume due to net efflux from the cell or tissue. A stage of quasi-equilibrium is then reached at which there is zero net volume flux followed by a recovery in volume as the influx of water and solute exceeds the efflux.

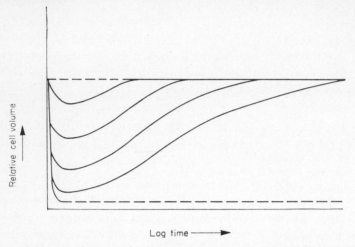

FIG. 6.1. Schematic diagram of volume changes in an initially turgid cell immersed in aqueous solutions. It is assumed that the permeability of the membrane to the solute increases from zero (bottom broken line) to a degree indistinguishable from that of the solvent (upper broken line) (after Slatyer, 1966b).

As the permeability of the membrane to the solute decreases, the point of quasi-equilibrium occurs at lower cell volume (for the same initial $\Delta\pi$) and volume recovery is slower. Had the cell been completely impermeable to the solute, the final situation would have been that depicted by the lower broken line.

At the quasi-equilibrium stage of minimum cell or tissue volume, $J_v = 0$. We then see, from Eq. (6.16), that

$$L_p\Delta P = -L_{pD}\Delta\pi_s$$

so that

$$\sigma = \frac{-L_{pD}}{L_p} = \frac{\Delta P}{\Delta\pi_s} \tag{6.18}$$

where σ, the reflection or selectivity coefficient, is the ratio of the apparent osmotic pressure to the potential osmotic pressure.

Equation (6.16) can now be re-written as

$$J_v = L_p\,(\Delta P - \sigma\Delta\pi_s) \tag{6.19}$$

It can be seen that when the membrane is impermeable to the solute, $\sigma = 1, L_p = -L_{pD}$ and, as in the ideal osmometer, $J_v = L_p\,(\Delta P - \Delta\pi_s)$. Similarly, when the membrane is completely non-selective, $\sigma = 0$ and $J_v = L_p\Delta P$. At intermediate states, as depicted in Fig. 6.1, $1 \geqslant \sigma \geqslant 0$.

In addition to volume flow, the solute flow J_s usually is measured instead of the exchange flow J_D (see Eqns. 6.13 and 6.14). Now J_s becomes (Kedem and Katchalsky, 1958):

$$J_s = J_v(1 - \sigma)\bar{c}_s + \omega\Delta\pi_s \tag{6.20}$$

In this expression, the solute flow is seen to be a function of the volume flow J_v and the force $\Delta\pi_s$. The new coefficient ω is introduced in place of the J_D of Eq. (6.17). Using the substitution $\Delta\pi_s = RT\Delta c_s$ at $J_v = 0$, Eq. (6.20) becomes

$$\omega RT = J_s/\Delta c_s \tag{6.21}$$

and ωRT is seen to be the conventional solute permeability coefficient, k_s, (in cm sec^{-1}), where RT is in erg^{-1} mole^{-1} and Δc_s in mole cm^{-3}.

These three coefficients L_p, ω and σ defined as

$$L_p = \left(\frac{J_v}{\Delta P}\right)_{\Delta\pi_s=0} \qquad \omega = \left(\frac{J_s}{\Delta\pi_s}\right)_{\Delta P=0} \quad \text{and} \quad \sigma = \left(\frac{\Delta P}{\Delta\pi_s}\right)_{J_v=0}$$

are therefore the three independent coefficients needed to provide a correct description of osmotic phenomena with the forces and flows used in the present system.

In real cells the situation is complicated by the presence of the vacuolar solutes to which, for practical purposes, the cytoplasmic membranes are impermeable. Equation (6.19) can be modified to account for this factor by writing:

$$J_v = L_p(\Delta P - \sigma\Delta\pi_s - \Delta\pi_m) \tag{6.22}$$

where $\Delta\pi_m$ represents the theoretical difference in osmotic pressure caused by the vacuolar solutes to which the membranes are impermeable. Under these conditions, L_p is defined as

$$L_p = \left(\frac{J_v}{\Delta P - \Delta\pi_m}\right)_{\Delta\pi_s=0}$$

2. *Characteristics of the reflection coefficient*

Although a number of other complications also arise, some of which will be dealt with below, the equations already presented indicate the importance of the reflection coefficient in studies of water and solute exchanges across membranes. L_p and ω have already been related to the conventional water and solute permeability coefficients, under specified conditions. Because σ is less easily understood, but is of considerable importance, it is of value to consider it in a little more detail and, in particular, to delimit its probable range, given values of L_p and ω.

If it is assumed that the solute and solvent interact in their passage through the membrane, the interaction induces in each a velocity component in the direction of the force acting on the other. It can be appreciated that the extent to which this interaction takes place will be least when solute and solvent follow different pathways through the membrane and greatest when a common pathway exists. To illustrate the effect of this interaction on the reflection coefficient, consider first that the membrane separates pure water from a solution and is impermeable to the solute, as in the classical osmometer. When there is water potential equilibrium on both sides of the membrane

$$\mu_w^o - \mu_w^i = \overline{V}_w \Delta \Psi = 0 = \overline{V}_w \Delta P + RT \Delta \ln N_w \qquad (6.23)$$

so that

$$\overline{V}_w \Delta P = -RT \Delta \ln N_w \cong RT \Delta c_s$$

Since the water outside the membrane is at atmospheric pressure and contains no solutes, the internal pressure, $P = -RT c_s = -\pi$, the osmotic pressure inside an osmometer, as shown previously. If it is now supposed that the membrane is permeable to the solute but that the water and solute cross the membrane independently, $\Delta \Psi$ remains at zero and there is no net flux of water but there is a volume flux equal to the solute flux, $J_v = J_s \overline{V}_s = \overline{V}_s k_s \Delta c_s$, where $k_s = \omega RT$ as shown previously. Therefore, in order to obtain a situation where $J_v = 0$, $\Delta \Psi$ must be reduced so that a water flow $J_w \overline{V}_w = L_p \Delta \Psi$, from the water side, balances the flow $J_s \overline{V}_s$ from the solution side. This is achieved by reducing ΔP and the new value, $\Delta P'$ is given by equating the two flows

$$-L_p \Delta \Psi = L_p (RT \Delta c_s - \Delta P') = \overline{V}_s k_s \Delta c_s \qquad (6.24)$$

$$\Delta P' = RT \Delta c_s \left(\frac{1 - k_s \overline{V}_s}{RTL_p} \right)$$

and since $\sigma = \Delta P / \Delta \pi = \Delta P / RT \Delta c_s$

$$\sigma = 1 - \frac{k_s \overline{V}_s}{RTL_p} \qquad (6.25)$$

If it is now assumed that not only is the membrane permeable to solute and water but that they do interact in their passage through the membrane, the net volume flow to the water side, due to solute flow, is greater than in the previous case because the solute flowing towards the pure water side drags water with it. Hence, the hydrostatic pressure has to be reduced still more than in the previous case so

$$\sigma < 1 - \frac{k_s \overline{V}_s}{RTL_p} \qquad (6.26)$$

This important result, derived by both Kedem and Katchalsky (1958) and Dainty (1963a), has several important implications. Firstly, $\sigma = 1$ may be regarded as a good test of semi-permeability of a membrane, since $\sigma \rightarrow 1$ only when k_s (or ω) $\rightarrow 0$. Secondly, since $\sigma \rightarrow 0$ for very coarse membranes, for given L_p and ω, σ is limited by

$$0 \leqslant \sigma \leqslant 1 - \frac{k_s \overline{V}_s}{RTL_p} \qquad (6.27)$$

Thirdly, the value of σ may enable a decision to be made on the mechanism of solute transfer. When σ is equal to or only slightly less than $(1 - k_s \overline{V}_s / RTL_p)$, the indication is for independent passage; whereas when $\sigma \ll (1 - k_s \overline{V}_s / RTL_p)$, the indication is for a linked flow of solute and water through pores in the membrane.

Measurements of σ for any particular solute and membrane are conveniently made when $J_v = 0$. At this stage, Eq. (6.22) becomes

$$(\Delta P - \Delta \pi_m) = \sigma \Delta \pi_s \qquad (6.28)$$

If the experiment is conducted so that the cell is at incipient plasmolysis when $J_v = 0$ (by immersing cells in a series of graded osmotic solutions and determining the concentration of the vacuolar sap in that solution which just causes plasmolysis), ΔP is also assumed to equal zero. If there is negligible solute entry by the time plasmolysis occurs, $\Delta \pi_s = \pi_s^o$ and, assuming the membrane is effectively impermeable to vacuolar solutes, $\Delta \pi_m = \pi_m^i$, hence

$$\pi_m^i = \sigma \pi_s^o \qquad (6.29)$$

The implications of Eq. (6.29) to measurements of permeability and cell and tissue water relations generally are described later in this chapter. Suffice it to say at this stage that many measurements in the literature are based on the assumption that $\sigma = 1$ and $\pi_m^i = \pi_s^o$ and failure to take σ into account may have significantly affected the results in some cases (Slatyer, 1966b).

3. *Effect of electrolytes*

When the system contains electrolyte solutions on both sides of the membrane, not only may different concentrations and pressures develop but also different electrical potentials, so that now three forces are involved, together with three conjugated flows. The dissipation function can therefore be written (cf. Eq. 6.7)

$$\frac{1}{A} \cdot \frac{TdS}{dt} = J_w\mu_w + J_w\Delta\mu_s + I\epsilon \tag{6.30}$$

where I is the flow of electrical current through the membrane (amp cm^{-2}), ϵ is the electromotive force (volts) and μ_s is the chemical part of the electrochemical potential ($= RT \ln \phi_s c_s$ where ϕ_s is the activity coefficient and c_s the concentration).

The three phenomenological equations now required include six independent Onsager coefficients and make the detailed study of the system rather complex. In addition to the three coefficients already derived, (L_p, ω and σ), another set of three electrical coefficients are required. Under conditions where $J_v = 0$, and there is equal salt concentration on both sides of the membrane ($\Delta\pi_s = 0$), the coefficients are (Katchalsky and Kedem, 1962)

$$\kappa = \left(\frac{I}{\epsilon}\right) \quad \nu = izF\left(\frac{J_s}{I}\right) \quad P_\epsilon = \left(\frac{\Delta P}{\epsilon}\right)$$

where κ is the specific conductance (ohms^{-1} cm^{-2}) measured as the flow of current per unit electromotive force, ν is the transport number (of the ion which does not interact with the reversible electrode) measured as the flow of solute per unit of current, where z is the valency of the ion, i is the number of ions per salt molecule and F the Faraday (96,500 coulomb). P_ϵ is the electro-osmotic pressure developed at zero flow per unit potential.

With these coefficients, Katchalsky and Kedem (1962) have written the three phenomenological equations as (cf. Eqns. 6.16 and 6.29)

$$J_v = L_p\Delta P - \sigma L_p\Delta\pi_s - \frac{P_\epsilon L_p}{\kappa} I \tag{6.31}$$

$$J_s = \bar{c}_s(1 - \sigma)J_v + \omega\Delta\pi_s + \frac{\nu}{izF} I \tag{6.32}$$

$$I = -P_\epsilon J_v + \frac{\kappa\nu}{izF} \frac{\Delta\pi_s}{c_s} + \kappa\epsilon \tag{6.33}$$

for the flow of volume, solute, and electricity, respectively. When there is no flow of electricity $I = 0$, and Eqns (6.31 and 6.32) are identical with Eqns (6.16 and 6.20) for volume and solute flow, respectively.

These expressions, or others derived on a somewhat different basis which also describe the flows of water, solutes and electricity, permit a sensitive and quantitative description of many membrane transport phenomena. By comparison with water and heat flux phenomena in soils, where problems arise because of non-linearity in the $\Psi(K)$ relationship (see Chapter 4), the equations appear to be effectively linear for a number of membrane transport systems, and the transfer coefficients constant, at better than first approximation level. It should be remembered, with special reference to water exchanges in plant cells, that for a number of solutes, the reflection coefficient of most plant cell membranes is $\sigma \cong 1$ and negligible error is introduced by the use of Eq. (6.3) rather than Eq. (6.22). However for other solutes, $\sigma < 1$ and the uncritical use of Eq. (6.3) can introduce important errors.

B. Osmotic Water Movement and Related Phenomena

Although very little has been said specifically about osmosis, it is apparent that many of the phenomena discussed above are associated with osmotic water movement.

Osmosis has been defined as diffusion across a membrane, generally with the proviso that it occurs in response to a gradient in the chemical potential of the diffusing substance. Osmotic water movement is therefore regarded as a wholly passive phenomenon and any water movement not readily explainable on this basis has sometimes been thought to be active, in the sense that it occurs in the absence of, or in the opposite diredtion to, a water potential gradient. In this section brief accounts of present viewpoints on the nature of osmotic water movement, and possible mechanisms of non-osmotic movement, are given.

1. Nature of osmotic water movement

In recent years there has been renewed interest in the actual mechanism of osmotic water movement. This has been motivated in part by observations that ΔP and $\Delta \pi$ are equivalent in their effects on water flow across membranes (Mauro, 1957; Robbins and Mauro, 1960) and partly by evidence that k_s for solutes such as D_2O, DHO and THO is less than the dimensionally equivalent value of $RTL_p/\overline{V}_w$ obtained from volume flux experiments, in which levels of ΔP or $\Delta \pi$ were imposed across membranes (Prescott and Zeuthen, 1953; Durbin Frank and Solomon, 1956; Villegas, Barton and Solomon, 1958). Much of the latter evidence is still controversial (see p. 193, also Dainty, 1963a), but a general model has gradually evolved for osmotic water movement which involves bulk flow through water filled pores (Robbins and Mauro, 1960; Ray, 1960; Dainty, 1963a). The following notes summarize this viewpoint.

Equations (5.17) and (6.22) for an ideal vacuolate plant cell ($\sigma = 1$, $\omega = 0$):

$$\Psi = P - \pi \tag{5.16}$$

$$J_v = L_p(\Delta P - \Delta \pi) \tag{6.22}$$

imply that the cell membrane is truly semi-permeable and that an osmotic pressure difference, $\Delta\pi$, produces the same kind of flow as a hydrostatic pressure difference, ΔP, using the same coefficient L_p. While some workers have argued that the flow caused by the water potential difference across the membrane must be diffusional in nature (Chinard, 1952), others, (see for example, Durbin, Frank and Solomon, 1956; Solomon, 1960) assume that it involves a bulk flow of water through water-filled pores in the membrane. Mauro (1960), Ray (1960) and Dainty (1963a) have pointed out that when $\sigma = 1$, if differential permeability involves the exclusion of solute from water filled pores (assuming that pores do exist), the pores must be filled with water alone. This implies that there must be a sharp transition in water concentration at the pore aperture adjoining the bulk solution.

Since, when $\Psi^o = \Psi^i$,

$$P^o + \frac{RT}{\overline{V}_w} \ln N_w^o = P^i + \frac{RT}{\overline{V}_w} \ln N_w^i \tag{6.34}$$

there must be an equally sharp change of P at the pore aperture. This situation is depicted in Fig. 6.2 (a) redrawn from Dainty (1963a). When there is pure water each side of the membrane, the mole fraction terms disappear from Eq. (6.34) and, during flow ($\Delta\Psi \neq 0$), the difference in P^o and P^i should be reflected in a gradient of P within the water filled pores, such as is depicted in Fig. 6.2 (b). By comparison, when there is no difference between the bulk values of P^o and P^i, water flow, under the same value of $\Delta\Psi$, should be reflected in pressure and water mole fraction relationships as depicted in Fig. 6.2 (c).

It is assumed, therefore, that, within the pore, the pressure profile is exactly the same as in the pure water case and this indicates that mole fraction differences produce flow by the same type of mechanism as does a hydrostatic pressure difference, in both cases hydrostatic pressure gradients providing the operative forces. The reason for the sudden pressure change at the pore aperture has been pictured by both Ray (1960) and Dainty (1963a) as being due to the greater concentration of water at the pore aperture than in the adjoining solution. Random activity of each water molecule results in a tendency to jump into any spaces which develop due to similar activity among its neighbours. Since there will be more spaces developed on the less water-concentrated solution side, more molecules jump from water to solution and the spaces so developed on the water side are filled by water molecules from further along the pore. The net creation of vacancies leads to a decrease in density of the water which can cause a decrease in pressure.

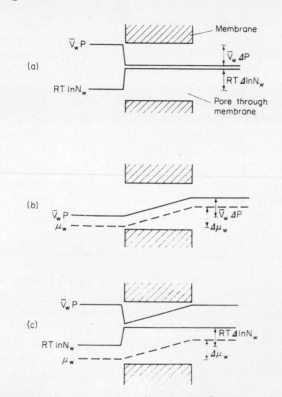

FIG. 6.2. Diagrammatic representation of pressure and concentration changes across a membrane: (a) equilibrium situation in which $\bar{V}_w \Delta P$ balances $RT \Delta \ln N_w$ (b) pressure induced flow with pure water each side (c) concentration induced flow with equal pressure on each side (after Dainty, 1963a).

This explanation appears satisfactory, if membranes do contain pores, but the subject is still controversial (see next section). Moreover, it should be remembered that, at the pore aperture, the net exchange of molecules may still have the characteristics of a diffusion process and it is this exchange which activates the whole process. However, if pores exist and if more rapid rates of flow occur across membranes than can be expected on a simple diffusional basis, then the mechanism just described appears to account satisfactorily for it.

2. *Non-osmotic water movement*

Assuming that non-osmotic, or active, water movement across a membrane implies $J_w \neq 0$ when $\Delta\mu_w = 0$, two mechanisms by which such water transport could occur would be electro-osmosis and the passive drag of water molecules by solute flow.

The importance of electro-osmotic flow in plant physiology has been debated for many years and has recently been re-investigated by Dainty (1963a) and Dainty, Croghan and Fensom (1963). As can be appreciated from Eqns (6.31)–(6.33), a flow of ions can be associated with a flow of electricity, and in this situation the equations for both volume flow (6.31) and flow of electricity (6.33) include an electro-osmotic coefficient, P_ϵ. Electro-osmosis therefore provides an obvious mechanism by which metabolic energy associated, for example, with an ion pump, could be linked to water movement. Assuming that $\Delta\pi_s = 0$ and $J_v = 0$, the electro-osmotic pressure represented by any excess hydrostatic pressure required to balance an inwardly directed electro-osmotic pump is obtained from Eq. (6.31):

$$P_\epsilon = \frac{\Delta P}{\epsilon}$$

where $\epsilon = \kappa I$.

Various values of $\Delta P/\epsilon$ have been calculated by Dainty (1963a) and Dainty, Croghan and Fensom (1963) using different models, the highest of which give estimates of the order of 10^3 bar volt $^{-1}$, so that even with driving potentials of $0{\cdot}01$–$0{\cdot}1$ volt, pressures of 10–100 bars could be expected. Despite the magnitude of these estimates, the actual effect is likely to be very small in most plant cells because of the relatively high permeability of the cell membranes to water. Since most passive water flow is by pathways other than those used by ions, water tends to leak back through these other high permeability channels and reduce markedly the effectiveness of any postulated electro-osmotic pump.

It was estimated by the above workers that these high permeability channels have a total "resistance" about 10^{-5} times the resistance in the electro-osmotic channels. Consequently, the effective electro-osmotic pressure is about 10^{-5} times the estimated figures. If $\Delta P/\epsilon$ reduces to 10^{-2} bar volt $^{-1}$ and the maximum driving potential is 10^{-1} volts, the maximum pressure developed would be a negligible 10^{-3} bars. Although these figures are for large coenocytic algal cells, it is unlikely that the value for other cells will be greater by the 3 or 4 orders of magnitude required, and the logical conclusion is that electro-osmosis is not likely to be of importance for water transport in plant cells, except in specialized cases where L_p is extremely low.

The alternative mechanism, of a passive drag of water molecules by solute molecules or ions, has already been considered in the case where there is a net solute flux. In this case, it was included in the reflection coefficient, σ, and was the factor which reduced σ below the value, $\sigma = [1 - (k_s \bar{V}_s / RTL_p)]$ (Eqns. 6.25 and 6.26). If conditions for active transport of water are made more stringent by the definition that it is flow occurring $(J_w \neq 0)$, when $J_s = 0$, $\Delta\mu_w = 0$, the observed water drag can still be passive if a membrane is envisaged with two independent parallel pathways through which different solutes pass in opposite directions, each dragging along different amounts of water (Kedem, 1965). Such a situation could lead to a net water flux while the net solute flux was zero, and would meet the criteria for active transport.

Another possible mechanism of active transport could involve the hydration or dehydration, with associated swelling and shrinking, of protein molecules in response to changes in pH, or ion composition and distribution (Goldacre, 1952). In such cases, however, an active ion pump is involved, with associated flow of ions or electricity, and the degree to which the associated water flux can be regarded as active is somewhat questionable.

In general, it appears that evidence of active water transport has sometimes really detected passive water movement following active ion transport, or passive water drag related to the solute flux. In the latter case, it is immaterial whether the solute flux has been caused by a concentration gradient or by active ion transport. In other cases it is probable that use of Eq. (6.3) instead of (6.22) has produced invalid evidence, due to neglect of the reflection coefficient. Electro-osmosis appears to offer the most likely direct mechanism for active water transport; whether the amounts of water so moved are significant depends largely on the hydraulic conductivity, L_p, of the membrane under study.

C. Volume Changes in Osmotic Cell Systems

If a plant cell is placed in an aqueous solution, at atmospheric pressure, containing a non-permeating solute, the change in volume can be described by the equation (cf. Eq. 6.19):

$$-\frac{dV}{dt} = L_p A(P - \Delta\pi) \tag{6.35}$$

where V and A are the volume and surface area of the cell, respectively,

P is the turgor pressure and $\Delta\pi = (\pi^0 - \pi^i)$ the difference in the osmotic pressure outside and inside the cell.

By assuming (1) that the order of magnitude of $k_w A$ does not change and so may be regarded as equal $k_w A_0$ [where the subscript $_0$ indicates the value of any parameter at zero turgor pressure, $P = 0$]; (2) that π^i is proportional to solute concentration, so that $\pi^i = \pi_0^i/(1+\mathbf{v})$ where $\mathbf{v} = V/V_0 - 1$, the relative departure of cell volume from V_0; (3) that the change in V is proportional to change of P, so that, for the $P(V)$ function, $P = \varepsilon\mathbf{v}$, where ε is the elastic modulus, corresponding to Broyer's (1952) "coefficient of enlargement"; and (4) that $\pi_0^i/(1+\mathbf{v}) \cong \pi_0^i(1-\mathbf{v})$, since $\mathbf{v}$ is generally rather smaller than unity, Philip (1958c) developed Eq. (6.35) and obtained

$$\frac{d\mathbf{v}}{dt} = \frac{L_p A_0}{V_0} [\pi_0^i - (\epsilon + \pi_0^i)\mathbf{v}] \qquad (6.36)$$

This equation can be integrated for the case where the cell is transferred from a solution of osmotic pressure $\pi^0 = \pi_0^i$ to one of $\pi^0 = 0$ to give

$$\mathbf{v} = \frac{\pi_0^i}{(\epsilon + \pi_0^i)} \{1 - \exp[-(\epsilon + \pi_0^i)\frac{L_p A_0}{V_0} t]\} \qquad (6.37)$$

This expression gives the rate of volume change regardless of the original and final values of π^0, as long as $P > 0$. From this expression a half-time for the approach to the new equilibrium volume is given by

$$t^{\frac{1}{2}} = \frac{(\ln 2)V_0}{L_p A_0(\epsilon + \pi_0^i)} = \frac{0\cdot693 V_0}{L_p A_0(\epsilon + \pi_0^i)} \qquad (6.38)$$

To the degree that the foregoing assumptions are valid, this analysis indicates that, for given cell dimensions and L_p, both the elastic properties of the cell wall and the internal osmotic pressure influence the rate of swelling and shrinking. Of the assumptions, that of linearity in the $P(V)$ relationship may not be valid in some cases (Broyer, 1947), and, if plastic stretching of the wall takes place during swelling, the increase of volume is one of growth rather than dynamic water exchange (Barrs and Weatherley, 1962). It is also assumed that cell volume equals vacuolar volume, but a simple arithmetical correction can account for this effect in many instances.

An important application of Eq. (6.38) is found in situations where cell wall water has been considered to buffer the protoplast against desiccation (Gaff and Carr, 1961; Carr and Gaff, 1962). In such cases, it is implied that substantial and prolonged reductions (of the order of hours) in cell wall water content may occur without compensating reductions in protoplast volume. However, inserting appropriate figures for V_0/A_0 $(3 \times 10^{-4}$ cm) (cell radius assumed to be 10 μ), Lp $(0 \cdot 5 \times 10^{-6}$ cm sec^{-1} bar^{-1}) and $(\epsilon + \pi_0^i)$ (50 bars) into Eq. (6.38) yields an estimate of half-time of the order of only 10 sec. Even if these figures are an order of magnitude in error, it suggests that any loss of water from the leaf which does not involve a flow of water out of individual cells must be very transient in nature.

Philip (1958c) has extended this approach to considerations of cell volume changes where the external solute enters the cell. When a turgid cell is placed in an aqueous solution of osmotic pressure, π^0, this gives

$$\frac{d\mathbf{v}}{dt} = \frac{L_p A_0}{V_0} \left[\pi_0^i - \pi^0 \exp \left(-\frac{k_s A_0 t}{V_0} \right) - (\epsilon + \pi_0^i) \, \mathbf{v} \right] \qquad (6.39)$$

subject to the initial condition $t = 0$, $\mathbf{v} = \pi_0^i/(\epsilon + \pi_0^i)$. This expression indicates that the most important quantity governing the time dependence of cell behaviour is $k_s A_0/V_0$. The other significant parameter of the system $(\epsilon + \pi_0^i) L_p/k_s$, influences the relative rate of approach to the quasi-equilibrium situation which develops when the initial excess of volume efflux over influx is eliminated. In Fig. 6.3, $\mathbf{v}$ is plotted against $k_s A_0 t/V_0$ for different values of $(\epsilon + \pi_0^i) L_p/k_s$ to illustrate the three phases of equilibration. There is first a rapid reduction

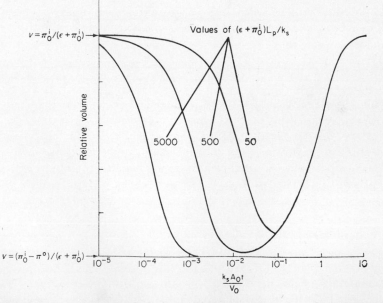

Fig. 6.3. Behaviour of an initially turgid cell placed in an aqueous solution containing permeating solute. For explanation see text (after Philip, 1958c).

associated with rapid water efflux from the cell, [assuming $(\epsilon + \pi_0^i)$ $L_p \gg k_s$], merging into a quasi-equilibrium situation where $J_v = 0$, which in turn merges into a slow drift back to the original cell volume, the rate of which is determined largely by k_s. Recovery is, of course, much quicker at lower values of $(\epsilon + \pi_0^i) L_p / k_s$ but this is not apparent from the diagram because of the dimensionless nature of $k_s A_0 t / V_0$.

The degree to which Eq. (6.38) can be applied to real situations is limited unless $(\epsilon + \pi_0^i) L_p \gg k_s$, since Philip (1958c) did not take account of the reflection coefficient and assigned the theoretical osmotic pressure to the external solution in developing the model. In many cases, the above requirement is met, and $\sigma \cong 1$, but in situations where $\sigma < 1$ (e.g. Collander, 1949, 1950; Dainty and Ginzburg, 1964a) marked deviations would occur. In the extreme case of, say, tritiated water $(\sigma \cong 0)$ no change of volume would occur (see upper broken line in Fig. 6.2).

2. WATER PERMEABILITY OF PLANT CELLS AND TISSUES

The measurement of the water permeability of single cells and of tissue segments presents a number of difficulties largely associated with the small size of most individual cells and the series-parallel nature of the internal resistances to water movement in tissues. In this section, factors affecting permeability will first be discussed, followed by a consideration of measurement techniques and problems associated with them.

A. Factors Affecting Permeability

The water permeability of individual cells is not determined solely by the permeability of the plasma membranes, although in most situations they constitute a large component of the total resistance to water flow. Obvious exceptions occur in cells where walls have undergone extensive secondary thickening, with associated lignification or cutinization, when the walls themselves may constitute the major barrier.

1. Water permeability of cell membranes

Across any one membrane, three sites of resistance arise; at each interface between solution and membrane, and within the membrane (Davson and Danielli, 1952). An idealized structure of a typical membrane has been drawn by Briggs, Hope and Robertson (1961) and is reproduced in Fig. 6.4. This structure is similar to that of Danielli

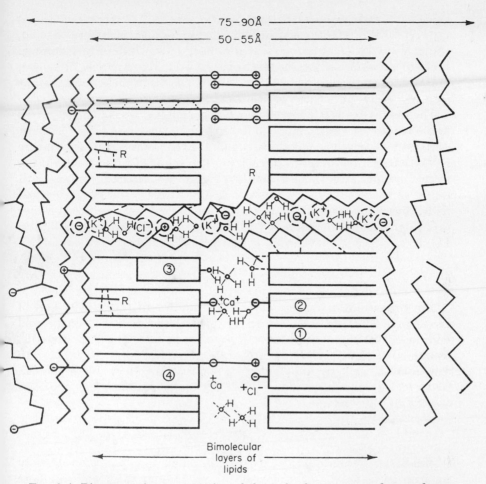

FIG. 6.4. Diagrammatic representation of the molecular structure of a membrane with aqueous pores: molecules (1) = triglycerides; molecules (2) = phosphatidic acid; molecules (3) = cholesterol; molecules (4) = lecithin or cephalin; (R) = non-polar side groups of the proteins (after Briggs, Hope and Robertson, 1961).

and Davson (1952) and Danielli (1954), but also shows a water filled pore along which water may move, encountering relatively little resistance.

Such a structure can be expected to show considerable stability, since numerous Van der Waal's bonds would exist between–CH_2–groups of neighbouring lipid chains. The electrostatic linkages between charged groups in lecithin and cephalin are particularly strong (Briggs, Hope and Robertson, 1961). In addition, hydrogen bonds would exist between – OH groups through water molecules, and ionic linkages may be formed between adjacent–COO$^-$ groups in phosphatidic acid, through bivalent cations. The membrane would also be expected to possess elasticity since the surface protein chains would fold and unfold in a limited way. Depending on the degree of cross-linkage between the amino acid chains, significant volume changes may also occur, associated with changes in pH and in the ion exchange complex, as discussed in Chapter 5.

A water molecule passing through such a membrane would encounter a fluctuating level of resistance due to the successive Van der Waal's forces encountered between adjacent–CH_2–groups. The diffusing molecule would only be able to cross the lipid layer when it possessed the minimum kinetic energy needed to break these bonds and separate the lipid molecules. It is possible that a "pore" is created in this manner (la Mer, 1962) and, rather than being a semi-permanent structural feature of the membrane, is transient only, the pore closing again when the kinetic energy of the hydrogen bonds between successive diffusing water molecules drops below that of the surrounding lipid molecules.

The magnitude of the barrier to water movement represented by the lipid layer is determined by the activation energy, which in turn is reflected in the Q_{10} of the water transport process, and can be measured by determining the effect of temperature on rate of transfer. In Fig. 6.5, data of this type obtained by Kuiper (1963) show two components of the temperature effect, one with a high Q_{10} ($Q_{10} \cong 4 \cdot 0$) which occurred at low temperature, the other with a low Q_{10} ($Q_{10} \cong 1 \cdot 2$) which occurred above a certain critical temperature. At low temperatures, it is probable that only transient pore development occurs, and single water molecules or transient files, rather than continuous files, may constitute most of the diffusing molecules. With increasing temperature, however, the extent and permanance of the pores may increase until, above the critical temperature, the activation energy would be such that water-filled pores could be effectively maintained. At this stage the low Q_{10} value is similar to that for the viscosity of water, which implies a minimal resistance to water transport.

The activation energy for water transport across different membranes varies widely, probably depending largely on the characteristics of the lipid layer, but also on the configuration and elasticity of the protein layers. It is well known that water uptake can be influenced to different degrees by different types and concentrations of ions (at concentrations too low to significantly affect $\Delta \pi$) and these effects are likely to be mediated by changes in protein layer characteristics.

Kuiper (1963) obtained data of this type for several ions (Fig. 6.6) which showed that Na$^+$ exerts a much stronger effect than K$^+$. He concluded that the specific effect of a cation on the permeability of the cell membranes depends on the degree of cross-linkage between the

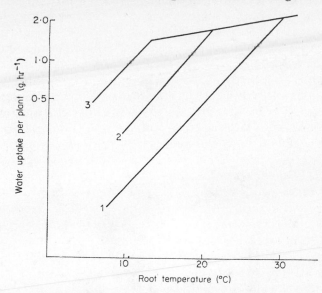

FIG. 6.5. Effect of root temperature on water uptake by bean plants. Plants grown at (1) 25° and under favourable aeration conditions; (2) and (3) at lower temperatures and less favourable aeration conditions (after Kuiper, 1963).

amino acid chains of the protein layers. In a membrane consisting of proteins with a low degree of cross-linkage, an increase in elasticity, and in rates of water uptake per unit of imposed pressure difference (J_v/L_p), can be expected with an increase in hydration number of the ions and with an increase in hydrated ionic volume. Since Li$^+$ has the highest hydration number of the ions depicted in Fig. 6.6, Kuiper (1963) considers this partial reversal of permeability as compared with hydration number indicates a moderate degree of cross-linkage in the membranes of the bean roots which were examined.

2. Other factors affecting cell permeability

The effect of temperature on permeability has already been mentioned, and its influence on the activation energy of the water transport process distinguished from that on the viscosity of water. CO_2 and other metabolic inhibitors have also frequently been observed to reduce permeability (see, for example, Kramer, 1959; Bennet-Clark, 1959; Slatyer, 1960c). The mode of action of these factors has generally been thought to be via a reduction in the flow of respiratory energy needed to maintain active ion exchange processes within the cell, protoplasmic streaming, and the structural integrity of the membranes themselves. However, the actual changes in the membrane are still not understood,

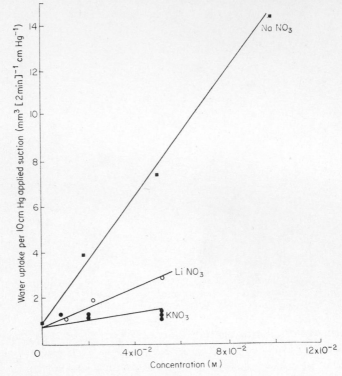

Fig. 6.6. The effect of Na^+, Li^+ and K^+ ions on the ratio of water uptake by bean roots per unit increment in suction applied to the cut stem. The values on the ordinate refer to increments in water uptake per 10 cm Hg increment in applied suction (after Kuiper, 1963).

although Kuiper (1963) has raised the possibility that protein elasticity and configuration may be affected.

The effects of inhibitors on permeability can be very pronounced but are generally reversible if the treatment is not prolonged. An example of this is given in Fig. 6.7 from Glinka and Reinhold (1964) in which dehydrated sunflower hypocotyl segments were placed in water through which either air or CO_2 was bubbled. The rate of water influx was much depressed in the CO_2 treatment but, on replacement of the CO_2 treatment with aerated water, it increased to the same level as in the standard aerated water treatment.

The effect of narcotics sometimes depends on the concentration at which they are applied. Lepeschkin (1932) noted that narcotics at low concentrations decreased permeability to dyes while, at higher concentrations, permeability was sometimes increased. Glinka and Reinhold (1964) observed a similar result with chloroform at $2 \times 10^{-2} M$ compared with $5 \times 10^{-2} M$. It is probable that low concentrations act through a

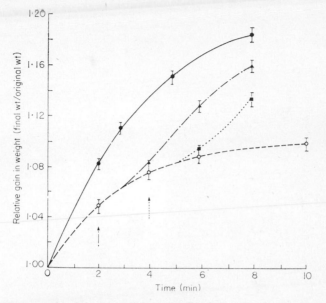

Fig. 6.7. An example of the reversibility of a CO_2 induced reduction in water uptake by sunflower hypocotyl segments. Aerated water treatment (●); CO_2 bubbled treatment (○); segments transferred from CO_2 to aerated water after 2 min (▲), after 4 min (■). Vertical lines indicate standard errors (after Glinka and Reinhold, 1964).

reversible reduction in metabolic activity, but that high concentrations cause pronounced structural changes in the membrane which increases permeability.

In this regard, compounds which tend to separate the membrane lipid layers, such as benzine vapour (Currier, 1951), ether and chloroform (Chibnall, 1923; van Overbeek and Blondeau, 1954) may cause a marked increase in membrane permeability when applied at appropriate concentrations. With this in mind, Kuiper (1964b) reasoned that deliberate incorporation of molecules with unsaturated hydrocarbon chains into the lipid layers should also increase the water permeability. He therefore investigated the effect of several alkenylsuccinic acids (hydrocarbon chain lengths from 6 to 18) and found that a $5 \times 10^{-4} M$ solution of decenylsuccinic acid (chain length 6) increased water uptake by bean roots nearly 8 times compared with the control, while the

analogous saturated compound, decylsuccinic acid, depressed uptake
by 55%. He concluded that the double bond in the hydrocarbon
chain was essential for the increase in water permeability.

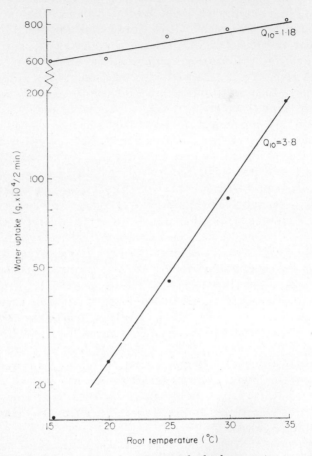

FIG. 6.8. Effect of temperature on water uptake by bean roots exposed to 5×10^{-4} M decenylsuccinic acid for 2 hr (O) compared with controls in normal aerated water (●) (after Kuiper, 1964b).

The effect of decenylsuccinic acid on water uptake, over a range of temperatures, is shown in Fig. 6.8. The appropriate Q_{10} values for uptake by the treated roots compared with the control were found to be 1·18: 3·8. The low value, comparable to those observed with water transfer through physical systems (see also Fig. 6.5), demonstrates that the membrane has been changed from a phase characterized by a high potential energy barrier to water transport to one in which only the effect of the viscosity of water is apparent. It is, of course, probable that membranes suffer some permanent damage by such treatment.

An effect of light on water permeability has seldom been clearly demonstrated, even though light induced volume changes have been demonstrated in chloroplasts (Itoh, Izawa and Shibota, 1963; Belsky, Siegenthaler and Packer, 1965; Packer, Siegenthaler and Nobel, 1965), and light effects on membrane potentials have been demonstrated in giant algal cells (Briggs, Hope and Robertson, 1961). In general, the effects appear to be attributable to ATP synthesis and ion transport followed by passive water movement along water potential gradients. However, if ion movement is associated with a change in protein configuration and elasticity, it is probable that some direct effects of light on water permeability may also occur.

The effect of water stress on cell permeability has not clearly been evaluated. It is probably true that, if dehydration is severe enough to interfere with metabolism, reduction in permeability can be expected in the same way as with other inhibitors. This will certainly be so if protoplasmic viscosity is increased (Brauner, 1930; Levitt, Scarth and Gibbs, 1936; Aykin, 1946). Data from plasmolysis experiments, however, have indicated an increase rather than a decrease in permeability following plasmolysis (Bogen, 1940, 1941; Myers, 1951).

Stocker (1960) has proposed a two stage reaction of protoplasm to water stress, in which an initial reaction phase, associated with reduction in viscosity and increased permeability, is followed by a restitution phase, associated with increase in viscosity and decline in permeability. During the reaction phase, he envisages a weakening of the protein-water bonding mechanisms, which is followed, during the restitution phase, by re-bonding in an altered conformation. It is possible that the two sets of conflicting evidence, just cited, can be reconciled on this basis.

Much more research is required in this field, and short-term effects must be separated from long-term effects which cause anatomical, as well as physiological, changes. In the meantime, it can be assumed that one effect of water stress will be via its influence on metabolism, and hence will be similar to that of other inhibitors. At more severe stress levels direct effects of dehydration on membrane structure can also be expected to have a pronounced influence (Kramer, 1950).

B. *Measurement of Cell and Tissue Permeability*

1. *Techniques for individual cells*

For most plant cells two main techniques have been developed for measuring water permeability. The first, which measures L_p (or k_w)

involves the use of an expression based on Eq. (6.3), or (6.22) if $\sigma < 1$, and the measurement of changes in protoplast or tissue volume with known, or estimated, water potential differences inside and outside the cell. The other technique, which measures a k_s for self-diffusion of water, (termed k_d), involves the use of an expression based on Eq. (6.1) and the use of isotopes of water, without volume changes being involved.

In the first case the hydraulic conductivity, L_p, is obtained, in units such as cm sec^{-1} bar^{-1}. Since the rate of influx or efflux follows an exponential decay related to $(\epsilon + \pi_0^i)$ (see Eq. 6.37), the dynamics of the process are represented, not by Eq. (6.3) which describes instantaneous or steady state flux, but by an expression of the form

$$\Psi^i = \Psi^o \left[1 - \exp \frac{(-L_p A \,(\epsilon + \pi_0^i)t)}{V} \right] \qquad (6.40)$$

With this expression the value of the exponent can be obtained under known conditions of the other parameters, and L_p can be properly evaluated.

In practice, isolated individual cells, or tissue strips in which the exposed layer of cells can be bathed with osmotic solutions and the volume change observed with a microscope, are generally used. Generally the cells or tissues are either equilibrated in pure water and then placed into solutions of known osmotic pressure, and hence of known water potential, Ψ^o, or plasmolysed and then placed in pure water. Volume changes are then observed microscopically and plotted against time. The solution is rapidly stirred during the determinations and the external solution concentration is not assumed to change. Internal water potential, Ψ^i, is estimated from the volume measurements and prior $\Psi^i(V)$ calibrations.

Values obtained with this procedure are given by Davson and Danielli (1952), Bennet-Clark (1959) and Stadelmann (1963), and indicate a range of over an order of magnitude from different experiments, from about $0 \cdot 5 \times 10^{-7}$ to $2 \cdot 0 \times 10^{-6}$ cm sec^{-1} bar^{-1}. An average figure of approximately $0 \cdot 5 \times 10^{-6}$ appears reasonable.

Sources of variability arise from real differences in permeability such as those considered above, but specific measurement errors may be introduced if either the change of volume, the cross sectional area across which movement takes place, or the water potential difference is not measured or estimated accurately. For example, it is generally desired to measure the permeability of a "membrane", comprising the cytoplasm itself and its bounding membranes. Vacuolar volume change is therefore required in order to measure J_v (or dV/dt) and errors will be introduced if cell or protoplast volume change is measured, since the wall or cytoplasm itself may change in volume during the determination. Similarly, estimates of the area across which flow is occurring, A, must refer to a mean cytoplasmic area, not total cell surface area (Crafts, Currier and Stocking, 1949; Mercer, 1955).

The second technique effectively measures the solute permeability coefficient, k_s, (referred to as k_d when isotopic water exchange is under consideration) in units of cm sec $^{-1}$, when the time course of diffusion of labelled water into or out of a cell or tissue segment is monitored. The internal concentration of isotope inside the cell changes exponentially, being proportional to the prevailing concentration difference. Instead of Eq. (6.1) the dynamics of the process are represented by an equation of the form

$$c_s^i = c_s^o \left[1 - \exp\left(-\frac{AK_s}{V}t \right) \right] \tag{6.41}$$

For some purposes the half-time of the process is preferred, given by

$$t^{1/2} = (\ln 2)\,\frac{V}{k_s A} = 0.693\,\frac{V}{k_s A} \tag{6.42}$$

In general, samples of cells or tissue strips are first equilibrated with the isotope solution being used, and are then placed in containers of pure water. Rapid stirring takes place, as before, and aliquots sampled from the external solution are taken at points on the time course of isotope efflux. Internal isotope concentration is measured on subsamples of tissue simultaneously with the water aliquots. Other sources of error, common to most techniques will be discussed separately in the next sections.

If osmotic water movement, involving L_p, is purely diffusional, and no pores exist in the membrane, $RT\,L_p/\overline{V}_w$ should equal k_d. However, in most cases (see for example Ray, 1960; Dainty, 1963a) measurements indicate that $RT\,L_p/\overline{V}_w > k_d$, order of magnitude differences not being uncommon. As stated previously, this has been regarded as evidence that osmotic water movement may be a bulk flow of water through water-filled pores in membranes, as distinct from a diffusional mechanism. However, specific sources of errors also exist in k_s measurements. For example, estimation of vacuolar isotope concentration presents difficulties in small cells where sampling methods almost invariably include non-vacuolar components. Also there is good evidence that in some cases, rate of isotope equilibration may be seriously rate limited by resistances to diffusion beyond the membrane surfaces (Dainty, 1963a).

A third method of measurement, which probably constitutes the most effective means of L_p determination, is that of transcellular osmosis, in which a flux of water is induced through a cell, the ends of which are immersed in solutions of different osmotic pressure. Because of size limitations, only giant algal cells have been used for studies of this type (Kamiya and Tazawa, 1956; Kamiya and Kuroda, 1956; Dainty and Hope, 1959; Dainty and Ginzburg, 1964b). Without exception, the values obtained for L_p are much higher than those obtained by other methods on other cells, approximating $1-2 \times 10^{-5}$ cm sec^{-1} bar^{-1}. [In such cells, partly because of the high permeability, Dainty and Hope (1959) first found clear evidence that associated determinations of k_d using D_2O were distorted by resistances external to the membranes]. The method is most appropriate for L_p determinations and the discrepancy between the L_p values so obtained and those quoted earlier is almost certainly due in part to the inadequacy of the techniques for small cells. However, much of the difference may be real, and caused by differences in membrane structure. Another contributing factor may well be the absence of suberization on the walls of algal cells, whereas most other cells, with surfaces exposed to intercellular air, appear to show some suberization (Scott, Schroeder and Turrell, 1948; Scott, 1950).

2. *Techniques for tissue segments*

The techniques used for tissue segments are generally the same as for individual cells, involving volume change or isotope exchange procedures. However, in tissue segments the measured permeability may frequently bear little relation to the membrane permeability of the individual cells which comprise it. This is partly because the measured volume or isotope concentration changes refer to the entire tissue, and so include changes in non-vacuolar compartments. It is also partly because the effective cross-sectional area through which water exchange occurs may be much smaller than the area of the surfaces bathed by the external solution. The problem, in tissue studies, of diffusion into individual cells being rate-limited by diffusion beyond the membranes is also exaggerated in tissue segments.

With leaf discs, for example, Weatherley (1955) and Slatyer (1958) noted that rate of water exchange varied proportionally with tissue circumference, and not tissue area, confirming that virtually all the water exchange took place through the cut surfaces of the discs, rather than through the cuticle. Similar results were obtained for *Avena* coleoptiles by Ordin and Bonner (1956). Philip (1958d) investigated this problem more generally, regarding a tissue segment as isotropic, and analysing the manner in which the half-times for water exchange varied with tissue dimensions. In Fig. 6.9, predicted cumulative uptake

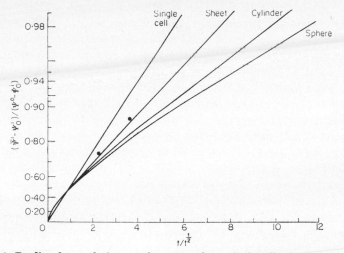

FIG. 6.9. Predicted cumulative uptake curves for a single cell, plane sheet, cylinder and sphere. t is the time (sec) and $t^{\frac{1}{2}}$ the half-time for the exchange process. Ψ^o, Ψ Ψ^i denote the water potential in the external solution, in the tissue at time zero, and the mean value in the tissue respectively. For isotope exchange considerations the values c_s^o, c'_{so} and $\bar{c}_s^i$ can be used representing the external isotope concentration, internal isotope concentration at time zero, and mean internal concentration, respectively. The points are from the data of Buffel (1952) (after Philip, 1958d).

curves for single cells, plane sheets, cylinders and spheres are plotted, together with data of Buffel (1952) in which isotopic water exchanges were measured on cylindrical *Avena* coleoptiles segments 4 mm long. The results suggest that rather than the exchange occurring through the cylindrical walls of the tissue segment, it occurred principally through the cut ends. The same analysis, applied to Ordin and Bonner's (1956) data, indicated a similar result.

The major problem in tissue permeability studies, however, is probably caused by internal geometry. In many cases flow of water into individual cells is rate limited by diffusion up to the effective membranes. This problem, common to almost all permeability studies-will now be considered in some detail. In concluding this section, however, it should be emphasized that tissue permeability studies of the type just described may yield little information concerning flow of water through tissues in the transpiration stream (Raney and Vaadia, 1965b). Under such conditions, the amount of water flowing through any one of the various pathways available is inversely proportional to the resistance encountered. Consequently, there may be negligible flow from vacuole to vacuole, the main pathways, in most tissues, being external to the protoplasts (Strugger and Peveling, 1961; Russell and Woolley, 1961; Weatherley, 1963).

3. *Effect of membrane boundary layers*

It is probably true to say that the factor most neglected in permeability measurements, yet most likely to affect the results in many cases, is the presence of an "unstirred layer" or diffusional boundary layer adjoining the membrane surfaces. Because water and solutes can only move across such a zone by molecular diffusion, it effectively increases the apparent thickness and resistance of the membrane, and does so to an increasing degree as the membrane permeability value approaches that for free diffusion of the solute species being studied.

Permeability measurements implicitly assume that the aqueous solutions, on both sides of the membrane under study, are so well stirred that the concentration at the interface is identical with that in the bulk solutions. This is quite impossible, even though effective stirring can probably reduce boundary layer thickness considerably and, in consequence, permeability values necessarily refer to the permeability of the boundary layer-membrane-boundary layer system rather than to the membrane alone.

Although the need for stirring has been intuitively recognized by many research workers, and boundary layer theory has been studied extensively in interfacial phenomena (see for example, Davies and Rideal, 1961) the quantitative evaluation of boundary layer effects in permeability studies is a recent development, due largely to the work of Dainty (1963a) (see also Dainty and Hope, 1959; Dick, 1959).

The thickness of a boundary layer, d, is not an actual thickness but rather an effective thickness assuming the concentration gradient across it (dc/dz) is uniform and equal to that at the actual interface, so that it is obtained from the expression

$$\frac{dc_s}{dz} = \frac{(c_s - c_s^m)}{d} \tag{6.42}$$

where c_s and c_s^m are the bulk solution and membrane interface concentrations, of the particular solute, respectively. The effective concentration profile for a solute moving from one side of the membrane to the other, therefore, is similar to that shown diagrammatically in Fig. 6.10 and the permeability coefficient generally is given as

$$k_d = J_s/(c_s^o - c_s^i) \tag{6.43}$$

whereas the true membrane permeability coefficient is given by

$$k_d^m = J_s/(c_s^{om} - c_s^{im}) \qquad (6.44)$$

where the concentration difference is measured across the membrane alone. Since $k_d = D/d$ where D (cm^2 sec^{-1}) is the diffusion coefficient

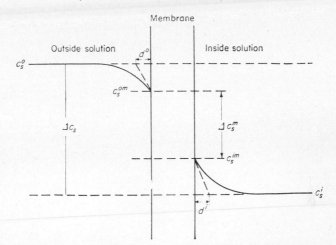

Fig. 6.10. Concentration profile for a permeating solute in the solutions adjacent to membrane. d^o and d^i are the thicknesses of the unstirred layers as defined by Eq. (6.38). (after Dainty, 1963a).

of the solute in solution, the two permeability values are related by the expression

$$\frac{1}{k_d} = \frac{1}{k_d^m} + \frac{d^o}{D^o} + \frac{d^i}{D^i} \qquad (6.45)$$

where d^o and d^i are the effective thicknesses of the external and internal boundary layers, and D^o and D^i are the diffusion coefficients for the external and internal solutions, respectively.

Typical values for boundary layer thicknesses range from 20–500 μ (Bircumshaw and Riddiford, 1952; Dainty and Hope, 1959) depending on size of surface and rate of stirring, but wider ranges undoubtedly exist and, especially with tissue slices and minimal stirring, effective thicknesses of >1 mm can be expected. Since, for self-diffusion of water, $D \simeq 10^{-5}$ cm^2 sec^{-1}, the permeability coefficients of boundary layers 10 μ and 1000 μ thick are 10^{-2} and 10^{-4} cm sec^{-1}, respectively. Clearly, measured permeabilities which approach these values are likely to be increasingly influenced, and the process finally rate limited, by the permeability of the unstirred layers themselves.

In this connection it is of interest that the D values Philip (1958d) calculated from the *Avena* coleoptile data of Buffel (1952) and Ordin and Bonner (1956) gave $D \cong 10^{-5}$ cm^2 sec^{-1}. It is therefore not surprising that Ordin and Bonner concluded that there was little difference in the permeability of any region except the cuticle, and it seems probable that, in this case, diffusion of water was primarily rate-limited by diffusion up to the effective membrane surfaces (both inside and outside the tissue) rather than by the surfaces themselves. Dainty and Hope (1959) have also shown that this is the case for water permeability values of the long internodal cells of *Chara* measured with D$_2$O, where $L_p RT/\overline{V}_w \cong 2 \times 10^{-2}$ cm sec^{-1}. At the other extreme, where measured water permeability values are lower than $k_d \cong 10^{-4}$ cm sec^{-1} ($L_p \cong 10^{-7}$ cm sec^{-1} bar^{-1}), relatively little error can be expected, even without active stirring.

In small cells of higher plants, errors introduced by boundary layer effects appear to be due mainly to the unstirred layer outside the protoplasts, because of the unknown but obvious effects of protoplasmic streaming inside the cell, and the small diameter of the average higher plant cell ($< 50 \mu$). This tendency would be accentuated in tissue segments, which may be 1 mm thick and in which the free space constitutes an extensive unstirred layer. In such situations, the boundary layer within the tissue but external to the vacuoles is probably of greater significance than the layer external to the tissue itself. In large algal cells, such as *Chara*, however, internal cell size is sufficiently large for the diffusion within the cells to be an important rate-limiting factor in D$_2$O exchange (Dainty and Hope, 1959).

The effect of unstirred layers may be much different when k_d is being measured with DHO, D$_2$O or THO and there is no volume flux, than when L_p is being measured and a volume flux occurs. In the former case, Eq. (6.45) describes the relative importance of k_d^m, d^o and d^i in determining k_d. In the latter case, however, the volume flow of water is caused by an osmotic pressure or hydrostatic pressure difference.

In the case of water influx from external solution of pure water into a plasmolysed cell, no external boundary layer arises, except in the cell wall. During water efflux from a turgid cell into an osmotic solution, it could be expected that external "bulk solution" concentration might be significantly reduced by the water efflux through the membrane, leading to boundary layer development with the lowest solute concentration next to the membrane. While this undoubtedly occurs, Dainty (1963a) has shown that this is not likely to significantly reduce the assumed "bulk solution" concentration, as long as boundary layer does not exceed 50 μ, because back diffusion of the external solute towards the membrane opposes the convective outward flow. A figure of $50\,\mu$ appears reasonable unless tissue slices are being used, in which case the effective thickness could be significantly higher.

Internal boundary layers presumably also develop because of the phenomenon just described, although to a smaller degree than in the isotope exchange measurements. Consequently, while both volume flow and isotopic exchange methods of water permeability measurement can be influenced by unstirred layers, those techniques involving isotopic water exchange are likely to be influenced to a much greater degree. The effect of unstirred layers on the observed permeabilities will increase as the true permeability increases, and as the distance from the membrane to the bulk solution increases. Measurements on individual cells of low permeability will therefore tend to be the most accurate, and those on tissue slices the least accurate.

An important implication of the foregoing discussion concerns the evidence as to the presence or absence of water filled pores in membranes and hence concerns the previous remarks concerning osmotic flow. The evidence for the presence of pores is based, in the main, on the fact that when water permeability is measured using J_v and L_p (Eq. 6.22), the values of the "osmotic" permeability coefficient, $RT\,L_p/\overline{V}_w$, usually appear to be significantly higher than the dimensionally comparable values of the "diffusional" permeability coefficient, k_d. A number of examples of this phenomenon are given by Prescott and Zeuthen (1953), Durbin, Frank and Solomon (1956) and Ray (1960). However, Dainty (1963a) contends that in every case the substantially greater effect of unstirred layers in the latter determinations casts considerable doubt on the validity of the comparisons. These criticisms are soundly based and, while a demonstrated discrepancy could be regarded as good evidence for the presence of water filled pores, definitive experiments are extremely difficult to devise, and the question is still open.

4. *Effect of solute entry and the reflection coefficient, σ*

Measurements of "osmotic" water permeability using techniques such as transcellular osmosis or plasmometry, and based on Eq. (6.22) generally, presuppose that the membrane is impermeable to the solute used to vary $\Delta\Psi$ and hence that $\sigma = 1$. The equation therefore simplifies to the original Eqns (6.2) to (6.5). Although, in many cases, the solute permeability coefficient, k_s, is so much smaller than the water permeability coefficient k_w (or L_p) that this source of error is negligible; it is now well established, for example, that sucrose, the most commonly used solute for water permeability measurements, is not excluded from many cells and that its rates of uptake may be quite substantial in some cases (Weatherley, 1954; 1955; Pennell and Weatherley, 1958; Slatyer,

1961, 1966b). Furthermore, there is increasing evidence that mannitol, long regarded as completely excluded from most cells, may be absorbed to a significant extent (Burström, 1963a; Groenewegen and Mills, 1960; Slatyer, 1966b).

With both solutes absorption appears to be less rapid with storage tissue or algal cells than with leaf tissue. Because leaf tissue has not been extensively used for permeability studies this may explain, in part, why the possible effect of $\sigma < 1$ in leaves has not been emphasized more. However, from the point of view of plant water relations generally, leaf tissue is of considerable importance and it is of value to examine the possible significance of solute entry, not only with respect to the permeability measurements themselves, but also to related measurements such as those of Ψ and π.

Referring again to Eq. (6.22), it can be assumed that, since the experiments are normally conducted with $P^o = 0$ and $\pi_m^o = 0$, $\Delta P = P^i$ and $\Delta \pi_m = \pi_m^i$. Because the amount of solute entry is not negligible at the time the measurements are made it cannot be assumed that $\Delta \pi_s = \pi_s^o$, and π_s^i must be accounted for. However, the permeability coefficient for outward movement of solute from the vacuole is very low and it seems valid to assign $\sigma = 1$ to the membrane with respect to the solute which has already entered the vacuole. Equation (6.22) therefore becomes, for volume efflux:

$$J_v = L_p[(P^i - \pi_m^i - \pi_s^i) + \sigma \pi_s^o] \qquad (6.46)$$

where the effective driving force is given by the two terms inside the square brackets. In some cases, the amount of solute entry will be negligible by the time the measurements are made, in which case π_s^i can be eliminated from the expression. The full term inside the round brackets, however, is the internal water potential, Ψ^i, and can be satisfactorily measured by a vapour equilibration technique regardless of whether π_s^i is detectable or not.

It is apparent that the driving force will be overestimated by neglect of σ, and L_p will consequently be underestimated. The degree of error will increase as σ declines from 1. In consequence, it becomes important to evaluate σ so that the correct driving force can be assigned and L_p accurately determined.

Very few measurements of σ for plant cells exist, as yet (see Dainty and Ginzburg, (1964a) for some determinations of *Nitella* and *Chara*). It is conveniently measured at the quasi-equilibrium stage of Fig. 6.1, when $J_v = 0$. At this stage, using Eq. (6.46)

$$\sigma = -\frac{(P^i - \pi_m^i - \pi_s^i)}{\pi_s^o} = -\frac{(P^i - \pi^i)}{\pi^o} = -\frac{\Psi^i}{\pi^o} \qquad (6.47)$$

where $\pi^i = (\pi^i_m + \pi^i_s)$. If the measurements are also made at incipient plasmolysis, so that $P^i = 0$

$$\sigma = \frac{(\pi^i_m + \pi^i_s)}{\pi^o_s} = \frac{\pi^i}{\pi^o} \qquad (6.48)$$

(cf. Eq. 6.29). The significance of these equations is enhanced considerably when it is realized that liquid exchange measurements of cell and tissue water potential, $\Psi^i = (P^i - \pi^i)$, involve equilibration of tissue with a range of osmotic solutions with the objective of determining that concentration in which $J_v = 0$ (see Chapter 5). In such cases it seems that the measured value of Ψ^i does not equal $-\pi^o$ but instead $-\sigma\pi^o$.

Estimates of σ for several types of tissue were made, on this basis, by Slatyer (1966b), and values of $\sigma = 0.7-0.8$ and $\sigma = 0.8-0.9$ were obtained for sucrose and mannitol, respectively. Earlier published data on discrepancies between plasmolytic and cryoscopic osmotic pressure determinations indicated similar values. Although other factors could have contributed to these results, particularly since tissue segments were used rather than single cells, it seems probable that the assumption of $\sigma = 1$ for solutes such as sucrose and mannitol, when used with leaf tissue, may introduce significant errors both to permeability determinations and to liquid exchange measurements of Ψ^i and π^i. It seems most desirable that σ should be specifically measured in conjunction with all water permeability studies, either using the technique based on measurements at $J_v = 0$ or other procedures (Dainty and Ginzburg, 1964a).

C. Active Water Uptake

From time to time there have been suggestions that water may be absorbed by plant cells against a water potential gradient and that respiration may provide the necessary energy (Bennet-Clark, 1959). The main supporting evidence has come from measurements of internal and external osmotic pressures (Bennet-Clark, Greenwood and Barker, 1936; Bennet-Clark and Bexon, 1940; Mason and Phillis, 1939; Currier, 1944), which have indicated discrepancies between observations made with the plasmolytic method and on expressed sap. Other support has been based on evidence of auxin induced water uptake against a water potential gradient (Reinders, 1938, 1942; Hackett and Thimann, 1952, 1953; Bonner, Bandurski and Millerd, 1953).

The first type of evidence has been strongly criticized on the basis of the methods used (Levitt, 1947; Mercer, 1955). It also seems probable that neglect of σ may have contributed to the observed discrepancies, since, with reference to Eq. (6.48), expressed sap measurements should provide a measure of π^i and plasmolytic measurements an estimate of $\sigma\pi^o$.

The other type of evidence also seems capable of re-interpretation. In general, it was noted that auxin significantly promoted water absorption and that the effect was inhibited by respiratory inhibitors. Some of the early investigators attributed the auxin effect to enhanced accumulation of osmotically active solutes in the vacuole (Reinders, 1938, 1942; Commoner, Fogel and Muller, 1943) but this was not confirmed in subsequent work (van Overbeek, 1944; Hackett, 1952; Brauner and Hasman, 1952; Ordin, Applewhite and Bonner, 1956). The primary effect appears to be through increased permeability and plasticity of the cell wall, which serves to reduce turgor pressure, P, and hence Ψ, and so induce water movement into the cell with associated enlargment.

Another phenomenon, which is sometimes regarded as evidence of active uptake, is the apparent polarity of water movement which has been demonstrated in some cells and tissues and which is reflected in an apparently higher value of k_w for water influx than for water efflux (Kamiya and Tazawa, 1956; Dainty and Hope, 1959). Dainty (1963b) has shown that the results of the second group of workers, obtained with *Chara*, could be satisfactorily accounted for by unstirred layer effects, so that when a calculated effective driving force ($\Psi^{om} - \Psi^{im}$) was used instead of the apparent driving force given by ($\Psi^o - \Psi^i$) the two permeability coefficients were not significantly different. The results of Kamiya and Tazawa, however, when adjusted in this way still indicated an apparent polar permeability to water, with k_w (influx) $> k_w$ (efflux).

In summarizing the remarks which have been made at several points in this chapter concerning active uptake, it seems valid to state that there is no evidence at present available which demonstrates significant water uptake against a concentration gradient. Water movement is, however, frequently observed in conjunction with other flows of matter or energy such as solutes, heat or electricity. This type of uptake is frequently regarded as "active" although it follows the normal thermodynamic relationships already discussed. In this context "active" uptake of water into cells, associated with ion accumulation or changes in turgor pressure due to changes in wall and permeability characteristics, is a commonly observed phenomenon. Electro-osmosis appears to come closest to the concept of a non-osmotic pump. It is undoubtedly a real and important phenomenon, but its role in water movement appears to be insignificant in most situations. This is because of the short-circuiting effect of the normal high permeability pathways along which water moves passively.

CHAPTER 7

Water Movement Through the Plant

Water movement through the soil to the root surfaces has already been considered in Chapter 4 and water movement from the leaf surfaces to the air will be considered in Chapter 8. In the present chapter, liquid movement through the plant is considered, with emphasis given first to flow across each segment of the transpiration pathway, and then to transport through the system as a whole.

This subject has been comprehensively reviewed in recent years and, for further detail, the reader is referred to papers by Philip (1957a, 1966), Gardner (1960b, 1965), Slatyer (1960c), Slatyer and Denmead (1964), Slatyer and Gardner (1965), Visser (1964) and Cowan (1965).

1. WATER MOVEMENT ACROSS THE ROOT

Although small amounts of water can be absorbed by the above ground parts of plants, the root system effectively constitutes the region of entry for virtually all water and minerals absorbed by higher plants. Considerable diversity exists in the character and spatial distribution of root systems of different species, and of any one species in relation to soil type, soil water and nutrient status and competition from roots of nearby plants. Background information on these aspects of root systems and root growth can be found in Kramer (1949), Shaw (1952) and Kramer and Kozlowski (1960). In this section, attention is directed to the absorbing zones of roots, the pathways for water movement and the factors affecting water uptake. Some aspects of active absorption and root pressure phenomena are also considered.

A. Water Absorbing Zones of Roots

The actively elongating meristematic terminal zone of each root is composed of numerous, small, actively dividing, tightly packed cells, almost completely filled with cytoplasm. As the root extends through the soil, cell division at any one point ceases as the root progressively differentiates into three main tissue phases, the epidermis, the cortex and the stele, containing the vascular system (see Fig. 7.1). As the age of the zone under consideration increases, and its distance behind the continually extending root tip lengthens, cutinization and suberization of the epidermis occurs, root hairs tend to disappear and secondary growth gradually commences. If an exodermis exists, the cell walls tend to thicken and also become suberized.

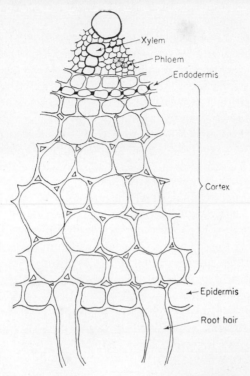

FIG. 7.1. Diagrammatic transverse section of young root in the zone of most rapid absorption (after Kramer, 1959).

Secondary growth occurs commonly in dicotyledons and gymnosperms, but less commonly in monocotyledons, particularly annuals. It is associated with an increase in the diameter of the stele as cambial activity increases and, as a result, the cortex and epidermis are ruptured and break off. Frequently a cork cambium develops and the roots become covered with a layer of corky tissue. In roots of woody plants, the cortex disappears completely, successive cork cambia develop and the arrangement of tissues in the root resembles that in the stem.

Anatomical and physiological investigations of the absorbing zones of roots (Kramer, 1956a; Esau, 1960; Brouwer, 1965) have shown that the zone of most rapid water absorption lies behind the meristematic region of the root tip itself, and ahead of the region where suberization develops. In consequence, the zone tends to move through the soil with growth of the root system, being located in a region characterized more by anatomical development and physiological age, rather than any other factor. The zone has been located from 1·5 cm to 20 cm from the root tip, but is usually found 5–10 cm behind the tip (Sierp and Brewig, 1935; Hayward and Spurr, 1943; Brouwer, 1953, 1954; Raney and Vaadia, 1965b). As can be expected, the length of the region depends largely on the rate of root extension; in slow growing roots it is very restricted and, when root growth ceases, for example as a result of water stress and low temperatures, few unsuberized roots can be found (McQuilkan, 1935; Reed, 1939).

The zone of rapid uptake coincides with the normal zone of root hair development. Although the water permeability coefficient, L_p, of root hairs appears to be no greater than that of unsuberized epidermal cells (Rosene, 1943, 1954), they may increase the effective absorbing surface by up to an order of magnitude (Dittmer, 1937; Evans, 1938) and hence are of considerable importance under conditions when the absorbing surface of the roots appear to be limiting absorption. Root hairs may persist for some time in roots which do not exhibit secondary growth, but their permeability is decreased by cutinization and lignification (Rosene and Walthall, 1954).

Beyond the zone of rapid uptake, rate of absorption decreases rapidly as suberization develops. An example of this phenomenon, from Brouwer (1954) is shown in Fig. 7.2. In this instance, average figures for 2·5 cm segments of bean root were obtained, excluding the root tip proper, and show a rapid decline at distances greater than 5 cm from the root tip. If secondary growth occurs, absorption is generally reduced still further. In monocotyledons not showing secondary growth, however, there may be a low and progressive decline from root tip to the base of the main stem (Hayward, Blair and Skalling, 1942). Even so, considerable quantities of water can be absorbed through suberized roots, as has been shown by several workers on a range of different species (Chapman and Parker, 1942; Hayward, Blair and Skalling, 1942; Addoms, 1946).

The role of mycorrhizae in water absorption is still unclear, but the additional absorbing area so exposed may induce an effect somewhat similar to that of root hairs, but which persists in older sections of roots than those which normally retain root hairs. Root-grafting also affects net water uptake by any one root system. Although somewhat beyond the scope of the present discussion, it can be an important factor determining the water relations of plant communities (la Rue, 1952; Bormann, 1957, 1961).

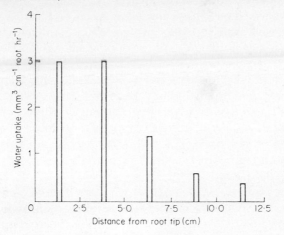

FIG. 7.2. Water uptake by different zones of a bean root at constant applied suction ($\Delta P = 1\cdot3$ bars). The root was examined in $2\cdot5$ cm segments, but the zone nearest the tip excluded the root tip proper (after Brouwer, 1954).

B. The Water Pathway Across the Root

1. Anatomical features

The pathway of water movement from the root surface to the stele, in the zone of rapid absorption, involves transport across the epidermis, the root cortex and the endodermis. The outer epidermal cell walls, in this region, may be slightly cutinized but appear to be highly permeable to water (Rosene, 1954). Many epidermal cells may develop root hairs, which are tubular extensions of the cells themselves, also of high permeability, as mentioned above. The root cortex is generally composed of 5–15 layers of loosely-packed parenchyma cells with conspicuous intercellular spaces. Water may move through this zone through the protoplasts, from cell to cell, or outside the protoplasts, via the cell walls. Both pathways are almost certainly utilized, the proportion of flow via each one varying inversely with the resistance encountered.

The outermost layers of the cortex may differentiate into an exodermis of tightly-packed cells lying directly beneath the epidermis, but these cells do not become suberized at this stage. The innermost layer of the cortex, however, which forms the endodermis, contains suberin in a band-like layer which extends around each cell within the radial and transverse walls (see Fig. 7.3). This band, called the Casparian strip, constitutes part of the primary wall, rather than a simple wall thickening, and the suberin is deposited continuously across the middle lamella. Furthermore, the protoplast is firmly attached to the Casparian strip.

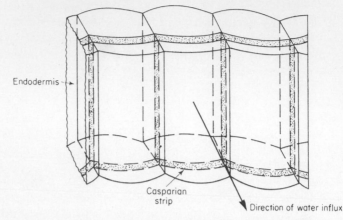

FIG. 7.3. Diagram of three endodermal cells showing Casparian strips in transverse and radial walls but not in tangential walls (after Esau, 1960).

In consequence, the cell wall pathway for water, and ion, movement is effectively blocked and transport across the endodermal cytoplasm is necessary when Casparian strip development is effectively continuous. In some cases, particularly if passage cells are located in the endodermis opposite xylem vessels, suberization is poorly developed and, at points where branch root initials are forming, the effectiveness of the endodermal barrier may also be considerably reduced. Overall, however, the endodermis constitutes an important barrier to the free diffusion of ions. It is also probably the main source of resistance to water entry in young roots.

Within the vascular cylinder there appears to be little resistance to water flow from endodermis to xylem tracheids or vessels. The cylinder consists of one or more layers of pericycle, and the vascular tissues. The pericycle generally consists of parenchyma and, in the zone of rapid absorption, the cells are thin walled without obvious regions of low permeability.

The pathway of water movement in suberized roots, in which secondary growth has occurred, is across the suberized outer tissue, consisting of a peripheral bark layer and layer of cork cambium, across or around the secondary phloem, and across the vascular cambium. None of these layers may constitute a single resistance barrier as great as the endodermis, but the outer suberized layers are heavily impregnated and can be expected to have very low permeability. There are, however, breaks in the tissue and there is some lenticel development, both possibly providing areas of low resistance.

2. Driving force for water transport

It has generally been assumed that the driving force for water flow across the root is the gradient of water potential $d\Psi/dx$, and that Eq. (6.3) adequately describes the observed water flux. However, there is obviously rapid transport of ions across the root and it has sometimes been suggested that a significant part of this ion flux may be passive (Hylmö, 1953, 1955, 1958; Epstein, 1956a, 1956b; Kramer, 1957). Walter (1955) has assumed that this is so and that, in solution cultures, free diffusion of the externally applied osmotic solutes results in a wave of solutes passing into the plant, causing rapid and complete internal osmotic adjustment.

Should the "free space" (Briggs, 1957) extend from epidermis to stele, free diffusion of solutes could be expected, but in practice there is indisputable evidence that most vascular solutes are effectively retained when a root system is placed in pure water (see for example, Long, 1943; Slatyer, 1961). There is also clear evidence that the concentration of the xylem sap, and of the individual ions which comprise it, may be substantially greater or substantially less than in the rooting medium (Broyer and Hoagland, 1943; Russell and Shorrocks, 1959; Russell and Barber, 1960; Scholander et al., 1962). There is in the root, therefore, a barrier which most effectively prevents ion efflux and which exhibits a high degree of discrimination in the degree to which any ion is absorbed and accumulated in the vascular system. In consequence, the free space appears to be discontinuous in the root and the logical site for this barrier appears to be the endodermis.

This is not to say that there is no passive ion transport across the root, in fact there is good evidence of passive uptake of both electrolytes and non-electrolytes (Groenewegen and Mills, 1959; Slatyer, 1961; Jackson and Weatherley, 1962a, 1962b; Lopushinsky and Kramer, 1961). While it appears that the reflection coefficient, σ, of the root to solute efflux is effectively unity; for solute influx, the situation where σ is significantly less than unity, may exist, under some circumstances.

Possible evidence for this view can be found in those experiments in which internal osmotic adjustment occurs, when plants in normal water culture are transferred to osmotic substrates. In such cases (Walter, 1955; Bernstein, 1961, 1963; Slatyer, 1961) the final internal concentration of the osmotic solute is frequently almost identical with the substrate concentration and the rate of solute entry can be very rapid. Other evidence comes from experiments in which an increase in the pressure difference imposed across a detopped root system causes a marked increase in salt transport (Lopushinsky and Kramer, 1961; Jackson and Weatherley, 1962a, 1962b). Experiments which show a dependence of ion uptake on transpiration rate have also been used as evidence for passive ion uptake, but Russell and Shorrocks (1959) have shown that this relationship only appears to develop when both high transpiration rates and high salt concentrations exist. Under these conditions, Russell and Shorrocks suggest that the rapid removal of ions as they reach the stele could develop a stronger sink for both passive and active transport. In some respects, both the pressure-stimulated ion flow and the transpiration-stimulated ion flow, show the characteristics of a metabolically facilitated passive transfer (Jackson and Weatherley, 1962b), but the nature of such an energy linked transport mechanism is still a matter of speculation.

In summary, it appears that although there is rapid solute uptake by roots, there is little direct evidence, as yet, that σ is less than unity for most solutes. Consequently, it appears to be valid to assume that the water potential gradient across the root $d\Psi/dx$ is the effective driving force for water transport, at least at first approximation level. However, measurements of σ are to be encouraged, both for their own intrinsic value in studies of water and solute uptake and also because they will indicate the degree to which error might be introduced by the foregoing assumption.

C. Factors Affecting Root Permeability

The foregoing discussion has indicated that the primary barrier to ion diffusion in the root is probably at the endodermis. The primary barrier for water transport is almost certainly also at this point, since resistance of flow of water through the cell wall pathway of the root cortex is likely to be relatively small, by analogy with water flow through meosphyll (Weatherley, 1963), from evidence with fluorescent dyes (Strugger, 1949) and from calculated relative permeabilities of cell walls and cytoplasm (Russell and Woolley, 1961).

Typical values for the water permeability coefficient, L_p, of root segments are 0.1–2.2×10^{-6} cm sec^{-1} bar^{-1} for bean (*Vicia faba*) roots (Brouwer, 1954); 0.2–0.8×10^{-6} for oat coleoptiles (Ray and Ruesink, 1963) and 0.6×10^{-6} for maize roots (House and Findlay, 1966). These values are of similar order of magnitude to those obtained for single cells (see Chapter 6) also suggesting that a single cell barrier of somewhat similar permeability characteristics is probably involved.

Some of the factors influencing root permeability have already been discussed, both in this Chapter and in the general discussion of Chapter 6. For present purposes they can conveniently be divided into two groups: those associated with the genetic characteristics of the plant and the environmental conditions under which it has been grown, which determine the basic morphology and anatomy of the root and the ultrastructure of the root cells, and so create the actual water pathway, and those which influence the permeability of the pathway so formed. Because the root is continually ageing, both interact continuously and short-term environmental influences rapidly result in structural changes in the developing root.

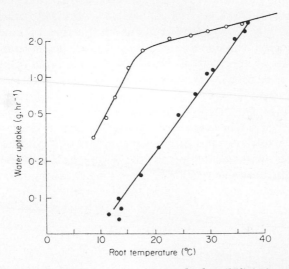

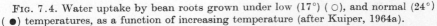

Fig. 7.4. Water uptake by bean roots grown under low (17°) (○), and normal (24°) (●) temperatures, as a function of increasing temperature (after Kuiper, 1964a).

A good example of the former group is shown in Fig. 7.4, where water uptake of root systems grown under two different temperatures was compared over a range of root temperatures. The roots of those plants grown in low temperature substrates absorbed more water at low temperatures than roots of plants grown under normal temperatures. At low temperatures it is possible that altered membrane structure of the endodermal cytoplasm may have resulted in a lower resistance to water flow. It is of interest, however, that the slopes of the temperature response curves (i.e. the Q_{10} values or activation energies) are similar for both treatments. This probably indicates a similar degree of metabolic control over the transport process despite the different growth conditions. The evidence that the critical temperature [above which changes in uptake are caused only by changes in the viscosity of water (cf. Fig. 6.5)] is lower in the low temperature treatment probably indicates a different membrane structure. The long-term effect of water stress is likely also to affect structure; indeed, if root growth is significantly reduced through any cause, suberization of the epidermis will, in itself, lead to a rapid decline in root permeability, apart from changes in membrane characteristics.

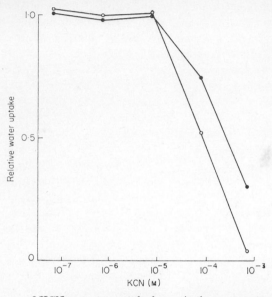

FIG. 7.5. Influence of KCN on water uptake by entire bean roots (●) and part of the root system (○) (after Brouwer, 1965).

The permeability of root to water is strongly influenced by the level of metabolic activity. As stated in Chapter 6, a continuous flow of respiratory energy is required to maintain membrane structure and function and the mode of action of most of the shorter term factors which affect the capacity of the root to transmit water appears to be largely via their effect on metabolism. Thus, limiting aeration, low or high temperature, or high CO_2 levels, appear to reduce permeability in much the same manner as specific metabolic inhibitors such as azide or KCN (van Overbeek, 1942; Rosene, 1944; Brouwer, 1954; Mees and Weatherley, 1957b) (see Fig. 7.5). Here again, however, there is a strong interaction with the genetic and longer-term pre-conditioning factors. As mentioned above, roots grown at low temperatures had relatively high permeability values at low temperatures compared with normal roots. In general, the effect of low temperatures on cold climate species is much less than on warm climate species (Brown, 1939; Kramer, 1942; Kozlowski, 1943). Similarly, with aeration, water plants can be relatively unaffected by conditions which seriously affect other species (Parker, 1950; Hunt, 1951; Kramer, 1951; Russell, 1952). Of course, with temperature effects, there is also a direct effect on permeability caused by changes in the viscosity of water. Under some conditions, this effect can dominate the pattern of water uptake (Kramer, 1949; Kuiper, 1963, 1964a).

As can be expected, the effects of poor aeration are frequently linked with temperature, since the oxygen requirements for respiration increase rapidly with temperature. Poor aeration often involves combined effects of low oxygen and high CO_2 concentrations, although the two gases have somewhat different modes of action (Glinka and Reinhold, 1962, 1964). The direct inhibiting effect of high CO_2 concentration appears to be very rapid (Kramer, 1940b; Chang and Loomis, 1945), but oxygen deficiency is generally slower acting and may take hours or even days to become apparent (Whitney, 1942; Hoagland and Broyer, 1942; Rosene and Bartlett, 1950; Rosene, 1950; Mees and Weatherley, 1957b). This delay is related to the degree of oxygen deficiency, since total absence of oxygen in small containers can be inhibiting in several hours (Mees and Weatherley, 1957b). Kramer (1949) suggested that the longer delay required for oxygen deficiency to become noticeable may be due to slow continued oxygen supply by intercellular diffusion, or through the vascular system. He also suggested that, if one of the main effects of low oxygen is a build up of toxic end-products of anaerobic respiration, some time may elapse before this accumulation becomes important. It is of interest that even high concentrations of CO_2 can be tolerated in some situations as long as oxygen concentration is high (Whitney, 1942; Glinka and Reinhold, 1962). In this regard, (Russell 1952) concluded that, under field conditions, high CO_2 is likely to be of minor significance unless oxygen levels are depressed.

With most of these inhibiting factors, the initial effects, or those associated with low dose rates, appear to impair membrane function only, being fully reversible even though they may reduce water flow through the root by more than an order of magnitude. With prolonged treatment and high dose rates, however, the treatment is not reversible, membrane structure also appears to be affected and, under some conditions, increases in permeability occur (Hoagland and Broyer, 1942; Kramer and Jackson, 1954; Brouwer, 1954; Glinka and Reinhold, 1962). In extreme cases, treatments which cause the death of the roots may also cause increases in permeability (Renner, 1929; Kramer,

1940a). All these responses are consistent with the view that active metabolism is essential to maintain normal permeability levels. It is of interest that, in general, effects on ion uptake are apparent at lower inhibitor concentrations and are more pronounced at any one concentration than on water uptake (van Andel, 1953; Brouwer, 1954).

The effect of water stress on L_p is unclear. In general, it can be expected that reduced cytoplasmic hydration, if prolonged, can be expected to increase cytoplasmic viscosity and decrease the permeability of both the cytoplasmic membranes and the protoplasm itself (Levitt, Scarth and Gibbs, 1936; Aykin, 1946). However, the degree of stress needed to cause this change may not be reached even in wilted plants, since the activity of the water in the cytoplasm is still high at this stage.

Initial effects of stress may result in a transient reduction in protoplasmic viscosity and an associated increase in permeability (Stocker, 1960). Moreover, some reduction in turgor may reduce the compression which exists between adjacent cell walls, and increase the permeability of the cortical cell wall pathway. Ordin and Gairon (1961), for example, noted an increased rate of isotopic water exchange in stressed tissue, which may have been due to this phenomenon. Kramer (1950) concluded that pronounced water stress causes an immediate reduction in permeability due to the factors mentioned above and, if prolonged, causes changes in cytoplasm and membrane structure, leading to irreversible effects. In many respects, therefore, water stress can be regarded as similar in its mode of action to other metabolic inhibitors, but it may have little effect, at least on cytoplasmic permeability, until stress is fairly severe.

As distinct from the effect of water stress on permeability, pressure differences across the root have been observed to have a pronounced effect on flow (Brewig, 1936; Brouwer, 1953, 1954; Mees and Weatherley, 1957a, 1957b; Kuiper, 1963). With entire root systems of tomato, for example, Mees and Weatherley (1957b) observed a five-fold increase in L_p as ΔP was increased by about 2 bars. This effect appears to be caused by two factors: a straight out increase in the permeability of the absorbing surfaces and an increase in the effective absorbing zone of the root or roots. The results of Brouwer (1954) given in Fig. 7.6 for the same root system as used in Fig. 7.2, illustrate this effect, showing the increase in permeability in all zones examined, as ΔP was increased from 1·3 bars to 2·5 bars, and a progressive rise in permeability towards the base of the root, where the permeability had previously been lowest.

Various explanations have been used to account for this phenomenon; that used by most authors assumes that the water flow is through water-filled pores and that, as ΔP is increased, pores which previously were not conducting water are brought into play. It can be expected that the ratio of $J_v/\Delta P$ will increase as new pores are brought into the transport system but, once all pores are open, the slope would be uniform. Brouwer (1965) considered that existing pores may also be widened by the treatment. If pores do exist, and if they are widened sufficiently by pressure differences, the results of apparent facilitated transport of solutes across the root could be explained on the basis of a passive solute (or bulk solution) flow through enlarged pores.

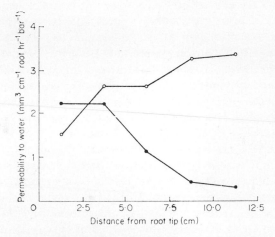

FIG. 7.6. Water permeability of bean roots, measured in 2·5 cm zones, with pressure differences of $\Delta P = 1·3$ bars (●) and $\Delta P = 2·5$ bars (○). The system is the same as used for Fig. 7.2 (after Brouwer, 1954).

It is possible that the phenomenon is an experimental artifact, associated with pressure-induced changes in protein hydration (Klotz, 1958) which may affect protoplasmic viscosity. If so, the effect may only be transient, and it is noteworthy in this connection that Mees and Weatherley (1957a, 1957b) noted a reduction in permeability if a pressure treatment was prolonged. If, however, the effect is real, the possibility of this phenomenon occurring in naturally transpiring plants must be considered, since large ΔP values undoubtedly develop across the root in response to rapid increases in transpiration rate or the imposition of osmotic substrates. Should it take place it could perhaps explain the observed rapid osmotic adjustments which have been observed, σ dropping to values of $\sigma \ll 1$ during the period of adjustment. It could be envisaged that, as equilibration approaches and ΔP declines, the pores would shrink to normal dimensions and active transport would again account for most ion uptake. At the present time this possibility is quite speculative but it may warrant study.

D. Root Pressure and Active Absorption

Root pressure is a well-known phenomenon, particularly evident in the exudation of xylem sap from wounds and cut stems of non-transpiring plants, and indicating a positive pressure in the xylem, rather than the negative xylem pressures which are associated with transpiration. It frequently reaches values of $P = 1 - 2$ bars and may fulfil an important role in refilling ruptured water columns in xylem vessels under some conditions (see next section). Active absorption of water by root systems is the mechanism by which root pressure is assumed to develop. However, the evidence for active water absorption, in the sense of non-osmotic absorption, is weak and, in almost all cases, the osmotic influx of water follows active ion transport into the xylem (Kramer, 1959; Slatyer, 1960c).

The evidence for non-osmotic water uptake by root systems has been based mainly on exudation experiments in which it has sometimes been noted that the osmotic concentrations of the root medium (π_s^0), required to prevent exudation, exceeds that of the exudate (π^i) (van Overbeek, 1942; Broyer, 1951). Observations of autonomic rhythms in rate of exudation (Grossenbacher, 1939; Hagan, 1949) have also been used as evidence, as have the observations of enhanced exudation following auxin application (Skoog, Broyer and Grossenbacher, 1938; Lundegardh, 1949).

In all these cases, however, there is no real evidence of a phenomenon other than that of passive water movement following active ion transport into the xylem. Moreover, in relation to the $\pi_s^0 > \pi^i$ evidence, Hodges and Vaadia (1964a, 1964b) have recently shown that, with chloride transport, a steady state of efflux is not established until exchange sites for chloride throughout the system are satisfied, and Raney and Vaadia (1965c) have shown that internal equilibration of THO, applied externally with chloride, is very slow. Both these factors indicate that the concentration of the emerging exudate may not effectively average the xylem sap concentration unless steady state efflux is prolonged, a requirement not usually met in the experiments cited previously. There is also the point made by Slatyer (1966b) that, if $\sigma < 1$ for the external solute, at zero flow $\sigma\pi_s^0 = \pi^i$ so that the situation $\pi_s^0 > \pi^i$ can be expected.

2. MOVEMENT OF WATER IN THE VASCULAR SYSTEM

Water moves in the vascular system from the xylem terminals in the root to those in the leaf, a distance of over 100 m in some species. Some characteristics of vascular tissues have already been discussed (Chapter 5). In this section attention will be devoted to the factors associated with longitudinal water transport through the plant. Useful additional background material can be found in papers by Greenidge (1957) and Zimmermann (1964b, 1965).

A. Pathway Charocteristics

Longitudinal water movement through the plant takes place primarily in the xylem, but the whole cross-sectional area of the root or stem is available for flow, except in dense woody regions, and some longitudinal water movement presumably occurs external to the vascular system, the proportion varying inversely with the relative resistance to flow.

Some idea of the flow in these two regions can be observed from the work of Huber and Höfler (1930) who estimated that water permeability coefficients of woody stems of conifers were of the order of 10^{-1} to 10^{-2} cm sec^{-1} bar^{-1}, approximately 5 orders of magnitude greater than the values quoted previously for radial water movement into roots, or for single cells. Assuming that the cell walls are approximately 50 times as permeable as protoplasts (Russell and Woolley, 1961; Weatherley, 1963), these figures still suggest a value for vascular permeability approximately 3 orders of magnitude greater than for the pathway external to the xylem. If the ratio of the cross sectional areas of the xylem elements and the longitudinal wall pathway ranges between 0·1 to 0·01, the proportion of flow external to the xylem elements would range from 1 to 10%, respectively, of the total flow. Since it is doubtful if the ratio is ever as low as 0·01 and may, in many cases, exceed 0·1, the significance of flow through the wall portion of the stem pathway, at least under most conditions, can consequently be seen to be negligible.

The characteristics of the pathway have been investigated by following the path of isotopically labelled water applied to the root medium (Biddulph, Nakayama and Cory, 1961). This type of study shows that the water in the xylem is rapidly replaced by the labelled water, confirming bulk flow of water in the vessels and tracheids. Associated with the passage of labelled water through the vascular tissue, there is also a progressive replacement of the water in the remainder of the outer stem tissue. This process resembles flow of an isotope through a porous-walled tube, with a reversible exchange of water molecules occurring between the moving stream and the porous walls.

B. Characteristics of the Water Flux

Although the anatomy of the vascular elements varies widely between species and between major plant groups such as angiosperms and gymnosperms, the relatively large diameter of the conducting elements and the high conductivity of the cross-walls, pits and other structures in the xylem, mean that differences in permeability from plant to plant are unlikely to affect water transport through the system as a whole.

It has long been assumed that liquid flow through the vascular system is a bulk flow of sap caused by hydrostatic pressure differences, which are themselves induced by the gradient of water potential from roots to leaves. It is assumed that negative pressures develop in the leaf xylem as transpiration reduces leaf water content and Ψ_{leaf}, and (since π_{xylem} is generally small) the pressures which develop are equal in magnitude to Ψ_{xylem} (see Eq. 5.3) so that there is water potential equilibrium and liquid phase continuity between the vascular elements and the surrounding tissue. The negative pressure is transmitted, by cohesion between water molecules, through a continuous liquid system to the root, where it influences liquid flow across the root cortex by its effect on $\Delta\Psi$ and ΔP. Although the water is under tension, cavitation is prevented by cohesion between adjacent water molecules and adhesion between the water molecules and the vessel walls.

The cohesion theory for water flow through stems was first advanced many years ago (see Dixon, 1914). In most respects it adequately explains why substantial negative pressures can develop and hence why water can rise in the stems above the height equivalent to zero atmospheric pressure ($\simeq$10 m) (see Huber, 1956; Greenidge, 1957; Zimmerman, 1965). It has been criticized, however, on the basis that flow continues even when a high proportion of the larger vessels may contain air at any one time (Preston, 1958); and that deep overlapping lateral incisions into the stem do not stop water flow, even though it may be reduced and the resistance to flow increased (Elazari–Volcani, 1936; Scholander, Ruud and Leivestad, 1957). However, only a small proportion of the vessels appear to be required, at any one time, to supply enough water to satisfy the transpirational requirement (Preston, 1958) so, when overlapping cuts are made, water can move laterally around them, maintaining liquid phase continuity (Preston, 1952; Postlethwait and Rogers, 1958). Because the cross walls, pits, etc., have very low air entry values, air bubbles introduced by cutting presumably do not spread.

One problem which is still not satisfactorily explained, however, is the mechanism by which vessels in tall trees refill once they have been ruptured. In herbaceous plants and in low trees it is probable that nocturnal recovery of turgor leads to refilling, and positive root pressures which can cause guttation could provide the necessary driving force. These pressures develop in the absence of tension in the vascular system and commonly occur at night with plants rooted in moist soil. It can be expected that, when the soil is moist, $\Psi_{soil} \cong 0$. Since the osmotic pressure in the xylem is of the order of $\pi = 1$–2 bars, water influx should lead to positive pressures of 1–2 bars, enough to raise water to a height of 10–20 metres, and more than adequate to refill vessels in all plants except tall trees. The occurrence of guttation, the visible manifestation of positive root and stem pressures and of turgor restoration, could presumably be good evidence that vessels have been refilled.

However, in trees which are higher than the equivalent height of rise due to turgor recovery and stem pressure, it is possible that the vessels, once drained, do not refill. In such cases it may be the continued growth of new cambial tissue which, during the growing season, develops new water filled vessels which play their part in the conducting system until they, too, rupture and drain. Gibbs (1958) has provided some interesting data which bear on this point, by showing that, even in the absence of water stress, the water content of stems of tall trees decreased during the summer (possibly as more and more vessels drained and did not refill) reaching a minimum level in the autumn and increasing in the following spring, as new water-filled tissue was laid down.

Although root pressure, or the laying down of new tissue, may be possible mechanisms by which refilling is achieved or avoided, some evidence suggests that refilling may occur by other means. Scholander, Love and Känwisher (1955) severed a vine stem, allowing air to enter all xylem elements and found that, on replacement in water, absorption was resumed at the same rate in short vine stems, suggesting rapid re-establishment of the water columns. In longer stems (> 10 m) absorption also was resumed, but at a slower rate, suggesting that many, but not all vessels were again active. Scholander, Hemmingsen and Garey (1961) froze a section of liana stem in dry ice, presumably causing cavitation in all vessels. However, on thawing, normal absorption resumed, suggesting that the gas bubbles were re-dissolved. Similar observations have been made on trees which normally freeze in the winter (Scholander, 1958; Lybeck, 1959), and on trees in which stem sections have been frozen experimentally (Johnston, 1959; Zimmermann, 1964a). If gases can be re-dissolved when column rupture occurs mechanically, or through water stress, most of the remaining problems in applying the cohesion theory to sap ascent are satisfactorily answered. Even without this matter being cleared up, however, it is clear that water does move through transpiring plants in continuous liquid columns under tension. The vascular elements comprise the effective pathway for flow, and the pressure gradient in the pathway is the effective driving force.

Measurements of pressure gradients along the stem have usually been made with indirect methods but suggest that gradients range from about 0.1×10^{-2}–0.5×10^{-2} bar cm^{-1} (Huber, 1932; Huber and Schmidt, 1936; Eaton, 1941; Stocking, 1945; Scholander, Hammel, Bradstreet and Hemmingsen, 1965). As might be expected, lower gradients occur in plants under conditions favouring low transpiration and low water stress, those of Scholander *et al.* (1965) for example, which were made on Redwood and Douglas fir, yielding values at the low end of this range. Since the basic gradient of pressure with height is 0.1×10^{-2} bar cm^{-1}, values of this order confirm the high conductivity of the tissue for water. Zimmermann (1964b) suggests a normal gradient in actively transpiring trees well supplied with water as 0.15×10^{-2} bar cm^{-1}.

The magnitude of tension developed in the xylem parallels the water potential values in the adjacent tissue. Presumably, therefore, values equivalent to $\Psi = -100$ bars, or lower, can exist, and dendrometer measurements by MacDougal, Overton and Smith (1929) indicate values of this magnitude. At one stage it was thought that cohesive forces of this order could not be maintained in practice, even though, theoretically, they could be expected. Recently, however, measurements in artificial systems have shown cohesion at P values lower than $P = -100$ bars (Wakeshima, 1961; Zimmermann, 1964b).

In contrast to the root system, permeability of the stem appears to be little affected by metabolic inhibitors, although, because of application difficulties, low temperature is the factor most commonly investigated. One early study (Handley, 1939) indicated that low temperature did cause a marked reduction in permeability, but later investigations with better temperature sensing equipment (Johnston, 1959; Zimmermann, 1964a) have shown conclusively that significant reductions in permeability do not commence until freezing commences. Maintenance of temperatures of 0–1°C, even for several days, did not cause wilting.

3. Movement of Water Through Leaves

Water moves in the liquid phase from the vascular terminals in the leaves to evaporation sites in the cell walls, and then in the vapour phase to the outside air. Transport in the gas phase will be included in the next section dealing with water movement through the soil-plant atmosphere system in the transpiration stream, and transpiration is separately considered, in detail, in Chapter 8. In this section, emphasis is given to the characteristics of the pathway for liquid flow. The absorption of atmospheric moisture, and its subsequent translocation, is considered later in the chapter.

A. Leaf Structure and Anatomy

From the stem water moves into the leaf via the petiole, or through the leaf sheaths, and then through either a reticulate or parallel network of vascular strands in the leaf blade itself. Esau (1960) has remarked that the principal characteristic of the vascular system of the leaf blade is the close spatial relationship between the mesophyll and the vascular strands, the latter forming an interconnected system in the mid-plane of the blade, parallel with the surface of the leaf.

Of the two types of venation, reticulate venation reveals a branching pattern with successively thinner veins diverging as branches from thicker ones. In parallel veined leaves, strands of relatively uniform size are oriented longitudinally side by side, gradually converging at the leaf apex. The main veins may vary in size, with smaller and larger ones alternating. The longitudinal veins are interconnected by much smaller veins. Thus, at the microscopic level, parallel veined leaves also have a reticulate arrangement of vascular strands.

There is considerable diversity in the two major vein patterns from species to species, although netted venation is most common in dicotyledons and parallel venation in monocotyledons. The number of conducting elements in each vein decreases progressively as distance from the main stem increases. The ultimate branches consist of single strands of elements, many thousands of which terminate in each square centimetre of leaf area, and the distribution system is so effective that few leaf cells are separated by more than two other cells from a vascular element. The vascular bundles are usually enclosed in one layer of compactly arranged cells which form the bundle sheath, so that no part of the vascular tissue is exposed to the intercellular air, the cells usually extending around the termination of any one vascular bundle. In many dicotyledons, the bundle sheaths are connected with the epidermis by panels of cells resembling bundle sheath cells. An exception to the foregoing exists in species which exhibit guttation, in which the xylem terminals in hydathode regions are exposed to the intercellular spaces of the epithem, a modified mesophyll which leads to the hydathode.

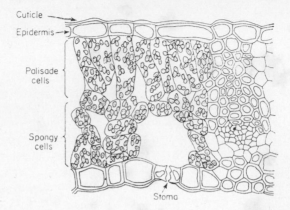

FIG. 7.7. Cross section of portion of a typical hypostomatous mesophyll leaf (after Meyer and Anderson, 1952).

Although leaves may markedly differ in size and thickness, the average thickness of the normal mesophyte leaf is of the order of 100–200 μ and, in cross section, reveals 5–10 layers of cells (see Fig. 7.7). The epidermis constitutes the outer layer, one cell thick, entirely sheathing the leaf, and consisting of a compact arrangement of cells, the outer walls of which are thickened and impregnated with cutin. (In some species, secondary wall thickening also occurs and may almost fill the lumen of the epidermal cells). Outside the wall proper, the incrustation is termed the cuticle. In many species the cuticle itself is covered with a wax "bloom" which may consist of several different waxes with differing characteristics.

The epidermis and cuticle together constitute a highly effective barrier to water transport (see next section and Chapter 8). The uniformity of the epidermal surface is, however, broken by two main types of structures, stomata and trichomes. The latter refer to all surface projections on the leaf, including hairs and scales. They generally arise from an epidermal cell and may be so frequent, in some cases, as to completely modify the appearance of the leaf. They can have an important indirect role in the water economy of the plant through their effect on the physical surface characteristics of the leaf. Little transpiration appears to occur through them, although this point needs quantitative confirmation.

The stomata, on the other hand, play a vital role in the gas exchanges between leaf and air, being the primary pathway for water vapour, CO_2 and oxygen. Stomata may be located on both surfaces. They occur more commonly on the lower surfaces, however, and occur at a frequency ranging from 50 to 500 per square millimetre. At this higher frequency, it can be appreciated that individual stoma are separated by only one or two other epidermal cells. The term, stoma (plural stomata), generally refers to a unit consisting of a pore and two surrounding guard cells. Each stoma may be surrounded by normal epidermal cells, or by companion or subsidiary cells, which differ in shape and structure from normal epidermal cells.

The guard cells of dicotyledons are commonly kidney-shaped. They are covered with a cuticle which extends over the surface lining the pore, and over the internal surfaces, although it is generally thicker on the outer surfaces. In monocotyledons, particularly grasses, the guard cells may be more elongated, and almost parallel to one another. In many species, especially those from extreme environments, the stomata are sunken below the level of the remainder of the epidermis. The typical appearance of these main types of stomata is shown in Fig. 7.8 for *Prunus*, *Oryza* and *Pinus* (Esau, 1960).

Within the leaf, apart from the vascular elements and their associated structures, the tissue consists of chlorophyll-containing parenchyma, the mesophyll. (The epidermal cells proper contain plastids but virtually no chlorophyll). It is this tissue in which almost all of the photosynthesis conducted by most green plants occurs, green stems and other tissues being relatively minor sites for photosynthesis.

Two main types of parenchyma are generally found in mesophyll tissue, the palisade parenchyma consisting of cells oriented perpendicularly to the surface of the leaf blade, and the spongy parenchyma, consisting of cells of various sizes, generally lying within, or below, layers of palisade cells. The spongy parenchyma forms a loose three-dimensional arrangement of cells, which can be appreciated from leaf cross-sections such as that shown in Fig. 7.7. The palisade appears much more compact in transverse cross-sections but, in paradermal sections, it is evident that a substantial part of the surface area of each palisade cell is also exposed to intercellular air.

It is most important to realise that this artifact (the apparent compact nature of mesophyll cell arrangement as seen in typical transverse cross-sections of leaves) has led to two misconceptions about internal leaf geometry, one concerning the ventilation of the palisade, the other concerning the nature of sub-stomatal cavities. In the first place,

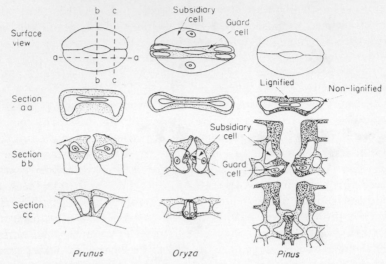

FIG. 7.8. Typical characteristics of stomata from a dicotyledonous mesophyte, grass and gymosperm. The upper row shows the surface view, the remaining rows show the sections indicated. In the left hand column is *Prunus*, centre column *Oryza*, and right hand column *Pinus* (after Esau, 1960).

the internal surface area of a leaf is usually a clear order of magnitude greater than the external surface area (Turrell, 1936, 1942) and the palisade generally has a larger, not smaller, internal exposed surface than the spongy parenchyma (Esau, 1965). Secondly, while there is generally a larger space directly below each stoma than between most meso-phyll cells, the appearance, in cross-sections, of a substomatal cavity lined by cells is quite misleading. The stomata are only ports by which the entire internal mesophyll surface can be ventilated. Therefore, while the path length for gaseous diffusion from cell surface to stomatal pore may be significantly longer and more tortuous from one part of the leaf than from another, it must always be realized that both CO_2 and water vapour exchange takes place across all of the exposed internal cell wall surfaces.

The leaf parenchyma conforms in most respects to the typical cell structure already given in Chapters 5 and 6, its most distinguishing feature being that it contains chloroplasts. In general, the cells range from 10–20 μ in diameter, with palisade cells tending to be more cylindrical in shape and the spongy cells more or less spherical. The longest dimension of the palisade may exceed 50 μ. The cell walls are generally thin and are usually cutinized at their outer extremities (Scott, Schroeder and Turrell, 1948; Scott, 1950; Fogg, 1947; Lewis, 1945, 1948). Thus the exposed internal surfaces appear to be hydrophobic to some degree and may appear visibly dry rather than moist, as has often been proposed. The possible effect of this phenomenon on transpiration is discussed in Chapter 8.

B. Pathways for Water Movement

There are three pathways for outward water movement from the vascular elements to the external surface of the leaf. The primary pathway is through the parenchyma to the mesophyll liquid-air interfaces and then, in the gaseous phase, through the stomata to the outside air. Associated with this transfer is the subsidiary parallel pathway through the parenchyma to the epidermal interfaces and then, in the vapour phase, via the cuticle to the outside air. The third pathway is the guttation pathway, only operative when transpiration is minimal, and which involves liquid transfer right to the surface. Because guttation transports a very small amount of water, under most conditions, compared with the normal transpiration pathway, it will briefly be dealt with first.

Guttation involves root pressure and hence only occurs where positive pressures build up in the conducting system, at times when transpiration is virtually zero. Positive pressure forces sap through the ends of the xylem elements in the hydathode region, into the intercellular spaces of the epithem. This pathway, beyond the xylem terminal, offers very little resistance to liquid flow. It is seldom more than a few cells diameters in length and terminates in a hydathode pore. The pore appears similar to a modified stoma but is inactive and does not close (Reams, 1953; Stevens, 1956). The exuded liquid is generally different in composition to xylem sap (Eaton, 1943; Curtis, 1944), suggesting some selectivity in ion transport through the xylem terminal, or some solute exchange in the epithem. The amount of guttation varies widely. It is insufficient to cause drip in many species, but daily amounts in excess of 1 litre have been reported. Further information on guttation, and on other exudation and secretion phenomena can be found in Stocking (1956b, 1956c) and Kramer (1959).

In the two main transpiration pathways, transport in the liquid phase to the evaporating surfaces appears to occur mainly in the cell walls. In the same way as in the root cortex, this is because of the much lower resistance the walls offer to water transport, compared with the cell to cell pathway via the vacuoles. The preponderance of flow through the walls has been demonstrated with fluorescent dyes and related techniques by Strugger (1949) and Strugger and Peveling (1961). Russell and Woolley (1961) calculated that the ratio of water flow in the walls compared with the vacuole-to-vacuole pathway would be about 50:1 and Weatherley (1963) has recently obtained a similar figure experimentally, by studying the time course of water uptake into the petioles of detached leaves in which transpiration was suddenly stopped. In consequence there appears to be little resistance to liquid flow through the parenchyma and the low permeability of the cuticular pathway, compared with the open-stomata value of the stomatal pathway, must be due to the high resistance to liquid and vapour flow across the cutinized outer epidermal walls and across the cuticle.

Because of extensive cutinization, water pathways through the outer epidermal walls, in most species, are very restricted and have very low hydraulic conductivity. This can be expected to apply particularly to those species, mainly from extreme environments, in which secondary thickening has virtually eliminated the lumen of the epidermal cells. In such cases removal of the outer cuticle would have little effect on overall permeability of the pathway. In other cases, such as clover leaves (Hall and Jones, 1961), grapes (Radler, 1965) and apples (Horrocks, 1964), removal of the soft external wax cuticle can increase cuticular transpiration by several times, suggesting that, if the evaporating surfaces are in the walls, the major barrier to water transport in the cuticle pathway may be in the vapour transport phase. However, it is difficult to assign specific liquid phase and vapour phase resistances to the cuticular pathway because the exact location of the liquid-air menisci which constitute the evaporating surfaces is not adequately known, and there is often intimate contact between the cutinized walls and the cuticle. It is probable, though, that there is liquid phase continuity almost to the outer wall surfaces and, in some cases, this may extend into the wax layer. It should also be pointed out that, even with increases in cuticular transpiration resulting from wax removal, the cuticular pathway would still offer substantially higher resistance to water transport than the stomatal pathway under open stomata conditions (see Chapter 8).

Resistance to liquid phase transport in the stomatal pathway, that is, from the vascular elements to the internal exposed cell wal surfaces, has generally been supposed to be very small (Milthorpe and Spencer, 1957; Slatyer, 1966a), despite some rather qualitative physiological evidence to the contrary and the anatomical evidence of internal suberization. Recently, however, experiments by Shimshi (1963a) have indicated that the effective water vapour pressure at the cell wall surfaces is significantly less than the saturation vapour pressure at the leaf temperature, which is the value to be expected if the surfaces are wet. This suggests that the suberization has effectively reduced pore size and the hydraulic conductivity of the outer wall layers to a stage where this zone represents the major source of resistance to liquid phase transport in the stomatal pathway. Even so the resistance is small compared with normal cuticular pathway resistance and appears to have no real controlling effect over transpiration.

The effect on transpiration of the various sources of resistance to flow in the vapour pathway, and of the effective vapour pressures at the external and internal surfaces, are discussed in detail in Chapter 8. At this stage it is only necessary to note that, in the liquid phase, most of the water transport in the mesophyll is through the cell walls in which water moves along water potential gradients encountering little resistance to flow. In the outer epidermal wall and cuticle, however, the hydraulic conductivity is very low and this zone constitutes a major resistance to flow. Across the outer surfaces of the internal exposed walls, where suberization has occurred, the hydraulic conductivity is also reduced but to a much smaller degree.

4. Overall Aspects of Water Movement

Water flow through the plant from soil to atmosphere encounters a series of resistances in each phase of the system. In this section emphasis will be placed on identifying the major sources of resistance to flow and the major processes controlling transport. The inter-relationships between water supply and water demand characteristics, in determining the observed flow, will also be considered. For more detail the reader can be referred to the extensive literature on this subject (for example, van den Honert, 1948; Philip, 1957a, 1966; Slatyer, 1960c; Slatyer and Denmead, 1964; Visser, 1964; Slatyer and Gardner, 1965; Cowan, 1965).

A. Factors Affecting Overall Flow

Water transport through the soil-plant-atmosphere continuum occurs in both the liquid and gaseous phases, which, under isothermal conditions, show a progressive drop in water potential from soil to atmosphere. The actual driving forces for transport across each compartment of the continuum, however, are not necessarily gradients of water potential, although this assumption can be made with respect to liquid phase transport in some compartments, and under most conditions, at better than first approximation level.

In moist soils liquid phase transport is the dominant process and, as was shown in Chapter 4, occurs along gradients of matric potential, or matric pressure, $d\tau/dx$, which are effectively gradients of water potential $d\Psi/dx$ in soils without steep salt concentration gradients. As soils dry and total flow declines rapidly, vapour transport becomes progressively more important, particularly under non-isothermal conditions, but the general assumption that most of the soil water flow to plant roots occurs in the liquid phase, along gradients of soil water potential, appears reasonable for present purposes.

In the plant, from the root surfaces to the liquid-air menisci at the evaporation sites in the shoots, flow is in the liquid phase. It is apparent, from the preceding sections of this chapter, that most of the water flow is, in fact, a flow of water and solutes, and that the main zone of resistance is in the roots (see also Kramer, 1938; Jensen, Taylor and Wiebe, 1961; Jensen and Taylor, 1961). Flow across the root cortex to the endodermis, and across the leaf mesophyll, via the walls, presumably occurs along gradients of matric potential, or matric pressure, $d\tau/dx$, and through the xylem along a hydrostatic pressure gradient dP/dx. In all these phases, a bulk solution is being transported, particularly in the xylem, but to a lesser extent in the walls because ion exchange and ion uptake phenomena will affect composition and concentration. In roots without secondary thickening the major source of resistance to liquid phase movement in the plant appears to be located at the endodermis. In roots, with secondary growth, it appears to be at the outer cambial and cork layers. Although there is some evidence for $\sigma < 1$ across the endodermis, the assumption that the effective driving force across these barriers is $d\Psi/dx$ appears reasonable.

Because the driving forces for water flow through the plant change as water passes from root cortex to endodermis, to xylem to mesophyll, it cannot be expected that liquid flow through the plant will be proportional to the water potential difference between root surface and leaf surface ($\Psi_{root} - \Psi_{leaf}$). This situation, however, could be approximated if flow across the endodermis, which does appear to be driven by the root cortex-xylem water potential difference, ($\Psi_{cortex} - \Psi_{xylem}$), encountered so much more resistance than elsewhere in the plant that ($\Psi_{cortex} - \Psi_{xylem}$) gave a reasonable estimate of ($\Psi_{root} - \Psi_{leaf}$). For plants other than tall trees, and in the absence of severe water stress, this requirement is probably met to a reasonable degree. However, doubling the water potential difference across the root ($\Psi_{cortex} - \Psi_{xylem}$) does not necessarily double the flow rate, because of the evidence previously presented which indicates a reduction in root L_p as ($\Psi_{cortex} - \Psi_{xylem}$) increases. While the major source of resistance to flow, under these conditions, probably remains in the root, so that ($\Psi_{cortex} - \Psi_{xylem}$) is still a reasonable approximation of ($\Psi_{root} - \Psi_{leaf}$), the assumption that flow rates through the plant will be proportional to ($\Psi_{root} - \Psi_{leaf}$) is not likely to be valid.

From the evaporating surfaces within the leaf, to the bulk air, water vapour flow occurs along gradients of water vapour concentration. The driving force therefore depends on the difference between the vapour pressure at the surfaces and in the bulk air, and the rate of flow is directly proportional to this value and inversely proportional to the resistances to vapour diffusion in the gaseous pathway (see Chapter 8).

In summary, therefore, it can be seen that flow from soil to root can be reasonably considered to be in the liquid phase, along gradients of water potential. Flow from root to leaf can also be treated in this manner, but only in very general and qualitative terms. Flow from leaf to air is in the vapour phase along gradients of water vapour concentration.

In order to place these flows on a common basis, with a view to identifying the major sources of control over water transport, van den Honert (1948) assumed that flow across each zone, soil-root, root-xylem, leaf-air, was proportional to the water potential difference, and inversely proportional to the resistance which developed across it. For steady state flow through the whole system he wrote an expression of the form

$$q = \frac{1}{A} \frac{dV}{dt} = \frac{\Psi_1 - \Psi_2}{r_{1,2}} = \frac{\Psi_2 - \Psi_3}{r_{2,3}} = \text{etc} \qquad (7.1)$$

By placing approximate values of $\Delta\Psi$ in the different segments which, for the leaf-air segment, involved converting relative vapour pressure data using the relationship $\Psi = RT \ln e/e^0$ (see Eq. 1.46), he showed that under almost all conditions in which normal plant growth occurs, the value $\Psi_{leaf} - \Psi_{air}$ represented the largest single drop in potential. This can be appreciated by taking typical values of Ψ_{soil}, Ψ_{root}, Ψ_{leaf} and Ψ_{air} as -1, -10, -15 and -1000 bars, respectively, when, in general, $\Psi_{leaf} - \Psi_{air}$ is seen to be greater, by more than an order of magnitude, than the combined $\Delta\Psi$ in all the other segments.

Clearly, this treatment can only be applied in the most general terms. Not only are the previous comments concerning liquid flow in the plant pertinent, but the use of $\Delta\Psi$ to describe the vapour phase driving force is conceptually incorrect (Ray, 1960; Rawlins, 1963). Even so, used for the purposes van den Honert envisaged (Philip, 1957a; Slatyer, 1960c) the analysis provides a useful analogy of overall water transport and leads to several valid and important conclusions.

These are firstly, that total flow is primarily controlled in the vapour phase (from the evaporating surfaces in the leaves to the bulk air). Secondly, the development of an increased resistance to flow elsewhere in the system, although initially reducing the flow across the zone in which it occurs, will only directly influence the total flow if the resistance becomes a significant fraction of the limiting resistance (put another way, if $\Delta\Psi_{zone}$ reaches a significant proportion of $\Psi_{leaf} - \Psi_{air}$). Otherwise the effect can only be indirect as a "filtration" resistance, serving to steepen $\Delta\Psi$ across the zone of resistance. Unless this in turn causes stomatal closure, or, in some other way increases the resistance from leaf to air, an effect on total flow may not occur or may only be transient, ceasing when $\Delta\Psi$ across the zone of resistance increases to re-establish the flow rate at the previous level. Thirdly, the location of the stomata in the vapour segment of the pathway highlights their importance as flow regulators, since they are sited in the only region where effective control can be exercised.

The application of these conclusions suggests that decreased root permeability due, say, to lack of aeration, will immediately influence water movement through the root, but transpiration will initially be unaffected. Continued transpiration will, however, reduce leaf water content and Ψ_{leaf} and, in turn, Ψ_{xylem} until $\Delta\Psi_{root}$ is increased to re-establish the flow rate across the root at the original rate. Only if the reduction in Ψ_{leaf} affects $\Psi_{leaf} - \Psi_{air}$ (strictly the leaf-air vapour concentration gradient) or $r_{leaf} + r_{air}$ (the sum of vapour diffusion resistances) will transpiration be affected. Since reduction of Ψ_{leaf} will only significantly reduce the vapour pressure at the evaporating surfaces under extreme conditions, the only way in which flow can be effectively reduced is by an increase in r_{air} and this is most commonly achieved by stomatal closure. The fact that the initial reduction in root permeability leads to a reduction of Ψ_{leaf}, which in turn causes stomatal closure, does not alter the fact that phenomena which affect root permeability have an indirect effect on transpiration and overall water transport.

These important conclusions remain valid in most cases, despite the knowledge that $\Delta\Psi$ is not the effective driving force across all zones and that the resistances to flow vary in characteristics and degree. Put another way, the thesis simply states that, unless a factor directly affects the vapour pressure at the evaporating surfaces in the leaf, or the total diffusion resistance of the leaf-air vapour pathway, its effect on flow is essentially indirect.

Even so, it must be remembered that, within the plant, the major source of resistance to transport appears to occur in the root and the factors which affect root resistance are, in most cases, the actuating influences which lead to changes in transpiration rate. While the effect of Ψ_{leaf} on the vapour pressure at the evaporating surfaces is generally small, its effect on stomatal aperture is rapid and pronounced.

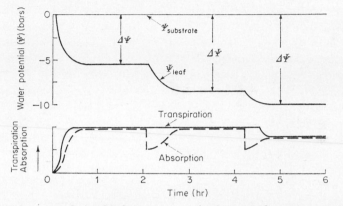

FIG. 7.9. Schematic relationship between transpiration rate, absorption rate, Ψ_{leaf} and ($\Psi_{substrate}-\Psi_{leaf}$) when root resistance is increased in two stages.

In Fig. 7.9 examples of these various responses are given in schematic form. It is assumed that, at time $t = 0$, a plant rooted in water culture in the dark ($\Psi_{root}\cong0$; $\Psi_{leaf}\cong0$) is illuminated under constant environmental conditions. Transpiration increases to a steady state value in approximately 30 min, and by $t = 1$ hr, a steady state situation exists, with transpiration and absorption proceeding at the same rates, constant $\Delta\Psi(= \Psi_{substrate} - \Psi_{leaf})$, and the lag of total absorption behind total transpiration represented by a steady state value of Ψ_{leaf}. Root permeability is then suddenly reduced by, say, cooling the water and the roots, and absorption drops to half its original value. The lag of total absorption behind transpiration increases, even though the rate of absorption commences to increase again, as decreasing Ψ_{leaf} increases $\Delta\Psi$. Finally, the rate of absorption again reaches the transpiration rate and the steady state system is re-established. Now, however, a much greater value of $\Delta\Psi$ is required to sustain the original rate of flow through the roots. It is to be noted, though, that transpiration rate has been unaffected throughout, the drop in Ψ_{leaf} being insufficient to cause stomatal closure or change in the vapour pressure at the leaf surface.

If it is now assumed that root permeability is reduced once more so that rate of absorption is again reduced by one half, the same sequence of events tends to recur. However, on this occasion, before $\Delta\Psi$ has increased sufficiently to re-establish the rate of absorption at the level of the transpiration rate, Ψ_{leaf} falls to a value which induces partial stomatal closure and causes a reduction in transpiration rate. It is assumed that the degree of closure is sufficient to prevent further decrease in Ψ_{leaf}, and transpiration rate drops to the level of the prevailing rate of absorption.

Although, in practice, stomatal closure induced in this manner usually overshoots (Slatyer, 1963; Begg et al., 1964) so that rate of absorption transiently exceeds transpiration, and there is some recovery in turgor and Ψ_{leaf} [sometimes resulting in re-opening of the stomata (Begg et al., 1964) or cycling in Ψ_{leaf} and stomatal aperture (Ehrler, Nakayama and van Bavel, 1965)]. Figure 7.9 provides a useful illustration of the indirect effects of root permeability on transpiration rate.

B. Water Supply and Water Demand

The previous discussion has emphasized the role played by Ψ_{leaf} and the stomatal apparatus, in regulating water flow through the soil-plant-atmosphere system. Clearly, however, both source strength, in the form of soil water potential and flow capacity to the roots, and sink strength, in the form of the potential transpiration, interact continuously in determining Ψ_{leaf}. For example, under conditions of extremely high potential transpiration, it is probable that Ψ_{leaf} will fall low enough to induce stomatal closure even when the soil is moist (Makkink and van Heemst, 1956) yet, if potential transpiration rates are extremely low, Ψ_{leaf} will probably remain above a critical value until very low values of Ψ_{soil} exist.

The main factors determining actual evaporation will then be the soil water potential Ψ_{soil}, the hydraulic conductivity of the soil, root volume and root density, internal resistance in the plant, the critical level of Ψ_{leaf} for stomatal closure (Ψ_{crit}), and the potential transpiration.

The first attempts to include these factors in an analysis of this problem were made by Philip (1957a) and Gardner (1960b). Subsequently, modified models have been developed by Visser (1964), Gardner (1965) and Cowan (1965). Cowan's model uses three main functions. The first describes soil water flow to the root in terms of the water potential of the soil mass, Ψ_{soil}, the water potential at the root surface, Ψ_{root}, the hydraulic conductivity of the soil, and root density, and yields generally similar results to those of Gardner (1960b) discussed in Chapter 4.

The other functions describe the influence of leaf water potential, and resistance to liquid flow in the plant, on stomatal aperture and hence on the capacity of the plant to transpire at the potential rate. Rate of water flow, and hence of transpiration, is assumed to be proportional to ($\Psi_{root} - \Psi_{leaf}$) and the sum of the internal plant resistances (Σr_{plant}) is assumed to be effectively constant (see relevant remarks in previous sections). A critical value of $\Psi_{leaf}(= \Psi_{crit})$ is adopted at which stomatal closure reduces transpiration to a degree which prevents further decline in Ψ_{leaf}. At Ψ_{leaf} values higher than Ψ_{crit}, it is assumed that transpiration proceeds at the potential rate. The value of the plant resistance is obtained from the rate of transpiration, E_{max}, at which $\Psi_{leaf} = \Psi_{crit}$ when $\Psi_{root} = 0$.

As can be appreciated from the foregoing discussion on driving forces and water permeabilities in the liquid pathway, some criticisms can be made of these assumptions, but in most respects they appear reasonable, at least at first approximation level and, in any case, can readily be modified as better experimental data became available. In the meantime they provide very interesting evidence of the interaction between soil, plant and atmospheric factors in determining the transpiration rate, as is shown in Figs 7.10–7.12 for a hypothetical crop rooted to a depth of 20 cm in Yolo light clay, the physical properties of which are depicted in Figs 4.5, 4.7 and 4.8. E_{max} is assumed to be $3 \cdot 0$ cm day $^{-1}$ and $\Psi_{crit} = -15$ bars.

In Fig. 7.10 the expected relationship between crop transpiration and soil water potential is given for three different levels of potential transpiration. The figure shows that, as the soil dries and Ψ_{soil} declines, there is initially no effect on transpiration, but when Ψ_{leaf} decreases to the stage $\Psi_{leaf} = \Psi_{crit}$, a progressive decline in transpiration commences. This stage is reached at higher values of Ψ_{soil} with higher prevailing rates of evaporative demand, since steeper gradients of Ψ_{soil} are required to sustain flow, resulting in lower levels of Ψ_{root}, and hence of Ψ_{leaf}, at any one value of Ψ_{soil} (see also Figs. 4.11–4.13). It can be appreciated that when potential transpiration rates, higher than those used in this example, occur, the actual daily transpiration may not at any stage equal the potential daily transpiration.

In Fig. 7.11 the effect of root density on the relationship between transpiration and Ψ_{soil} is depicted. In this case only the intermediate

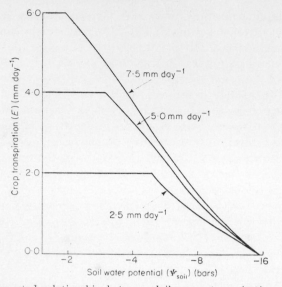

FIG. 7.10. Expected relationship between daily crop transpiration and soil water potential, with a crop of intermediate root density (4 cm³ soil cm⁻¹ root length) rooting to a depth of 20 cm. Three levels of potential transpiration are assumed as indicated on the figure (after Cowan, 1965).

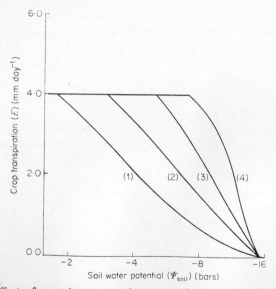

FIG. 7.11. Effect of root density on the expected relationship between daily crop transpiration and soil water potential. Same conditions as Fig. 7.10, but assuming an intermediate level of potential transpiration. Curve numbers (1)–(4) refer to root densities of 8, 4, 2 cm³ soil cm⁻¹ root length, and very dense rooting, respectively (after Cowan, 1965).

level of potential transpiration is used. As root density increases, it is apparent that potential rates of transpiration can be sustained until much lower values of Ψ_{soil} exist, smaller valuer of ($\Psi_{soil} - \Psi_{root}$) being adequate to sustain flow at the desired rate.

In Fig. 7.12 the same data are plotted against time, for a soil initially at a water content of $\theta = 0.2$ cm^3 water cm^{-3} soil. The marked effect of root density is again apparent and, in both this figure and Fig. 7.11, the response patterns range from one suggesting a progressive decrease in soil water availablity for transpiration, to one suggesting little effect on soil water availability until low Ψ_{soil} values exist.

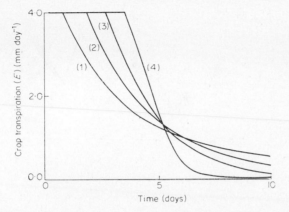

Fig. 7.12. Expected decrease of daily crop transpiration with time for the conditions of Fig. 7.11 (after Cowan, 1965).

Many field and laboratory studies, dealing with the effect of soil water status and evaporative conditions on evapotranspiration, have been conducted (see for example Makkink and van Heemst, 1956; Lemon, Glaser and Satterwhite, 1957; Scholte-Ubing, 1959; Bahrani and Taylor, 1961; Denmead and Shaw, 1962; Gardner and Ehlig, 1962, 1963; Ehlig and Gardner, 1964). Most of these provide confirmation of the main conclusions from the Cowan model, even though, under conditions where the soil surface is wet and not shaded by crop leaves, soil evaporation, not included by Cowan, can represent a substantial part of the total evapotranspiration (Penman and Long, 1960). Another important over-simplification, for plants under prolonged stress, is the assumption that transpiration ceases when $\Psi_{soil} = \Psi_{crit}$ (in this analysis, $\Psi_{crit} = -15$ bars). Clearly the soil is dried to much lower values of Ψ_{soil} under prolonged dry weather, as a result of soil evaporation and continued transpiration, much of which may be cuticular.

As an example of these differences, Fig. 7.13 from Denmead and Shaw (1962) shows the observed pattern of evapotranspiration of a corn crop, in relation to soil water content under a range of potential transpiration conditions. While there is qualitatively good agreement with Fig. 7.10, it is apparent that, in the field study, evapotranspiration of the low and moderate potential transpiration treatments was maintained at the potential rate until lower soil water contents developed than those predicted by the model, and the curves tail off as zero transpiration is approached, rather than intercepting the abscissa at

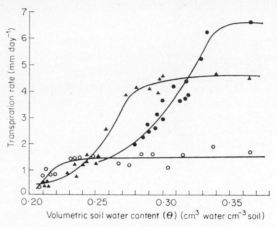

Fig. 7.13. Effect of decreasing soil water content of evapotranspiration of a corn crop, on days of high (●), moderate (▲), and low (○) potential transpiration (after Denmead and Shaw, 1962).

a value close to $\Psi_{soil} = -15$ bars. The first of these differences may have been due to Cowan's assumption of a sinusoidal fluctuation in daytime potential transpiration, whereas, in the field study, the days of low and moderate transpiration were cloudy and humid and there was probably no pronounced peak in the daily evaporative regime. The second is primarily due to the inapplicability of the assumption that transpiration ceases when $\Psi_{soil} = \Psi_{crit}$. Even so, a great deal of information is to be gained by studies of the type conducted by Cowan (1965), and, with progressive improvements in experimental techniques, even more realistic models will be developed.

5. Absorption of Atmospheric Water by Leaves and Shoots

In Chapter 2, some physical aspects of dew deposition were discussed. In this section the physiological evidence for, and significance of, absorption of dew and atmospheric water generally, by the aerial organs of plants will be examined. More detailed information can be found in reviews by Gessner (1956), Stone (1957a) and Slatyer (1960c).

Studies of water absorption by aerial organs have been conducted for many years. Most investigators have found evidence of some uptake, either from saturated or near saturated water vapour or from liquid water, but the amounts have generally been small and subsequent redistribution of water within the plant has been very slow. However, exceptions to this general pattern have been reported, particularly by Breazeale and co-workers (Breazeale, McGeorge and Breazeale, 1950, 1951; Breazeale and McGeorge, 1953a, 1953b) who have observed substantial amounts of water transfer, not just into the plant, but also into the soil or substrate in which the test plants were rooted. On the other hand, other investigators have failed to demonstrate uptake or transport (Janes, 1954; Hohn, 1954; Wiersma and Veihmeyer, 1954). In consequence it is of value to examine first the probable mechanisms involved in water uptake and then consider the significance of the phenomenon in nature.

A. Mechanisms Involved in Water Uptake by Leaves

Since most of the water vapour involved in transpiration passes through the stomata, it can be expected that the same diffusive resistances, in reverse, would apply to re-absorption of water vapour, as long as the stomata were open and the vapour pressure gradient was towards the leaf. However, the fact that relative vapour pressure at the exposed surfaces of the internal mesophyll cells is likely to be > 0.95 in most non-transpiring plants, even when wilted, and is rarely < 0.80 (Stone, Went and Young, 1950; Whiteman and Koller, 1964) means that the gradient is likely to be very small. Moreover, since it is likely to be towards the plant only at night when the stomata are generally closed and there is little air movement, both the internal and external resistances to water vapour diffusion will be high.

In consequence, the amount of water absorbed as vapour can be expected to be negligible and, although Slatyer (1956) observed "negative transport" of small quantities of water vapour from the air around the shoots to the air around the roots, under very closely controlled conditions, it is probable that most of the transfer was via the stomata and occurred during periods when they were open. A similar explanation may apply to the results of Stone, Went and Young (1950) who observed water vapour uptake by leaves in a small transparent chamber whenever the bulk vapour pressure exceeded the value at the leaf surfaces (calculated as Ψ_{leaf}), and observed transpiration whenever the gradient was in the other direction. In the vapour equilibration techniques for Ψ_{leaf} measurements, although water exchange is slow (Slatyer, 1958; Ehlig, 1962) it is more rapid than would occur through the cuticle alone and probably a good deal occurs through the cut edges of the leaf tissue used.

If dew, or applied water, is the main source of water absorbed, a prerequisite to uptake is the wettability of the cuticle. The ability of a liquid to wet a surface is a function of its contact angle on the surface; which in turn depends on the surface tension of the liquid and the nature of the surface. Fogg (1947) noted that marked differences existed in the contact angle of water on several species, and that the contact angle was influenced by age of leaf, and water content. Fogg attributed these differences primarily to variations in surface conformation, pubescence, and the composition of the cuticle.

Since contact angle is reduced by the addition of wetting agents, it could be expected that the addition of such materials would materially improve absorption, at least on some types of leaves. In this connection the addition of detergents to foliar sprays (Guest and Chapman, 1949; Cook and Boynton, 1952) has been found to increase nutrient uptake. Although no similar studies dealing directly with water absorption have been made, a similar result could be expected.

Whether or not the main pathway for water entry is via the cuticle or stomata is still in doubt. Since dew is primarily of nocturnal occurrence, and since the stomata of most species are normally closed for most of the night, it seems probable that uptake occurs primarily through the cuticle. Even if some uptake occurs via the stomata, its significance will be reduced in species with relatively high cuticular permeability to liquid flow. In this regard, Vaadia and Waisel (1963) showed that, in stressed sunflower plants, presumably with closed stomates but with relatively permeable cuticles, water uptake was only slightly less rapid than in non-stressed plants, presumably with open stomata. By comparison, in *Pinus halepensis* with a comparatively thick and impermeable cuticle, rate of uptake was much slower, even in non-stressed plants, than in sunflower and in the stressed plants was only about one-half of the non-stressed value (see Fig. 7.14).

Another possible pathway is via specialized epidermal cells (Zamfirescu, 1931; Meidner, 1954). Meidner noted that most of the water absorbed by *Chaetacme aristata* leaves occurred through specialized cells in the epidermis. He also noted that absorption by the upper leaf surface, where most of the cells are located, was much greater than on the under surface where stomatal frequency could be expected to be highest.

Regardless of the pathway for water entry, it is to be expected that initial rate of uptake would be inversely related to leaf water content and Ψ_{leaf}. Although this was found by Krause (1935) and Slatyer (1956), Eisenzopf (1952) observed a peak rate of absorption after 90 min immersion in water, following which rate of uptake decreased fairly rapidly. It seems probable that the increased rate for the first 90 min was caused by a progressive increase in cuticular permeability, as a result of increased cuticle hydration, or possibly by opening of stomata. The decreased absorption after this time was attributed by Eisenzopf to decreased permeability of the cuticle, although a more probable explanation is that the water potential gradient had been progressively reduced.

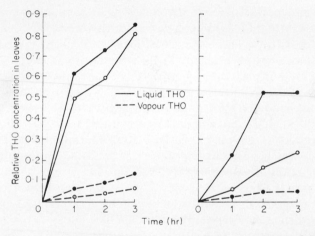

FIG. 7.14. Entry of tritiated water into leaves of Sunflower (left) and *P. halepensis* (right). Relative THO concentrations denote ratio of activity in leaf tissue compared with external medium. Plants not stressed (●), plants stressed (○) (after Vaadia and Waisel, 1963).

Transfer of the absorbed water through the plant can be expected to follow normal pathways and be subject to similar resistances and driving forces as normal transpiration (Jensen, Taylor and Wiebe, 1961; Jensen and Taylor, 1961). However, experiments have to date given conflicting results and it appears that in many cases the water potential gradients which are developed within the plant are insufficient to cause flow, particularly when the opposing influence of root and stem pressure, contributing to active absorption, is taken into account.

The most frequently cited experiments revealing negative transport through the whole atmosphere-plant-soil system are those of Breazeale,

McGeorge and Breazeale (1950, 1951) and Breazeale and McGeorge (1953a, 1953b). In addition to these studies Haines (1952, 1953) and Slatyer (1956) observed negative transport when roots were placed in empty flasks, and a steep vapour pressure gradient from leaf to root was established.

However, several other experiments conducted along similar lines have not shown transport. Thus with plants rooted in soil, even when the soil was as dry as the permanent wilting percentage, and an adequate water potential gradient should have been established, no accumulation of water in the soil was observed by some investigators, although the leaves of the plants under study regained turgor (Höhn, 1954; Janes, 1954). On the other hand Hagan (1949) observed rapid negative transport when plants in dry soil were detopped, and the stumps connected to potometers, and Bormann (1957) demonstrated sufficient cross-transfer of water through natural root graft systems to maintain growth of plants with no other water source.

Several factors probably combine to cause these conflicting results. In the first place, because absorption of water by leaves generally appears to be very slow, the quantities of water absorbed are usually quite small. Secondly, although transfer of this water to other branches has been observed (Dixon, 1914; Brierley, 1934, 1936) the rate of internal movement also appears to be fairly slow (Basler et al., 1961; Wilson and McKell, 1961). Consequently, the amount of water transferred to other branches and to the roots, may, in some cases, have been insufficient to enable recovery of turgor by more distant parts of the plant, and hence may have escaped detection.

Under extreme conditions significant resistance to water movement may also occur at the root surface. Death of small roots and root hairs in dry soil, together with rapid suberization, may markedly reduce the root surface available for transfer, and increase the resistance to movement across this zone. This could explain the failure in some experiments to observe negative transport, while at the same time recovery of turgor by the plant tops was observed. An alterntive explanation for the lack of transport into dry soil might be that, as soon as any exudation occurred into the soil directly adjacent to the root surface, the water potential in this immediate region would be increased to $\Psi_{soil} \cong 0$, and so the gradient from atmosphere to soil would be virtually eliminated, although a steep gradient would remain within the soil just away from the root. Unless complete saturation was approached, significant water movement from this zone into the soil mass would not be expected, because of the very slow rates of un-saturated water movement. As a result the water content in a narrow cylinder surrounding each root might be close to field capacity, but the bulk of the soil mass could remain dry, for short periods at least.

In summary, it can be stated that there appears to be little doubt that in some species, and under some conditions, significant amounts of water, probably similar to those involved in relatively slow transpiration, can be absorbed by the leaves and transferred through the plant, even into the soil. In most species, however, cuticular resistance appears to be so high that the total amounts absorbed in an overnight period are seldom adequate to more than partially re-saturate the leaf tissue and redistribution to other organs is not observed.

B. Significance of Aerial Absorption Under Natural Conditions

The presence of liquid water on the shoots of plants can influence the water economy of the plant in two ways, directly by actual entry of water into the plant, or indirectly by reducing transpiration. Both these factors tend to reduce the degree to which leaf water deficits develop, and so increase turgor and promote plant growth. Under natural conditions such phenomena could possibly occur as a result of wetting of the leaves by rain, dew or sprinkler irrigation. Greatest interest is centred around the possible utilization of dew, because the presence of rain or irrigation implies soil water abundance, and minimizes the importance of foliar absorption.

Although dew could never comprise more than a small proportion of the water requirement of a normally transpiring plant, it might possibly represent an important water source to plants under stress. In the first place it occurs at night, often while the absorption lag behind the previous day's transpiration is still being overcome. At such a time water intake could accelerate the restoration of leaf turgor, and hence promote growth. Furthermore, dew frequently remains on the leaves for several hours after sunrise, delaying the onset of renewed stress.

It seems that most of the observed beneficial effects of dew can be explained solely by the presence of surface water on the leaves. Thus Walter (1936) concluded that the primary effect of dew in the South West African deserts was in reducing transpiration. Duvdevani (1953) reported that plants exposed to dew developed great length and number of branches, and greater leaf area, than those not exposed. Although he considered that this was due in large measure to the effect of dew absorption by the plants, it is apparent that similar responses could be expected from a reduction in transpiration, giving rise to a smaller water deficit in the exposed plants.

If absorbed water is to make a real contribution to the water economy of the plant, it must be sufficient to cause noticeably greater growth or production than would otherwise be the case, or to enable plants to persist longer during protracted periods of dry weather. Direct evidence of the effect on growth is lacking, although Duvdevani's (1953, 1957) results of increased growth and production may well reflect some influence of absorbed water, in addition to the effects of reduced transpiration loss. Evidence for prolonged survival is much stronger (Stone and Fowells, 1955; Stone, 1957b). These authors have noted that artificial dew at night can prolong the life of seedlings of *Pinus* spp. rooted in dry soil, by as much as a month and a half. As soil water levels did not change during these experiments, transpiration losses appear to have come largely if not entirely from the applied water. These results thus provide evidence of a direct effect of applied water, either through resaturation of leaf tissues dehydrated during the day, or else by delaying the onset of fatal levels of dehydration. In either case normal metabolic functions were maintained far longer than would have been possible otherwise.

CHAPTER 8

Special Aspects of Transpiration

In Chapter 2 ecological aspects of evapotranspiration from the soil and plant surfaces of plant communities were considered, in terms of the overall energy and water balance of the ecosystem. General aspects of liquid flow through the plant have been considered in Chapter 7. This chapter will concentrate on the physics of the water and energy exchanges involved in leaf transpiration, on the physical and physiological factors which influence them, and on methods of measurement.

1. Energy and Water Exchanges Involved in Leaf Transpiration

Just as is the case with evapotranspiration from plant communities, transpiration from individual leaves is dependent on an energy supply to satisfy the latent heat demand, on a water vapour concentration gradient which provides the driving force for vapour flow, and on the diffusive resistances which are located in the vapour pathway. To a large extent, all of these three influences are interdependent, the latter two particularly so. As with evapotranspiration, it can be quite misleading to assume that to double the net radiation, doubles the transpiration rate, or to double the total diffusive resistance halves the rate. Instead, following a change in one of these factors, a completely new energy balance is established in which the new fluxes of radiation, heat and water vapour may bear no simple proportionality to the old. It is difficult therefore to discuss any one of the major groups of factors influencing transpiration without immediately introducing qualifications to cover changes in the others. In order to partition the subsequent discussion to some extent, consideration will therefore be given first to general aspects of the energy balance, and then to specific aspects of diffusion of water vapour out of the leaf. For background information, the reader is referred to Raschke (1960) and Gates (1962).

A. Radiant Energy Exchange Between Leaf and Environment

As distinct from the plant community surface which was generalized in Chapter 2 to the extent that radiation fluxes were only assumed to take place between the surface and the sky, an isolated plant leaf exchanges long- and short-wave radiation with the adjacent soil and plant surfaces, and so is in a more complex energy exchange environment. The appropriate energy balance equation, for such a leaf, can be written (cf. Eq. 2.1):

$$\mathbf{R}_n^u + \mathbf{R}_n^l = \alpha_s(\mathbf{R}_s^{iu} + \mathbf{R}_s^{il}) + \alpha_l(\mathbf{R}_l^{iu} + \mathbf{R}_l^{il}) - (\mathbf{R}_l^{ou} + \mathbf{R}_l^{ol}) \qquad (8.1)$$

where $(\mathbf{R}_n^u + \mathbf{R}_n^l)$ is the net radiant energy flux towards the upper (^{u}) and lower (^{l}) leaf surfaces. $\mathbf{R}_s^i$, $\mathbf{R}_l^i$ and $\mathbf{R}_l^o$ are the inward fluxes of short-wave radiation and inward and outward fluxes of long-wave radiation respectively, and α_s and α_l are the corresponding fractions of incident short- and long-wave radiation absorbed (absorption + transmission + reflection = 1). The first two bracketed terms on the right hand side of the equation therefore represent the energy absorbed by the leaf, and the third term the energy which is re-radiated.

Typical absorptivities for short-wave radiation of single leaves (see Chapter 2) appear to range from 0·5–0·8, with most albedos and transmission factors ranging over 0·1–0·4 (Billings and Morris, 1951; Kleschnin, 1960; van Wijk, 1963; Gates et al., 1965). However extensive measurements of α are lacking, and it is probable that with thick dull leaves values may be significantly higher than 0·8, while with very thin shiny leaves values significantly below 0·5 may be observed.

Absorptivity data for long-wave radiation are also scarce, although Gates and Tantraporn (1952) have presented data for a wide range of wavelengths in the infrared, confirming that leaves are almost completely "black" with respect to absorptivity and emissivity in the 3·0–25·0 μ range. Based on this work Knoerr and Gay (1965) have recently used a figure of 0·97 for absorptivity and emissivity of leaves for long-wave radiation, but for many purposes it can be assumed to be unity.

For the upper surface of an unshaded horizontally oriented leaf, virtually the only sources of incident short wave radiation are the sun and the sky, but if the leaf is inclined or partially shaded significant contributions may be received by transmission and reflection from other surfaces, particularly adjacent leaves and soil. By comparison, the major sources for lower surfaces of leaves are reflected rather than direct beams, except where a leaf is steeply inclined. The reflected and transmitted contributions to the effective beam incident on each surface may influence its albedo and transmission characteristics, particularly since some attenuation occurs during transmission and reflection, as a result of which a higher proportion of the energy may be received at longer wavelengths (Lemon, 1963). This effect can be evaluated by direct measurement, but will usually involve only a relatively small correction to the mean value of α.

The main source of incident long-wave radiation for the upper surface of an unshaded horizontally oriented leaf is the atmosphere (see Chapter 2) but again, if the leaf is not horizontal or is shaded, the main sources may be adjacent leaves and stems. The lower surfaces will receive long-wave radiation mainly from these sources, and also from the soil surface.

The net radiant load on a leaf due to all the energy exchanges described in Eq. (8.1) can be measured by determining the sum of the net radiant fluxes ($\mathbf{R}_n^u$) and below ($\mathbf{R}_n^l$) the leaf, using all-wave net radiometers (see for example Chapter 2 and Funk, 1959). However a certain net radiant load, so determined, may vary markedly in its components, due especially to changes in leaf temperature which determine the magnitude of the re-radiation term, ($\mathbf{R}_i^{ou} + \mathbf{R}_i^{ol}$). Also, while transpiration is influenced primarily by the net heat load, regardless of its composition, and by leaf temperature, photosynthesis, on the other hand, is considerably influenced by the spectral composition of the short-wave radiation. For reasons such as these, it is frequently important to evaluate each component separately.

This can be done in manner recently described by Knoerr and Gay (1965) in which the net radiant heat load was measured with all-wave net radiometers, as described above, and the short-wave exchanges by solarimeters. The difference, representing the net long-wave exchange was then partitioned into a net inward flux and an outward re-radiation (emission) term using the Stephan–Boltzmann law to compute re-radiation. This well known law states that radiation is emitted from any surface according to the expression

$$\mathbf{R} = \mathbf{e}\mathbf{q}T^4 \qquad (8.2)$$

where $\mathbf{R}$ is radiant flux in units such as cal cm^{-2} sec^{-1}, $\mathbf{e}$ is the emissivity, $\mathbf{q}$, the Stephan–Boltzmann constant ($= 1\cdot32 \times 10^{-12}$ cal cm^{-2} deg K^{-4} sec^{-1}) and T is surface temperature in deg K.

The net long-wave exchanges are often not separately considered in energy balance studies, partly because they can be sufficiently accounted for by all-wave net radiation measurements, and partly because the net long-wave fluxes may be relatively small. Nevertheless, the magnitudes of the separate inward and outward fluxes are quite considerable. This can be appreciated when Eq. (8.2) is used to calculate the re-radiation from a surface at $25°$ C ($298°$ K). Assuming $\mathbf{e} = 1\cdot0$, this gives $\mathbf{R} = 1\cdot04 \times 10^{-2}$ cal cm^{-2} sec^{-1} or $0\cdot62$ cal cm^{-2} min^{-1}, a value of similar magnitude to commonly measured values of net radiation above plant communities in summer. Because of the fourth-power dependence on temperature, this value will double with an increase in temperature to about 75°C.

It can therefore be seen that, under conditions where leaf temperature reaches appropriate levels, re-radiation alone could dissipate all the absorbed energy. At night this is not infrequently the case because the radiant load is relatively small, and emission at normal, or even below normal, temperatures can account for it. During the day, however, the required temperatures would almost certainly result in tissue damage or death. In practice, of course, it is unnecessary to speculate on the dissipation of all the incident energy in this manner because sensible and latent heat transfer, the latter representing the energy consumed in transpiration, also constitute effective energy dissipating mechanisms.

B. Dissipation of the Radiant Energy Load by Heat Transfer

Convective transfer of sensible heat and water vapour occurs from leaf to air whenever there is a gradient of temperature or water vapour concentration to provide the appropriate driving forces. Conversely, when the gradients are reversed, i.e. towards the leaf, the sensible and latent heat flows are also towards the leaf leading, in both cases, to a net increase in the total heat load on the leaf and, in the case of water vapour, sometimes to dew deposition. Commonly occurring situations when both fluxes are towards the leaf are at night; during the day the water vapour flow is rarely inwards, and the heat flow is only inwards when relatively warm ambient air is present. Under such conditions the effective heat load on the leaf exceeds that due to absorbed radiation alone, and re-radiation and transpiration constitute the only mechanisms for energy dissipation.

Energy is also consumed in photosynthesis and small quantities of heat can be stored in the leaf so that the equation for the whole energy balance (cf. Eq. 2.2) can be written:

$$(R_n^u + R_n^l) + (H^u + H^l) + l(E^u + E^l) + G + aA = 0 \qquad (8.3)$$

where $(R_n^u + R_n^l)$ is the net inward radiant flux at the upper (^{u}) and lower (^{l}) surfaces, respectively (cal cm^{-2} sec^{-1}) as in Eq. (8.1.), H and E are the net sensible and latent heat fluxes in cal cm^{-2} sec^{-1} and g cm^{-2} sec^{-1}, respectively, l is the latent heat of vaporization ($\simeq$590 cal g^{-1}), G is heat flux into the leaf (storage term) in cal cm^{-2} sec, A is net photosynthesis (g cm^{-2} sec^{-1}) and a is the chemical energy storage coefficient ($\simeq$3600 cal g^{-1}). All fluxes are positive inwards.

G can be neglected for all except very short ($\simeq$1 min) periods and A can also be neglected except when the sum of the first three bracketed terms in Eq. (8.3) becomes small. Although $\mathbf{aA}$ seldom exceeds more than a few percent of the incident short-wave radiation (see Chapter 2) short periods sometimes occur when, for example, $(\mathbf{R}_n + \mathbf{lE}) \cong 0$ and failure to take $\mathbf{aA}$ into account, under these conditions, can lead to significant percentage errors in estimates of $\mathbf{H}$ (Slatyer and Bierhuizen, 1964a). However, over periods longer than a few hours this is seldom important. In the subsequent discussion, therefore, attention is confined to the convective transfer of sensible heat, $\mathbf{H}$, and latent heat, $\mathbf{E}$, as the major processes, other than re-radiation, concerned with energy dissipation.

Two classes of convective transfer are normally recognized, free (or natural) convection in which the flow is created solely by density differences caused by temperature differences within the fluid (in this case, air) and forced convection in which the flow is caused by bulk air movement. In practice the former class seldom operates alone with leaves out-of-doors, because the air is rarely still. Forced convection therefore predominates, although sometimes both processes are important.

At the leaf-air interface, the air stream moving across the leaf is stationary, the velocity increasing with distance away from the surface, until eventually it is indistinguishable from the bulk air movement. The zone in which the velocity increases linearly with distance from the surface is called the laminar boundary layer (the unstirred layer of Chapter 6), and occurs at all surfaces within a moving fluid. In the case of heat transfer it represents a transition zone for fluid motion and temperature difference between the leaf surface and the air. Its characteristics are depicted in Fig. 8.1 in which it can be seen that an "effective" boundary layer thickness (cf. Chapter 6) can be defined as the distance across which a uniform concentration gradient, equal to that at the interface, would have to exist to give the same total drop in concentration, thus:

$$\frac{dT}{dz} = \frac{T_a - T_l}{d} \tag{8.4}$$

where d is the effective thickness of the temperature boundary layer, dT/dz is the slope of the linear portion of the temperature gradient nearest to the surface, and T_a and T_l are the temperatures of the bulk air and leaf surface, respectively. It will be appreciated that within this effective boundary layer there occurs a truly laminar sub-layer and then part of a transition zone leading to the fully turbulent bulk air beyond.

It is assumed that heat is transferred across the boundary layer by conduction (effectively diffusion down a concentration gradient) and is then removed by the convective motion of the bulk air. Consequently rate of heat transfer, $\mathbf{H}$, in cal cm^{-2} sec^{-1} will be proportional to the temperature difference $\Delta T (= T_a - T_l)$ between leaf and air and the

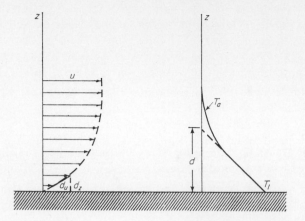

FIG. 8.1. Wind velocity and air temperature profiles near a plane surface where the wind speed in the free air is u, the air temperature T_a, and the surface temperature T_l. The boundary layer thickness next to the surface, d, is defined in terms of the temperature profile as in Eq. 8.1 (after Gates, 1962).

area of surface in contact with air and inversely proportional to the thickness of the boundary layer and can be expressed as

$$\mathbf{H} = \frac{l}{A}\frac{dQ}{dt} = \frac{k(T_l - T_a)}{d} \qquad (8.5)$$

where Q is the heat flow in calories, A is the cross sectional area of surface (cm^2), d is the thickness of the boundary layer (cm) and k is a proportionality coefficient, which in this case is the coefficient of thermal conductivity in air ($\simeq 6 \cdot 0 \times 10^{-5}$ cal cm^{-1} sec^{-1} deg C^{-1}). A heat transfer coefficient can now be introduced, k_h, and also a diffusive resistance to heat transfer across the boundary layer, r'', defined by

$$k_h = \frac{k}{d} \qquad (8.6)$$

and

$$r''_a = \frac{c_p \rho_a}{k_h} \qquad (8.7)$$

where k_h is in cal cm^{-2} sec^{-1} deg C^{-1} and r''_a is in sec cm^{-1}. The specific heat of air is c_p ($\simeq 0 \cdot 242$ cal g^{-1} deg C^{-1}) and ρ_a is the density of air ($\simeq 1 \cdot 19 \times 10^{-3}$ g cm^{-3}) at normal temperatures. r''_a is related to D_h, the thermal diffusivity of air ($\cong 0.22$ cm^2 sec^{-1}), by the expression $r_a'' = d/D_h$.

Equation (8.5) can therefore be re-written in the form

$$H = k_h(T_l - T_a) = c_p \rho_a \frac{(T_l - T_a)}{r_a''} \tag{8.8}$$

Expressions such as these are frequently used in heat transfer experiments. It should be mentioned that k_h, d and r_a'' are all averaged values because the thickness of the boundary layer increases with distance downwind across the leaf, while $(T_l - T_a)$ is also an averaged value because T_l tends to decrease downwind. As a result of both these factors, H itself must vary downwind, and must also be averaged. These variations are recognized phenomena in heat transfer studies (Hsu, 1963) and have been observed in studies with leaves by Raschke (1958, 1960). In most cases the changes are small, particularly with higher wind speeds, and mean values are quite effective in explaining observed results.

Because wind speed and leaf geometry are the primary factors affecting boundary layer characteristics, and hence heat conduction to or from an individual leaf, it is of value to consider the influence of these factors in a little more detail. They have been investigated by a number of workers over the years, not only with respect to heat transfer but also with water vapour transfer, since the boundary layer influences diffusion of both quantities similarly. Among these investigations may be mentioned those of Martin (1943), Raschke (1956, 1958, 1960) Kuiper (1961), Gates (1962, 1963), Tibbals et al. (1964), Knoerr and Gay (1965), and Gates, Tibbals and Kreith (1965). These studies, both empirical and fundamental, have shown that, for a flat broad leaf, the boundary layer is influenced by wind speed and downwind leaf dimension, and for a cylindrical leaf by wind speed and diameter.

Values of k_h for different geometric surfaces can be obtained from heat transfer considerations (Gates, 1962; Hsu, 1963). For a flat broad leaf

$$k_h = 9.55 \times 10^{-5} u^{0.5} b^{-0.5} \tag{8.9}$$

where u is wind speed in cm sec^{-1} and b is downwind width (cm). The constant 9.55×10^{-5} has the dimensions cal cm^{-2} sec$^{-0.5}$ deg C^{-1}. For a cylindrical leaf, with perpendicular air movement,

$$k_h = 1.03 \times 10^{-4} u^{0.33} (2r)^{-0.67} \tag{8.10}$$

where r is the radius of the cylinder in cm, and the constant has the dimensions given above. These values, which apply to forced convection, are thought to apply for wind speeds down to the order of 1 cm sec^{-1}, since both Kuiper (1961) and Slatyer and Bierhuizen (1964a) have shown the $u^{0.5}$ relationship to hold down to this range. These data appear to confirm that free convection seldom operates on this scale, except under artificial conditions. This is somewhat surprising for the under side of the leaf where it might be assumed that, with very slow wind speeds, a pool of relatively warm stagnant air could be developed and maintained under the leaf by free convection buoyancy effects. Under conditions of low transpiration (closed stomata) and high leaf-air temperature, differences of these effects may occur but with normal transpiration rates and moderate levels of energy input they do not appear to be significant. The general assumption that the boundary layer has the same characteristics on both upper and lower leaf surfaces, and hence that k_h is the same for both, may, hovever, not always be acceptable.

Real leaves are neither perfectly cylindrical nor perfectly flat and, partly for this reason, experimental observations seldom support Eqns (8.9) and (8.10) exactly. Both Martin (1943) and Raschke (1956, 1958) obtained expressions of the form

$$k_h = \text{(constant)} \, u^{0.5} \, b^{-0.3} \qquad (8.11)$$

However, the correction is small, and, regardless of whether an experimental or a theoretical approach is used, there are three important conclusions to be drawn from the preceding equations. Firstly, significant amounts of heat can be transferred by convection. Secondly, the smaller the downwind width of the leaf the greater the effectiveness of heat transfer (per unit area), the most effective shape being cylindrical. Thirdly, there is a marked influence of rate of air movement.

Figure 8.2, obtained from Eqns. (8.8) and (8.9) vividly demonstrates these conclusions. It can be appreciated that, at a wind speed of 200 cm sec^{-1} ($\simeq$5 mph) and $(T_l - T_a) = 5°C$, a heat loss of about 0.27 cal cm^{-2} min^{-1} can be expected from a leaf 10 cm wide, and that this figure rises to over 0.8 cal cm^{-2} min^{-1} with narrow leaves. (In computing total transfer per leaf, actual leaf area is twice the area of each surface).

As wind speed increases, the air flow across the leaf changes in its structure and characteristics from aerodynamically smooth to rough form (see Sutton, 1952 for a detailed account of this phenomenon). This decreases the thickness of the laminar sub-layer still further, makes the transition zone more abrupt, and enhances transfer across the boundary layer. In turn this changes the function of k_h versus wind speed to the form $u^{0.7-0.8}$ rather than $u^{0.5}$ (Raschke, 1956, 1960; Kuiper, 1961). From physical heat transfer consideration, however, rough flow may not always occur across leaves within crop canopies, since it requires wind speeds of the order of 15 cm sec^{-1} across a leaf as wide as 10 cm, and substantially higher wind speeds for narrower leaves. Leaf flutter may promote rough flow, or enhanced turbulence, at lower wind speeds, however.

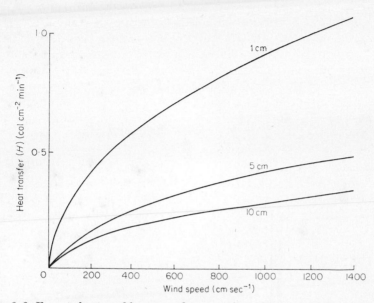

FIG. 8.2. Expected rates of heat transfer per unit surface area of flat plates 1 cm, 5 cm and 10 cm in breadth, in relation to wind speed. Surface-air temperature difference $(T_l - T_a)$ is assumed to be 5°. Equivalent amounts of heat transfer can be expected from each surface of a leaf [based on Eq. (8.9)].

Convective transfer of sensible heat thus constitutes the second important energy dissipation mechanism, re-radiation having already been discussed. The third major mechanism, latent heat transfer, can now be considered. Before discussing the specific aspects of water vapour transport, however, it is of value to examine Fig. 8.3 to determine the relative quantities of energy accounted for by re-radiation and sensible heat transfer. The remainder, neglecting heat storage in the leaf and photosynthesis, constitutes the energy supply available for and dissipated by, transpiration.

The diagram, based on the data of Fig. 8.2 (see also Gates, 1964), shows the expected relative contributions of re-radiation, sensible heat transfer and latent heat transfer to energy dissipation, with changing leaf temperature. It is assumed that air temperature is constant at 25°C and the rate of energy absorption by the surface is 1·4 cal cm^{-2}

min^{-1}. The diagram shows that, when $\Delta T = (T_l - T_a) = 0$, **H** is zero and re-radiation and transpiration effectively dissipate all the absorbed energy, the contribution of re-radiation being determined by Eq. (8.2).

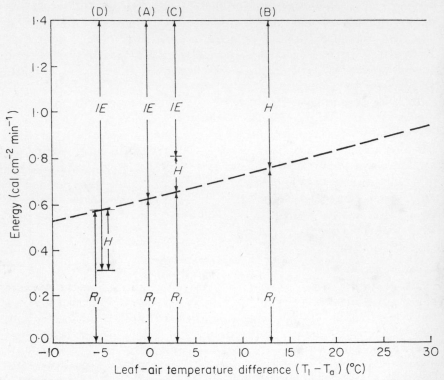

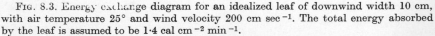

Fig. 8.3. Energy exchange diagram for an idealized leaf of downwind width 10 cm, with air temperature 25° and wind velocity 200 cm sec^{-1}. The total energy absorbed by the leaf is assumed to be 1·4 cal cm^{-2} min^{-1}.

This is represented by case (A). When transpiration is virtually zero (due to stomatal closure), increase in leaf temperature results in all the energy being dissipated by sensible heat transfer and re-radiation [case (B)]. In most cases, however, a dynamic equilibrium is established, as in case (C), in which all three mechanisms are involved, and the leaf temperature reaches an intermediate value. In case (D) a situation is envisaged in which there is a flow of sensible heat to a cooler leaf, i.e. the effective heat load is increased above the radiant energy load by sensible heat flow. Under these conditions transpiration tends to increase to dissipate the extra heat.

In these four cases the Bowen ratio (**H/lE**) ranges from 0 (case A) to a value approaching infinity (case B), to a value which is positive, but < 1 (case C), to a value which is negative (case D), hence its sign and magnitude are determined to a considerable extent by leaf temperature. The importance of leaf temperature in these energy exchanges can be readily appreciated since its absolute value directly determines the outgoing re-radiation flux, and, indirectly, through its effect on saturation vapour pressure, influences the latent heat flux. Furthermore, its value relative to the adjacent air provides the temperature difference necessary for sensible heat flux. It therefore mediates all three of the major energy dissipation mechanisms.

It must be emphasized, however, that leaf temperature cannot be regarded as controlling the partitioning of the energy balance. Its role is essentially passive, and there is a continuous and dynamic interaction between all the energy fluxes, with leaf temperature rising and falling as a result of these interactions to alter the relative magnitude of the different processes. Although leaf temperature varies widely it can be appreciated from the foregoing that $(T_l - T_a)$ seldom exceeds a range of $(T_l - T_a) = -2$ to $+15°C$ (Gates, 1963; Mellor, Salisbury and Raschke, 1964; Knoerr and Gay, 1965).

C. Latent Heat Transport

Transpiration from plant leaves involves water vapour transfer along a concentration gradient from the evaporating surfaces within the leaf to the natural leaf surface and then from leaf surface to bulk air. In this latter part of the pathway it is assumed that latent heat flow encounters the same boundary layer resistance, r_a, as sensible heat transfer, so that an equation similar to Eq. (8.8) can be written for latent heat transfer, assuming that the vapour pressure at the outer surface of the leaf is known.

$$\mathbf{E} = \frac{(c_l - c_a)}{r_a} = \frac{273}{PT} \rho_v \frac{(e_l - e_a)}{r_a} \qquad (8.12)$$

where $\mathbf{E}$ is the evaporation in g cm^{-2} sec^{-1}, $\Delta c = (c_l - c_a)$ is the difference between the water vapour concentration of the air (g cm^{-3}) at the leaf surface and in the bulk air, $\Delta e = (e_l - e_a)$ is the corresponding difference in vapour pressure, P is atmospheric pressure in the same units as e (say mm Hg), and ρ_v is the density of water vapour. $(273\rho_v/PT$ constitutes a conversion factor to change Δc to Δe. It has a value of approximately 10^{-6} so that 1 mm V.P. represents about 1 mg water vapour per litre of air). The expression can be converted to units of cal cm^{-2} sec^{-1} by multiplying both sides by 1 ($\simeq$590 cal g^{-1}).

The symbol r_a represents the diffusive resistance to water vapour transfer in air, related to the diffusion coefficient of water vapour in air

$D_w(\cong 0.24$ cm^2 sec^{-1}) by the relationship $r_a = d/D_w$ where d is the effective thickness of the boundary layer (cm). Since r_a'' is related to the thermal diffusivity of air, D_h ($\cong 0.22$ cm^2 sec^{-1}) by the similar relationship $r_a'' = d/D_h$, the ratio of the two diffusive resistances (assuming that l is the same for both) is given by

$$\frac{r_a''}{r_a} = \frac{D_w}{D_h} \cong 1.1 \tag{8.13}$$

In real leaves Eq. (8.12) can only be evaluated if the external surfaces are wet, when e_l is taken to be the saturation vapour pressure corresponding to the leaf temperature. In many cases such a surface can be closely approximated by pieces of wet blotting paper, cut to leaf dimensions and of similar energy absorption characteristics (Gaastra, 1959; Slatyer and Bierhuizen, 1964a).

In normal transpiration, however, the evaporating surfaces are inside the natural surface of the leaves, so an additional complex internal resistance, r_l, is added to the external resistance, r_a. In the next section the characteristics of r_l will be considered. For the present, however, it can be included in the basic Eq. (8.12) to give

$$\mathbf{E} = \frac{(c_w - c_a)}{(r_l + r_a)} = \frac{273\rho_v}{PT} \frac{(e_w - e_a)}{(r_l + r_a)} \tag{8.14}$$

where $\mathbf{E}$ is in g cm^{-2} sec, $(c_w - c_a)$ and $(e_w - e_a)$ are the differences in vapour concentration or vapour pressure, as the case may be, at the evaporating surfaces within the leaf and in the bulk air, respectively, and $(r_l + r_a)$ is the combined resistance to vapour diffusion within the leaf and in the boundary layer, with the effective length of the diffusion pathway given by $r = (d + d_l)/D$ where D is again the diffusion coefficient of water vapour in air ($\cong 0.24$ cm^2 sec^{-1}), d is the effective boundary layer thickness and d_l is an analogous length of the internal diffusive pathway. Equations (8.12) and (8.16) are general in form and provide an estimate of mean r_l from both surfaces of the leaf. When the stomata are mainly on one side of the leaf, r_l^u and r_l^1 may differ markedly. It can then be important to distinguish between them (Holmgren, Jarvis and Jarvis, 1965).

2. Characteristics of the Internal Water Vapour Pathway

In this section the anatomical and physical characteristics of the internal water vapour pathway will first be discussed, with reference to stomatal movement and the actual sites of evaporation within the leaf. The relative magnitudes of the internal diffusive resistances in the pathway will then be evaluated in relation to changes in stomatal aperture and external resistance. Finally techniques for measuring leaf transpiration rate will be examined. For general background information the reader is referred to Milthorpe and Spencer (1957), Heath (1959), Milthorpe (1960) and Slatyer (1966a).

A. Factors Affecting Stomatal Aperture

1. Controlling mechanisms

Detailed accounts of stomatal physiology are available in the literature (see for example Stalfelt, 1956; Heath, 1959; Ketellapper, 1963; Zelitch, 1963, 1965; Meidner and Mansfield, 1965). Accordingly, in this section only a brief account of stomatal function will be given.

It appears to be well established that increase in guard cell turgor relative to that of adjoining cells, causes opening of the stomatal pore, and decrease in turgor causes the pore to close. During swelling, the opening movement presumably occurs because the thin dorsal wall of each guard cell (away from the pore) lengthens considerably, and the relatively inextensible ventral wall (lining the pore) is drawn into a more semicircular shape (Heath, 1959). Good direct evidence for this hypothesis for guard cell movement comes from experiments in which either guard cells or adjoining cells have been punctured and the responses observed (Heath, 1938).

Swelling can occur because of an absolute increase in guard cell turgor, or because of an increase in guard cell turgor relative to that of the adjoining cells. The second phenomenon frequently arises when, during transient increases in evaporative demand or decrease in internal supply, the tendency for water loss to occur preferentially from the adjoining cells (which are less cutinized) results in a relative increase in turgor (Milthorpe and Spencer, 1957).

The primary factor controlling stomatal aperture appears to be intercellular space CO_2 concentration. Below a certain intercellular space, CO_2 concentration, which varies between species and appears to be affected by factors such as water stress, opening movements are initiated. Above this concentration closing movements are initiated. The degree of opening or closing which follows depends on the magnitude of the concentration change. Also, in the light, the reduction of CO_2 in photosynthesis means that a closing movement results in reduced internal CO_2 concentration, which in turn tends to cause re-opening. Thus there tends to be some overshoot and cycling in stomatal reactions which are gradually damped out to lead to a new steady state situation, in which stomatal aperture and internal CO_2 concentration are again in balance.

A model for these reactions has been proposed by Raschke (1965b) and is diagrammatically depicted in Fig. 8.4. Raschke envisages a system rather like an industrial control system in which the prevailing intercellular CO_2 concentration is first compared with a reference value. This generates an opening or closing signal, the strength of which depends on the disparity between the actual and reference values. The signed is amplified and passed into a servo-system which converts the signal into a mechanical force that changes guard cell turgor and hence stomatal aperture. As stated previously this affects internal CO_2 concentration, perhaps reversing the original signal so that cycling occurs before a steady state situation develops.

This model accounts in general terms for most observations, although evidence for the signal monitor stage is still meagre. The problem still remains, however, to identify specific mechanisms by which CO_2 concentration can so sensitively control stomatal aperture. Clearly a flow of energy is required (the amplifier stage) and, associated with the energy flow, a mechanism for inducing turgor change. The fact that Meidner and Mansfield (1965) and Zelitch (1965) have recently reviewed this subject, and drawn rather different conclusions, demonstrates how little quantitative knowledge yet exists.

It appears that changes in turgor could most likely be mediated by changes in guard cell osmotic pressure, or colloidal swelling pressure. The former effect could follow a gain or loss of ions, soluble carbohydrates or organic acids, the latter could follow the accumulation, or substitution, of ions or changes in pH. Evidence for the existence of all these processes has been found at different times by different workers (see Heath, 1959; Meidner and Mansfield, 1965) but the speed of the stomatal reactions, and the evidence that opening occurs in the dark when CO_2 is flushed through a leaf, eliminates some of them.

Two mechanisms which have received considerable attention involve the production of soluble carbohydrates from glycolic acid metabolism (Zelitch, 1963, 1965) and the production of organic acids following dark fixation of CO_2 (Heath, 1950; Meidner and Mansfield, 1965). As supporting evidence for his view, Zelitch has shown that stomatal opening is inhibited by α-hydroxy sulphonates, inhibitors of glycolic acid metabolism, and that this effect could be reversed by supplying additional glycolic acid to the tissue. However, the opening of stomata in CO_2 free air in the dark prevents this mechanism from being generally applicable even though it may contribute to opening movements during illumination, either in the manner just proposed, or if glycolic acid is

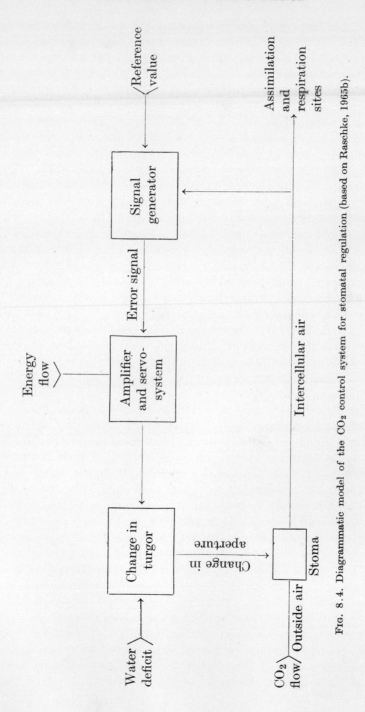

FIG. 8.4. Diagrammatic model of the CO_2 control system for stomatal regulation (based on Raschke, 1965b).

involved, in photophosphorylation reactions producing ATP (Butt and Peel, 1963) which could provide an energy source for an ion pump.

A mechanism involving dark fixation of CO_2 and organic acid accumulation in the guard cells could lead to direct effects on guard cell turgor via osmotic or pH changes. It is supported by evidence that dark fixation can occur in the light as well as the dark (Ranson and Thomas, 1960) and that it is sensitive to small changes in CO_2 concentration, within the physiological range (Bonner and Bonner, 1948). Furthermore Shaw and Maclachlan (1954a, 1954b) have shown that dark fixation does occur in guard cells, and Sayre (1926) and Scarth (1929), that pH is lower in closed than open guard cells.

At the present time, however, no general and fully satisfactory explanation exists and further research is urgently needed on this most important subject. It is possible that the mechanisms of turgor change involved in opening and closing movements may differ. For example, one could envisage a continuous ion efflux from guard cells tending to cause stomatal closure. This could be enhanced if increase in CO_2 concentration caused a reduction in σ (see Chapter 6). Opposed to this passive, CO_2 enhanced, efflux might be an active ion pump, utilizing normal salt respiration sources (Briggs, Hope and Robertson, 1961) which is progressively inhibited by increasing CO_2 concentration. The capacity of the pump could be such that rapid ion accumulation could occur yet, in the absence of an active ion influx, leakage could cause quite rapid closing movements.

Regardless of the actual mechanism of the CO_2 control system, it is clear that intercellular CO_2 concentration exercises sensitive and effective control over stomatal aperture and that a flow of energy is required, at least for opening movements, since metabolic inhibitors can effectively prevent opening (Zelitch, 1963). Low temperature, which can also be regarded as a metabolic suppressant, has a similar effect (Stalfelt, 1962). It is of interest that sodium azide, a well-known respiration and oxidative phosphorylation inhibitor, also inhibits closing movements, but at relatively high concentrations (5×10^{-4} M) (Walker and Zelitch, 1963). Zelitch (1963) considered this to be evidence that different processes were involved in opening and closing reactions, a view also supported by the evidence of Siegenthaler and Packer (1965). While this may be so, it may also be that some inhibitors act indirectly by primarily modifying the internal CO_2 concentration rather than by directly affecting the energy flow mechanism.

Some of the best evidence for the CO_2 control system is found in the stomatal behaviour of succulents exhibiting massive dark fixation of CO_2. In these species, stomata commonly open in the dark and close during the day (Nishida, 1963). Although other explanations for this behaviour have been advanced by Bruinsma (1958) and Nishida (1963), it appears probable that opening at night is a direct response to the low levels of intercellular CO_2 caused by dark fixation, and closing during the day is a response to the high levels of CO_2 associated with CO_2 evolution.

2. *Effect of light, temperature and water stress*

The effect of light on stomatal aperture appears to be mainly mediated by photosynthetic reduction of CO_2 concentration. Even though light has such pronounced effects on stomatal aperture under normal CO_2

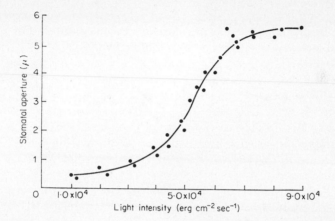

FIG. 8.5. Effect of light intensity on stomatal aperture (after Kuiper, 1961).

concentration (see Fig. 8.5), and even though photosynthetic production of soluble carbohydrates has been regarded, at times, as an important means of inducing guard cell turgor changes (see Heath, 1959), the fact that CO_2-free air induces widest stomatal opening clearly indicates an indirect role of photosynthesis.

There does, however, appear to be a significant direct effect of light on the degree of stomatal opening (Heath and Russell, 1954) showing that even when CO_2-free air was flushed through a leaf, responses to light intensity could be measured. Although the effects they observed were slight, such an effect could be associated with the increased permeability of cell membranes to active ion transport observed in the light (Briggs, Hope and Robertson, 1961) should such phenomena be involved in stomatal responses.

The main effect of temperature on stomatal aperture also appears to be mediated by internal CO_2 concentration, although prevention of opening by low temperature (Stålfelt, 1962) may be a direct result of suppressed metabolism. With high temperatures, enhanced respiration rates may lead to an increase in internal CO_2 concentration and Heath and Orchard (1957) have suggested this could be the cause of the midday stomatal closure which is sometimes observed. This explanation was supported by the work of Meidner and Heath (1959) which showed that midday closure could be prevented by supplying CO_2-free air (but this result may only have demonstrated an overriding of the water deficit control mechanism by the CO_2 control mechanism). A more direct effect of temperature may be on the rate of stomatal opening, since Meidner and Heath (1959) showed that this process had a Q_{10} of 2·2, suggesting control by a dark chemical reaction.

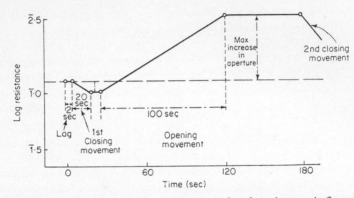

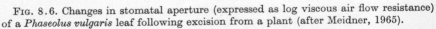

FIG. 8.6. Changes in stomatal aperture (expressed as log viscous air flow resistance) of a *Phaseolus vulgaris* leaf following excision from a plant (after Meidner, 1965).

Water deficits can clearly exert a direct effect on stomatal aperture, by their effect on relative and absolute turgor levels in guard cells and surrounding cells (Stålfelt, 1955, 1961; Allerup, 1961; Ehlig and Gardner, 1964; Meidner, 1965). Two types of reaction can be recognized, one being a transient change in stomatal aperture as a result of changes in guard cell turgor relative to that of adjacent cells, and the other a longer term change associated with severe stress.

The first type is well illustrated by the changes in stomatal aperture which occur when a transpiring shoot is detached, or placed in an osmotic solution. In Fig. 8.6, an example of this type of response obtained by Meidner (1965) is given for an illuminated bean leaf. The general hypothesis is that adjacent epidermal cells tend to exchange water more readily with both the vascular system and the atmosphere than the guard cells themselves. On excision, therefore, release of tension in the xylem elements ($\Psi \to 0$) led to a water flow into the adjacent cells, slightly increasing their turgor relative to that of the guard cells and causing partial stomatal closure. [This phase, which persisted for only about 20 sec in the experiment under consideration, can be prolonged if the excision is made under water when an enhanced water supply is established (Allerup, 1961; Meidner and Heath, 1963)]. The interrupted water supply, combined with the tendency for more rapid evaporation from the adjacent cells, then results in a decline in their turgor relative to the guard cells and opening occurs, sometimes to a pronounced degree. Finally, as the water deficit spreads through the leaf, turgor loss in both guard cells and adjacent cells leads to progressive stomatal closure.

The second type of response occurs at different values of Ψ_{leaf} for different species and for plants grown under different conditions (Stålfelt, 1961; Ehlig and Gardner, 1964). It appears that water deficit, in itself, may not affect stomatal aperture until a critical value is reached and then, as the water deficit increases, progressive decreases in stomatal aperture occur until almost complete closure exists. This type of response can frequently be confounded with the short term changes just described (Gregory et al., 1950; Milthorpe and Spencer, 1957). In many cases, even at quite severe water deficits, stomatal opening can be observed as water deficit increases, and closing as water deficit declines. The long-term response is also mediated to some degree by internal CO_2 concentration, since reduced photosynthesis caused by water stress, and perhaps enhanced respiration, lead to increased internal CO_2 levels which themselves influence aperture (Heath and Meidner, 1961; Heath and Mansfield, 1962). In consequence, specific Ψ_{leaf} values are seldom associated with specific degrees of stomatal aperture, even though a general relationship exists.

Overall, the water stress effect is probably similar to that of most other inhibitors and acts partly by causing changes in internal CO_2 concentration through an effect on respiration and photosynthesis, and partly by acting on the processes directly concerned with energy flow and guard cell turgor. Under field conditions, where ambient CO_2 levels are more or less constant at levels which, if they existed in the intercellular spaces, would cause stomatal closure, light intensity and water stress are the two environmental factors most responsible for observed changes in stomatal aperture. It is important to bear this in mind, even though the effects of both factors may be mediated to a significant degree by internal CO_2 concentration.

B. Sites of Evaporation Within the Leaf

There are two main sites of evaporation within the leaf; one appears to be located in the outer epidermal walls and the other in the walls of the mesophyll cells which are exposed to intercellular air spaces. The vapour then flows to the leaf surface through the cuticle in the first case and through the stationary air zone inside the leaf, and the stomatal pores, in the second case. When the stomata are open there is generally much less resistance in the latter pathway and a high proportion of the total vapour transport is along it. When the stomata are closed it is generally assumed that no flow occurs through them and the only pathway available is through the leaf cuticle. Since these cuticular and stomatal pathways are in parallel, the resistances they represent may be linked as follows:

$$\frac{1}{r_l} = \frac{1}{r_c} + \frac{1}{r_s} \tag{8.15}$$

where r_l is total internal resistance of the whole leaf and r_c and r_s are the resistances in the cuticular and stomatal pathways, respectively (in sec cm^{-1}). We may then consider r_s itself to comprise the following components

$$r_s = (r_w + r_i + r_p) \tag{8.16}$$

where r_w, r_i and r_p are the successive resistance in the mesophyll cell walls, intercellular spaces, and stomatal pore, respectively. Alternatively all the separate resistances may be included as a single more complicated expression in the form used by Milthorpe and Spencer (1957) and Milthorpe (1959).

The specific location of the evaporating surfaces is not yet known. In the cuticular pathway it appears logical to expect liquid phase continuity almost as far as the outer cutinized surfaces of the epidermal cell walls. Secondary deposition and thickening in these walls reduces void sizes to molecular dimensions (Esau, 1965), so that they would be expected to have a very low hydraulic conductivity, and any withdrawal of the liquid-air menisci into the voids would be prevented by surface tension. This can be appreciated from Eq. (1.9), which shows that water in voids smaller than 1 mμ diameter will not commence to drain until equilibrium relative humidities drop below 10% at normal temperatures. In some cases liquid-phase continuity may extend into the cuticle itself but, if any discontinuity occurs, it will probably be close to the wall-cuticle interface, since mechanical disruption and fracture of the cuticle would tend to occur along this boundary.

Even though liquid-phase continuity may extend into the cuticle, the hydraulic conductivity of the walls is probably so low that very steep water potential gradients develop across the zone of secondary thickening. Although evidence exists of several-fold increases in cuticular transpiration following cuticle removal (Hall and Jones, 1961; Horrocks, 1964; Radler, 1965), this effect does not necessarily indicate a high water permeability of the cutinized walls, but may only mean that resistance to water vapour diffusion in the cuticle itself constitutes a still greater barrier to water transport. Even with the cuticle removed, resistance in the cuticular pathway, in most species, far exceeds that in the stomatal pathway when stomata are open. The main exceptions would be in species adapted to very mesic habitats in which there is little epidermal wall thickening.

Because of doubt as to the exact location of the liquid-air interface in the cuticular water pathway, it is quite difficult to accurately partition the resistances into those which affect liquid-phase transport, for which the driving force is the gradient of water potential, and those which affect vapour phase transport, for which the driving force is the gradient of water vapour concentration. This in turn presents difficulties in assigning value to the water vapour concentration at the interface, and in evaluating the r_c term of Eq. (8.15).

In most cases, it has been assumed, for calculation purposes, that the water potential at the liquid-air interface is zero; that is, that the water vapour concentration at the interface is the saturation vapour concentration at the leaf temperature (see, for example, Slatyer and Bierhuizen, 1964a). While this may be satisfactory for some purposes, it clearly oversimplifies the real situation, and further experimental work is required on this subject. Since the $\Psi'(e/e^0)$ relationship is known, and since the water potential in the mesophyll (Ψ'_{leaf}) and the water vapour concentration in the bulk air (c_a) can be measured reasonably accurately, experiments in which cuticular transpiration is measured at different bulk humidity values, but with other conditions constant or measurable, should enable some estimate of the liquid- and vapour-phase resistances to be made.

In the stomatal pathway, the liquid-air menisci are located in the walls of the leaf mesophyll. Although most investigators have assumed that the menisci are effectively at the interface between the cell walls and the internal air (Gregory et al., 1950; Milthorpe and Spencer, 1957; Slatyer, 1966a), it was proposed many years ago that, during periods of water stress induced by rapid transpiration or inadequate soil water supply, the menisci may retreat into the interfibrillar spaces of the cell walls, thereby lengthening and increasing the tortuosity of the vapour pathway. It was considered that this could cause a significant wall resistance to vapour flow, r_w, to develop, and was termed "incipient drying" (Livingston, 1906; Lloyd, 1908; Knight, 1917).

The evidence for this view was based primarily on observations that transpiration appeared to fluctuate without detectable changes in stomatal aperture. Later investigators (Maskell, 1928; Milthorpe and Spencer, 1957; Slatyer, 1966a) have suggested that other factors could have caused the observed responses, and the use of gold sols in the transpiration stream (Gaff, Chambers and Markus, 1964) has shown water transport right to the cell wall-air interface. The size of the inter-fibrillar spaces also makes any withdrawal of the menisci unlikely. Because of suberization of the outer layers of the mesophyll cell walls (Scott, Schroeder and Turrell, 1948; Scott, 1950) the void dimensions are of the order of 1 mμ, so that they would not be expected to drain [see Eq. (1.9) and Table 1.III] until water potentials dropped to values which would cause death of the entire plant. Apart from this evidence, Slatyer (1966a) has also calculated, on the basis of the supply capacity of the leaf mesophyll and the high internal-external leaf surface ratio, that even under extreme evaporative conditions no withdrawal could be expected. In consequence, in the stomatal pathway, the evidence points to the evaporation sites being effectively located at the cell wall-internal air boundary or, if beneath the surface, at constant depth.

C. Water Vapour Concentration at the Evaporating Surfaces

As stated above it has generally been assumed that, at the evaporating surfaces in both the cuticular and stomatal pathways, the water potential $\Psi \cong 0$, so that the water vapour concentration at the surfaces is given by the saturation vapour pressure at the leaf temperature (Milthorpe, 1959; Slatyer, 1966a). It is conceded that leaf water potential may vary from $\Psi = 0$ to $\Psi = -50$ bars under the range of conditions normally encountered, but the equilibrium relative vapour pressure equivalent to the latter figure is still $e/e^0 = 96\%$ at normal temperatures, so that the error introduced into the $(c_w - c_a)$ term of Eq. (8.14) does not exceed about 8% as long as bulk humidity values are less than $e/e^0 = 0.50$.

It can be appreciated that with high atmospheric relative humidities, significant errors can be introduced by this assumption, especially in plants under severe stress. In xeromorphic species which can tolerate water potential values lower than, say, $\Psi_{leaf} = -100$ bars (Stone, Went and Young, 1950; Slatyer, 1960a; Whiteman and Koller, 1964) the assumption is untenable and measurements of leaf water potential are required in order to insert more appropriate values for c_w.

From the foregoing section, however, it is apparent that even measurements of leaf water potential may not adequately indicate the water potential at the evaporating surfaces, Ψ_{surf}. In the cuticular pathway, steep water potential gradients probably exist across the cutinized outer cell walls and the assumption that $\Psi_{surf} = \Psi_{leaf}$ results in the bulking of the vapour and liquid phase resistances into an equivalent cuticular vapour resistance. This could introduce significant errors in the calculation of the true r_c in cases where the vapour phase component was small, since reduction of atmospheric e/e^0 from, say, 0·90 to 0·80 would only double the vapour concentration gradient (assuming $\Psi_{leaf} \cong 0$) but may increase $d\Psi/dx$ across the walls by an order of magnitude.

In the stomatal pathway, although the hydraulic conductivity across the suberized outer wall layers is much higher than across the cutinized epidermal walls, there is increasing evidence that steep water potential gradients may exist across this zone, so that the assumption that $\Psi_{leaf} = \Psi_{surf}$ may also introduce errors in this pathway. At the present time quantitative data on this phenomenon are meagre although, Shimshi (1963a) has estimated $\Psi_{surf} \cong -80$ bars in corn leaves under low stress conditions. While subsequent experimentation may yield more reliable values, it is apparent that figures of this magnitude significantly reduce the effective leaf-air vapour concentration difference $(c_w - c_a)$ compared with that to be expected if it is assumed that $\Psi_{surf} \cong 0$. It can also be mentioned that, if the vapour component of the cell wall resistance is negligible (i.e. the liquid-air menisci are effectively at the surface), the magnitude of $\Psi_{leaf} - \Psi_{surf}$ can be expected to vary more or less with transpiration rate, since the rate of flow will be proportional to the water potential difference if the hydraulic conductivity of the pathway remains the same.

Apart from low hydraulic conductivity *per se*, Ψ_{surf} may be influenced by solute accumulation at the evaporating surfaces since, associated with the liquid flux to the surfaces, there must be an accumulation of solutes at the sites of evaporation until the rate of back diffusion of solutes balances the rate of arrival. Boon-Long (1941) showed with artificial systems that evaporation across a membrane enclosing an unstirred 1·0 M Sucrose solution could be substantially less than when water alone was enclosed, presumably because of solute accumulation at the surface and consequent reduction in Ψ_{surf} and c_{surf}. This explanation is supported by unpublished experiments of P. J. Kramer (personal communication) in which the Boon-Long study was repeated but with active stirring in the solution to minimize the otherwise considerable unstirred layer effect. This procedure effectively prevented any significant reduction in evaporation during the test period.

In the case of the mesophyll evaporation surfaces, the problem is essentially the same "convection opposed by diffusion" situation found with water and solute permeability across unstirred layers, referred to in Chapter 6. In this instance an appropriate expression is more difficult to develop because the diffusion is not free diffusion as in the external solution case treated previously, but involves diffusion back longitudinally through the walls towards the vascular elements and also involves some movement by diffusion and active transport into the vacuoles of the mesophyll cells. However, assuming that the volume of tissue available for back diffusion was equal to the free space, Slatyer (1966a) calculated that the probable increase in the osmotic pressure at the evaporation surfaces, compared with the xylem sap, was only 13%. This would have a negligible effect on c_w since xylem sap concentration is about 1–2 bars. While this analysis suggested little effect of solute accumulation, the assumptions made with respect to solute diffusivity in the zone of suberization may have been in error. Quantitative experimentation is required on this matter.

In concluding this section, it should be mentioned that until more information is available on the actual sites of evaporation in both the cuticular and stomatal pathways, the assignment of various fractions of the total resistance to the liquid or vapour phases is rather arbitrary. As can be seen from Eqns (8.14)–(8.16), if the assumption is made that $\Psi_{surf} = \Psi_{leaf}$, the effects of liquid phase resistance are bulked into r_c, in the case of the cuticular pathway, and r_w in the case of the stomatal pathway. If, on the other hand, the evaporation sites in the stomatal pathway are considered to be at the actual cell wall-air interface, then r_w is assumed to be zero. If r_w if finite, its effect is wrongly assigned to a hydraulic resistance influencing c_w.

It should also be remembered that, even if c_w tends to be reduced by the factors just described, a compensating increase could follow a small rise in leaf temperature.

D. Relative Magnitudes of Internal and External Diffusive Resistances

In this section the preceding anatomical and physiological features of the water vapour pathway will be considered in terms of the resistance they offer to the net water vapour flux out of the plant in the transpiration pathway. The degree to which leaf resistance, r_l, can be expected to control transpiration of single leaves and entire communities, under various external conditions, will be examined. Attention will also be given to the effect of transpiration suppressants on the internal and external resistances.

1. *External and internal diffusive resistances*

From Eqns. (8.6) and (8.9), the effect of wind speed and downwind leaf width on external resistance r_a can estimated. Taking a leaf such as cotton, about 10 cm wide, calculated values of r_a (per unit external leaf surface area) range from about $3 \cdot 0$ sec cm^{-1} at wind speeds of about 10 cm sec^{-1}, to about $1 \cdot 0$ sec cm^{-1} at 1 m sec^{-1} and to about $0 \cdot 3$ sec cm^{-1} at 10 m sec^{-1}. Comparable figures for a grass leaf 1 cm wide can be expected to be $1 \cdot 0$, $0 \cdot 3$ and $0 \cdot 1$ sec cm^{-1}. Very good corroborative data for these values have been obtained with a number of different plant species and wet blotting paper discs by Martin (1943), Raschke (1956), Kuiper (1961) and Slatyer and Bierhuizen (1964a). Consequently, although factors such as leaf hairiness can significantly increase the general level of r_a expected from a leaf of known geometry, the range of values for most leaves is from about $0 \cdot 1$ to 3 sec cm^{-1}. When, in response to water stress, a leaf folds or rolls around the surface containing most stomata, the values on that surface increase considerably. In the case of xeromorphic genera such as *Triodia* (Burbidge, 1946) in which the leaf completely folds up, it is probably more appropriate to regard the resultant leaf shape as a different leaf, in which some of the resistance in the enclosed air space is treated as an internal diffusive resistance.

External resistance is generally evaluated in a wind tunnel or assimilation chamber using surface-wetted leaves or pieces of wet filter paper cut to leaf dimensions and of similar surface characteristics. Subsequently, the internal leaf resistances are evaluated using normal leaves, or the same ones after the surface has dried (Kuiper, 1961; Slatyer and Bierhuizen, 1964a). It is generally assumed that, when the stomata are closed, stomatal resistance is infinitely high and the measured transpiration is due to vapour flow via the cuticle with the total diffusive resistance $(r_a + r_l)$ being represented by $(r_a + r_c)$. When the stomata are open r_s and r_c are in parallel (Eq. 8.15) and the proportion of vapour transport along each pathway is inversely proportional to the resistance encountered.

Typical estimates of cuticular resistance, r_c, (again expressed per unit external leaf surface area), range from < 20 sec cm^{-1} for shade plants to > 200 sec cm^{-1} for xerophytes (see Holmgren, Jarvis and Jarvis, 1965). For a number of crop plants values appear to range between 20–80 sec cm^{-1}. Kuiper (1961), for example, has obtained r_l values in the dark (presumably r_c) of about 20 sec cm^{-1} for tomato and bean; Slatyer and Bierhuizen (1964a) obtained 64·6 sec cm^{-1} for cotton.

Since r_c is determined largely by cuticular characteristics, it is probable that changing environmental conditions and age effects which cause changes in cuticle thickness and composition are likely to affect it (Mueller, Carr and Loomis, 1954; Silva Fernandes, Baker and Martin, 1964). Holmgren, Jarvis and Jarvis (1965) noted a marked effect of leaf temperature on cuticular resistance r_c declining as temperature increased from 17–22°C in three temperate zone species (see Fig. 8.7). A more pronounced effect on insect nymph cuticular resistance at temperatures corresponding to the melting points of cuticle waxes, has been observed by Beament (1961, 1965). Most investigators have considered r_c to be fairly constant under any one set of environmental conditions, but Slavik (1958) considered that, as water deficits developed, cuticular transpiration declined due to desiccation of the cuticle. His rates of cuticular transpiration relative to total transpiration were, however, very high ($\cong 30\%$) suggesting that some stomatal transpiration may have been included in this figure. More quantitative data on all these phenomena are required.

Stomatal resistance usually implies total stomatal diffusive resistances, r_s, rather than stomatal pore resistance, r_p. Estimates of r_s range from < 2 sec cm^{-1} (again on an external surface area basis) to infinity, since it is generally assumed that when stomata are closed all the water vapour flux is cuticular. Although stomatal closure may not be completely effective this assumption appears reasonable since Slatyer and Jarvis (1966) could not detect an N_2O flux through an amphistomatous cotton leaf when stomata were closed.

Typical open stomata values of r_s range from 0·5 sec cm^{-1} for wheat (Milthorpe, 1960) and 0·8 sec cm^{-1} for corn (Shimshi, 1963b) to 2·0 for cotton (Slatyer and Bierhuizen, 1964a), 3·2 for turnip (Gaastra, 1959) and > 4·0 for tomato and bean (Kuiper, 1961). Higher values occur in other species (Holmgren, Jarvis and Jarvis, 1965). This variation is due in part to differences in stomatal pore geometry and stomatal frequency, and also to differences in intercellular space resistance, r_i, discussed below. The effect of age can also be quite pronounced (Slatyer and Bierhuizen, 1964b), showing a three-fold increase in open stomata values of r_s for cotton leaves over a period of four weeks following the attainment of maximum leaf size. The main factors which lead to a reduction in aperture (below the open stomata value), under natural conditions, are reduced light intensity and water stress. A good example of the light intensity effect based on data of Kuiper (1961) [as presented by Rijtema (1965)] is given in Fig. 8.8.

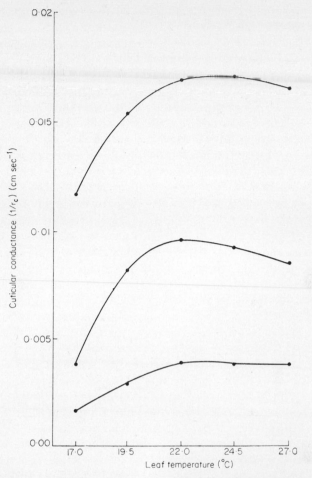

FIG. 8.7. Influence of leaf temperature on cuticular conductance for water vapour ($1/r_c$) on the adaxial surface of leaves of (a) *Lamium galeobdolon* (b) *Betula verrucosa* (c) *Acer platanoides* (after Holmgren, Jarvis and Jarvis, 1965).

Resistance to water vapour transport within the leaf, between the evaporation sites and the stomatal pores, varies with the effective path length from different cells. The value of r_s as a whole, therefore, includes a weighted average of r_l, since proportionally more evaporation takes place from the cells nearest to the stomatal pores. The greatest distance for travel occurs in thick leaves and in those leaves in which the stomata are located on only one surface. In such leaves highest values of r_i can be expected. For any one leaf it seems likely that r_i should be more or less constant, since it depends on internal geometry.

However, Meidner (1955) considered that, at least in hypostomatous leaves, increased desiccation caused an increase in r_i, presumably because shrinkage led to a more tortuous intercellular pathway.

Only one set of direct measurements of r_i are known. These values, obtained by Jarvis, Rose and Begg (1966) for a cotton leaf, using a diffusion porometer, indicated a value of $\cong 3 \cdot 0$ sec cm^{-1} from one surface to the other. This suggests that r_i was the largest single component of the open-stomata value of r_s ($\cong 4 \cdot 0$ sec cm^{-1} for both surfaces added in series). This result, while somewhat surprising, may explain in part why significant reduction in stomatal aperture is required before r_s is substantially increased. (In Figs 8.5 and 8.8, for example, it can be seen that reduction in light intensity from 8×10^4 to 5×10^4 erg cm^{-2} sec^{-1} reduced stomatal aperture by 50% but only caused a 25% rise in r_s).

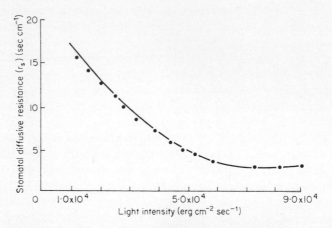

FIG. 8.8. Effect of light intensity on stomatal diffusive resistance, r_s, of a bean leaf. Light intensity values refer to the visible part of the spectrum; r_s values are based on the area circumscribed by the leaf outline (after Rijtema, 1965).

In comparison with these values for r_i, measurements with an air flow porometer, in which a bulk flow of air is induced through a leaf indicate almost negligible values of r_i (Heath, 1941; Milthorpe and Spencer, 1957). It has generally been assumed that the only process by which water vapour is transferred out of a leaf is that of diffusion, in which case the latter values may be of limited significance. If, however, bulk air movement occurs through a leaf under natural conditions, the estimates of r_s already quoted must be significantly reduced since the effective path length for water vapour diffusion could be dramatically affected under open stomata conditions.

This phenomenon requires detailed investigation but it seems probable that in amphistomatous leaves, small pressure differences across a leaf (of the order of 1–2 cm water-manometer pressure) could significantly enhance water vapour transport by markedly reducing r_s/Σ_r. It is equally probable that, under wind conditions in which leaf flutter develops, pressure differences of this magnitude could be expected, at least in large leaves. In hypostomatous leaves or very narrow leaves, however, the possibility of such enhanced flow under natural conditions is much reduced. Should such a phenomenon exist, it could lead to much lower vapour pressures within the intercellular spaces than under simple diffusion conditions and could provide an evolutionary justification for the development of internal suberization.

2. *Modification of internal resistance by transpiration suppressants*

In recent years the possibility of deliberately influencing r_s by the use of transpiration suppressants has created a considerable amount of interest and has much potential significance to plant production. For many years nurserymen have coated seedling leaves with various films (Miller, Nielson and Bandemer, 1937) in an attempt to reduce transpiration during transplanting, although the effects of these procedures has sometimes been deleterious, due in part to the marked resultant rise in leaf temperature (Thames, 1961). More recently Roberts (1961) attempted to suppress transpiration by use of cetyl alcohol, a compound developed for evaporation suppression from exposed water storages, which forms a monomolecular film on water surfaces. Later experiments have not always confirmed his initial success (Kriedeman, Neales and Ashton, 1964; Slatyer and Bierhuizen, 1964c). Subsequently plastic films (Gale, 1961; Gale and Poljakoff-Mayber, 1965; Angus and Bielorai, 1965) and compounds which metabolically induce stomatal closure, have been used.

Although it has been known for many years that a number of compounds, including metabolic inhibitors, will induce stomatal closure (see for example Heath, 1959; Ketellapper, 1963), the work of Zelitch (1961) and Zelitch and Waggoner (1962a, 1962b) has been mainly responsible for focusing interest in this area. These workers have found that phenyl mercuric compound and, more recently (Zelitch, 1964) alkenyl succinic acids, in concentrations of $10^{-4}–10^{-5}$ M can partly or completely close stomata with few apparent toxic effects and it has been demonstrated (Shimshi, 1963a, 1963b; Slatyer and Bierhuizen, 1964b, 1964c) that the effects may persist for several weeks.

The effect of such compounds on r_l has been investigated by the above workers and by Slatyer and Bierhuizen (1964b) who found that, with cotton leaves for example, 10^{-5} M phenyl mercuric acetate increased r_l (per unit external leaf surface area) from an open stomata value of 1·2 sec cm^{-1} to a value of 4·4 sec cm^{-1}, and a concentration of 10^{-4}, to 7·4 sec cm^{-1}. This represented a four-fold increase in r_l which, under the low wind speed conditions employed ($r_a = 3$·2 sec cm^{-1}) reduced transpiration by more than 60%. Almost complete closure was caused by a concentration of 10^{-2} M (see Fig. 8.9).

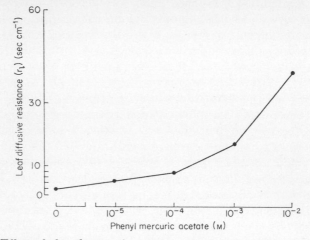

FIG. 8.9. Effect of phenyl mercuric acetate on internal leaf diffusive resistance, r_l, ($r_l \simeq r_s$) of a cotton leaf (after Slatyer and Bierhuizen, 1964b).

3. *Effect of r_l in controlling transpiration*

Early research on transpiration of single leaves and individual plants did not properly account for the effect of air movement, and hence of r_a, in determining the observed rates of transpiration. Partly because of this, early investigators obtained evidence of marked changes in transpiration with apparently constant stomatal aperture and conceived the idea of "incipient drying" (Livingston, 1906; Lloyd, 1908; Knight, 1917). Their ideas were also based on the work of Brown and Escombe (1900) which concluded, on the basis of measurements with perforated surfaces, that stomatal pore resistance would be unlikely to influence diffusion of water vapour until the stomata were almost closed. Unfortunately Brown and Escombe had assumed that pore resistance alone was operative in their system, and so neglected external resistance, r_a, in interpreting their experiments.

The important difference which is made by including r_a, is that under still air conditions, such as were used by Brown and Escombe, r_a is substantially greater than r_l so that even marked changes in r_l do not affect $(r_a + r_l)$ to any significant extent. Renner (1910) and Maskell (1928) first pointed out the effect of neglecting r_a and several other authors have subsequently elaborated on it (see for example, Penman and Schofield, 1951; Bange, 1953).

In order to investigate the relative magnitude of r_a and r_l, changes in both resistances have to be measured at constant or known $(c_w - c_a)$ (see Eq. 8.14). A good experiment of this type, conducted by Bange (1953), demonstrates well the reason for Brown and Escombe's conclusions, as well as the significance of wind speed and stomatal aperture in influencing r_a and r_l, respectively.

Taking first a still air situation (analagous to that used by Brown and Escombe, 1900) which included an r_a term of $\cong 2 \cdot 0$ sec cm^{-1} and a moving air situation with $r_a \cong 0 \cdot 1$ sec cm^{-1} (small leaf discs 2·8 cm diameter were used), Bange showed that for open stomata, r_s ($\cong r_l$) was approximately 1·0 sec cm^{-1}. Thus the total resistance $(r_a + r_l)$ in the open stomata case was 3·0 sec cm^{-1} in still air and 1·1 sec cm^{-1} in moving air. Reducing stomatal aperture to one-half its original value increased r_l to 1·5 sec cm^{-1}, so that the total resistance values $(r_a + r_l)$ increased to 3·5 and 1·6 sec cm^{-1}, respectively. If $(c_w - c_a)$ were unchanged this would be expected to reduce transpiration to only 3·0/3·5 = 0·85 of its original value in the still air situation, but to 1·1/1·6 = 0·69 in moving air. At one-tenth of the original aperture, the r_l value was 4·5 sec cm^{-1}, the total resistances 6·5 and 4·6 sec cm^{-1} and the ratios of final/original transpiration would be expected to be 3·0/6·5 = 0·46 and 1·1/4·6 = 0·24, respectively.

The influence of changes in stomatal aperture on transpiration obtained by Bange, are shown in Fig. 8.10 where it can be seen that the predicted changes are in close agreement with measured changes. In the still air case when, with open stomata r_s (or r_l) $\ll r_a$ there appears to be little stomatal control of transpiration until stomatal aperture is substantially reduced and r_s (or r_l) $\cong r_a$, whereas in the moving air situation, even with open stomata r_s (or r_l) $\gg r_a$, and there is effective control through the entire range of apertures. Because Brown and Escombe (1900) and Ting and Loomis (1963) failed to include r_a in their considerations, they attributed the relative absence of stomatal control in still air to r_s (or r_l) instead of to $(r_a + r_l)$; naturally while $r_l \ll r_a$ there can be little apparent effect of changes in r_l on total vapour transport.

As long as water vapour moves out of the leaf by diffusion, the situation $r_l \ll r_a$ will be unusual for single leaves under conditions of even moderate air movement. If, however, the foregoing suggestions concerning a bulk flow of air through leaves are confirmed, the situation

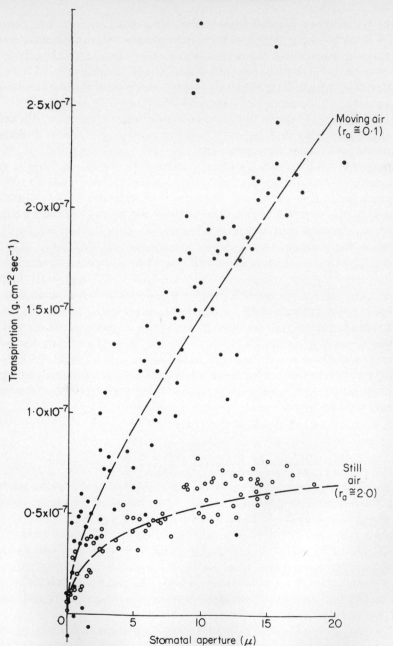

FIG. 8.10. Influence of stomatal aperture on transpiration of *Zebrina* leaves under moving and still air conditions. Dashed lines show predicted behaviour (after Bange, 1953).

would be of more common occurrence and marked stomatal closure would have to occur before significant changes in transpiration could be expected. Regardless of the validity or otherwise of these suggestions, however, it must be appreciated that is the magnitude of true r_l relative to r_a which determines the degree to which r_l must be increased before detectable changes in transpiration occur.

Under field conditions, the relative magnitudes of r_l and r_a for crops can differ markedly from those which apply to single leaves, and there is no simple relationship between crop and single leaf values. With reference to r_l values, this is because, under field conditions, there exists an array of leaves of complex structure and exposure, with a range of r_l values for individual leaves. Because leaf air index (LAI) normally exceeds 1, the area of leaf surface exceeds the area of ground surface, and it can be expected that the effective r_l value for the crop will be less than for a single leaf and will decline as LAI increases (the individual resistances being connected in parallel). However, as mentioned in Chapter 2, crop r_l values can be expected to change with factors other than stomatal aperture. External resistance, r_a, can also be expected to be influenced by factors other than wind speed (Monteith, 1965; Philip, 1966). In consequence, leaf diffusive resistance data cannot be simply extrapolated to field conditions. Perhaps the only generalization that can be made is that, for a crop, the ratio of r_l to r_a is likely to be smaller under most conditions, so that greater reduction in stomatal aperture will be required before equivalent reductions in transpiration occur.

3. MEASUREMENT OF SINGLE PLANT TRANSPIRATION

There are two main methods for the measurement of transpiration from individual leaves or plants, the first involving determination of the change in weight of the system when other sources of weight loss are eliminated or accounted for, and the second involving determinations of the rate of vapour loss. An energy balance method can also be used but is seldom necessary. Other methods, generally more indirect and less quantitative have not achieved wide use, and accounts of them can be found in texts by Crafts, Currier and Stocking (1949) and Meyer and Anderson (1952).

The first of these methods is analogous to the water balance method used for the whole plant community. Applied to an entire plant it reaches its most accurate and meaningful form when change of weight of plant, substrate and container is determined and evaporation from the soil (or other substrate) is minimized or otherwise accounted for. Alternatively, change in the volume of water in the substrate container can be measured by soil water determinations if the plant is rooted in soil (see for example, Ashton, 1956), or by determining the volume of water added to bring the culture solution level up to a given point, if the plant is in water culture.

Over short periods of several days, changes in water content within the plant can be neglected if the measurements are made at the same time each day, say, at sunrise, or can be estimated from plant water content determinations. Over longer periods changing fresh weight of the plant has to be accounted for, generally by periodic harvests of replicated groups of plants. These procedures are straightforward and do not need separate descriptions; details can be found in many papers dealing with transpiration, as for example Slatyer (1957b).

Extrapolation of transpiration so measured to transpiration from assemblages of plants under natural conditions should only be attempted when the plant is rooted in soil in what is essentially a micro-lysimeter located in the plant community itself (Pasquill, 1949). When the measurements are made in the laboratory or on a greenhouse bench it will be apparent, from what has been said in Chapter 2 and in the preceding parts of this chapter, that the environment may be so different that transpiration so measured bears little relationship to that out-of-doors.

The problem of extrapolation is more pronounced when transpiration of a single leaf is measured by detaching a leaf and measuring its loss of weight (the "cut shoot" method) (see Eckardt, 1960) or by placing it in a small potometer and measuring the change of weight or volume of the system (see Meyer and Anderson, 1952). While these techniques do provide a measure of transpiration from the leaf employed, under the environmental conditions imposed at the point of measurement (and subject to physiological changes in the tissue) the extrapolation to considerations of transpiration of even a single plant is fraught with difficulties. This is because the leaf cannot readily be oriented with respect to incident energy and air movement in a way that is typical of an array of leaves on a whole plant and its water supply is also influenced. It can be appreciated that any change in the energy load and wind structure immediately influences the energy balance of the leaf and hence transpiration.

It is not intended here to condemn the use of single leaf measurements for studies on single leaves, but to point out the serious sources of error which can be introduced if extrapolation to whole plants, or groups of plants, is attempted. This criticism applies particularly to the cut-leaf method which has, over the years, been used extensively to indicate relative transpiration rates of different species, or of the same species at different times of day, or seasons, (see for example, Oppenheimer and Mendel, 1939). Every effort should be made to discourage this procedure, which is not only subject to errors due to the altered environment in which the leaf is placed for weighing (so that the individual leaf is not in an environment typical of a leaf identically oriented on the plant) but also to extrapolation errors as applied to the whole plant, and to errors due to physiological changes in the tissue caused by cutting and associated with changes in stomatal aperture.

The last mentioned sources of error were investigated by Weinmann and le Roux (1946) for a range of species and were found to cause marked differences in most cases. Subsequently many workers (see for example, Andersson, Hertz and Rufelt, 1954; Decker and Wien, 1960; Brun, 1961, and references on p. 254) have shown that there is a brief surge in transpiration directly following cutting (2–5 min), associated with an increase in stomatal aperture. During this time transpiration may increase by 10–20% under normal conditions, and then decline progressively as water content and stomatal aperture decrease. Because of these effects the cut-shoot method, if used at all, must be employed with great caution.

It is sometimes observed that the cut-shoot method gives better results when the plant is under water stress and transpiration is slow. This can be expected to apply to all the methods when extrapolation is involved. It is mainly because, when r_l is very high ($r_l \cong r_c$), it dominates the magnitude of $(r_a + r_l)$ so that even the marked reduction of wind speed (and increase in r_a), which accompanies the weighing operation in the case of the cut-shoot method, does not unduly affect transpiration as long as $(c_w - c_a)$ is fairly constant. Also, such observations are usually made under arid conditions when bulk humidity is very low so that the effect on $(c_w - c_a)$ of change in leaf temperature (and c_w, see Eq. 8.14) during weighing is also reduced. These phenomena, while reducing the magnitude of the errors, obviously do not eliminate the sources of error.

The second main method for transpiration measurement, involving the determination of the vapour flow away from the plant, is analogous to the vapour flow methods described in Chapter 2 for plant communities. It involves enclosing a whole plant or portion of a plant and determining the rate of change in water content of the air passing through the chamber so formed, if the system is closed (Koller and Samish, 1964; Grieve and Went, 1965), or of the difference in water content of the air entering and leaving the chamber, if the system is open (Decker and Wien, 1960; Heath and Meidner, 1961; Slatyer and Bierhuizen, 1964a).

This approach enables a more sensitive monitoring of transpiration than the change of weight methods since, with suitable sensing instruments, it has a response time of less than 1 min. When used under controlled environmental conditions in the laboratory the open system approach undoubtedly provides the best means at present available for detailed studies of the transpiration process and details of suitable equipment for this purpose, particularly for use with single leaves, have been given by, amongst others, Heath and Meidner (1961), Björkman and Holmgren (1963) and Bierhuizen and Slatyer (1964a). However, if extrapolation is attempted, similar problems to those just considered arise due to the different environment inside and outside the chamber.

These problems still exist when the measurements are made out-of-doors, unless precautions are taken to simulate external conditions, particularly leaf temperature, bulk humidity, and wind speed [that is the factors affecting $(c_w - c_a)$ and r_a]. The possible magnitude of these effects can be appreciated when it is recalled that, if leaf temperature increases by 5°C, the value of $(c_w - c_a)$ may increase by, say, 50%. In addition, if air movement through the chamber is inadequate to control bulk humidity at the external level, $(c_w - c_a)$ will be reduced and, if the ventilating air stream is increased until bulk humidity and leaf temperature reach external levels, r_a may be reduced significantly and affect $(r_l + r_a)$. In addition, depending on the spectral transmission characteristics of the chamber walls, the radiation fluxes themselves may change. While all of these factors may be controllable and the measured transpiration may closely approximate natural transpiration, it is apparent that considerable thought must be given to the experimental layout before any measurements are made.

With most closed system methods, bulk humidity increases continuously as soon as the chamber is closed, as does leaf and air temperature. While a circulating fan may keep r_a at a level similar to that outside, it is apparent that $(c_w - c_a)$ changes continuously in a direction which depends on whether the increase in bulk vapour concentration, c_a, compensates for the increase in c_w caused by increasing leaf temperature. Extreme caution must therefore be used with this equipment. Koller and Samish (1964) avoided this problem by using a null-point compensation method, in which changes in the temperature, water vapour and CO_2 concentration of the air are continuously compensated for, the rate of compensation indicating the rate of the process under study. This procedure is very attractive for many purposes.

In summary, it can be stated that measurements of transpiration of individual plants or individual leaves refer only to the material with which they are made and the environmental conditions under which they are conducted. Subject to this proviso, the simplest method for an entire plant appears to be the measurement of weight loss from the whole system. If greater accuracy is required for single leaf studies the most sensitive technique is the open circuit vapour flow method or the null-point compensation method, depending on the objectives. Extrapolation, when required, must always be conducted cautiously and, whenever possible, the energy balance of the leaf and wind speed should be controlled. Use of detached plant parts, which may alter the water supply to the leaf even in the potometer method, restricts the extent of extrapolation and the cut-shoot technique should not be used at all. The evidence that the errors are reduced when r_l is high does not justify use of this method, or justify extrapolation of results from other methods, because the basic sources of error remain. Whenever possible, out-of-doors measurements should be made with the methods described in Chapter 2.

Methods of measuring internal diffusive resistance, sometimes a more significant measurement than transpiration itself, have been much improved in recent years. The basic instrument is a porometer, and two main types have been developed — an air flow instrument in which the pressure drop associated with bulk air flow through a leaf is monitored, and a diffusive flow instrument, in which rate of diffusion of water vapour, or another gas, is monitored under estimated or measured gradients of gas concentration.

The first type of instrument is well described by Heath (1959) who has been responsible for the development of many air flow porometers. Its main disadvantages are that a viscous air flow resistance is measured, whereas a diffusive resistance is generally required, and part of the leaf must be enclosed for a significant period of time while the measurement is being made. The first objection is progressively being removed as calibration procedures are developed to relate viscous to diffusive resistance (Waggoner, 1965; Jarvis, Rose and Begg, 1966). The latter has been largely overcome by the development of a simple yet rapid instrument by Alvim (1965) (see also Bierhuizen, Slatyer and Rose, 1965). Other faults in these instruments, associated with the use of pressure differences high enough to directly affect stomatal aperture, have recently been pointed out by Raschke (1965a) who has shown that pressure differences of only about 10 cm water should be used.

Measurements of transpiration from surface wetted leaves, together with measurements of leaf temperature and atmospheric water vapour concentration, c_a, can provide estimates of r_a (see Eq. 8.12). With these estimates of r_a, if the same air movement conditions are used, if there is no wall resistance r_w in the stomatal pathway, and if it can be assumed that c_w is the saturation vapour concentration at the leaf temperature, measurements of transpiration provide estimates of $(r_a + r_l)$ (see Eq. 8.14). Estimates of r_l can then be obtained by difference and, if cuticular resistance can be neglected or accounted for, estimates of r_s can be obtained.

While this series of assumptions must always be checked and appropriate additional measurements made where required, the procedure provides the basis of r_l and r_s measurements, (see, for example, Gaastra, 1959; Kuiper, 1961; Slatyer and Bierhuizen, 1964a; Holmgren, Jarvis and Jarvis, 1965). In most cases, however, the elaborate procedures necessary for controlling r_a preclude the field use of the apparatus described by the above workers. A satisfactory field instrument has, however, been developed by Wallihan (1964) and modified by van Bavel, Nakayama and Ehrler (1965). It is similar in many respects to the closed system of transpiration measurement described previously, but, because measurements are only made over a narrow range of c_a values and for a short period (< 1 min), few of the disadvantages of that procedure affect diffusive resistance measurements.

The instruments consist of a porometer cup which encloses a section of leaf and contains a solid state humidity sensor. The main advantage of this procedure is that r_a is constant and so independent control systems are not required. The time course of increase in vapour concentration inside the cup is used to compute both the rate of evaporation and the value of $(c_w - c_a)$, assuming again that c_w can be obtained from the leaf temperature. In the same way as before, r_a is measured with surface-wetted leaves or wet blotting paper, $(r_a + r_l)$ is measured with the leaf under study, and r_l obtained by difference.

When water vapour is used as the diffusing gas, problems arise in estimating r_s, because of the assumptions concerning r_w and c_w. These can be removed if another gas is used and Spanner (1953) and Slatyer and Jarvis (1966) have developed procedures which use hydrogen and nitrous oxide, respectively. Although these techniques include a measurement of r_i (as well as upper and lower surface r_p and r_a), they provide a useful method of measuring stomatal pore resistance under conditions where r_i and r_a are constant and measurable. The Slatyer and Jarvis (1966) method has the added advantage that continuous monitoring of r_s (or r_p) is possible, at least under laboratory conditions, in conjunction with simultaneous measurements of transpiration and photosynthesis (Jarvis and Slatyer, 1966a).

CHAPTER 9

Development and Significance of Internal Water Deficits

It was shown in Chapter 7 that liquid water flow through the soil-plant system tends to occur along gradients of decreasing water potential, even though the actual driving force across any one segment may not necessarily be the gradient of total water potential, $d\Psi/dx$, but may be instead, the gradient of a component potential. However, subject to appropriate qualifications, flow requires that Ψ_{plant} must be lower than Ψ_{soil} and the point of lowest potential is naturally in the leaves which constitute the sink for liquid flow.

In consequence, any factor which reduces Ψ_{soil} below zero tends to result in the development of internal water deficits in the leaves, and these deficits are increased not only by reductions in Ψ_{soil} but by the gradients of water potential within the soil and plant associated with flow. It is the purpose of this chapter to discuss the manner in which such deficits develop and their significance to physiological plant processes. For additional information on this subject the reader is referred to Kramer (1949), Crafts, Currier and Stocking (1949), Richards and Wadleigh (1952), Bernstein and Hayward (1958), Steward (1959), Slatyer (1960c), Vaadia, Raney and Hagan (1961), Kozlowski (1964) and Slavik (1965). Much useful information is also given in the Encyclopedia of Plant Physiology Volumes II and III (Ruhland, 1956b, 1956c).

1. Development of Internal Water Deficits

A. Progressive Changes in Ψ_{soil} and Ψ_{plant}

As transpiration, from a plant rooted in initially wet soil ($\Psi_{soil} \cong 0$), proceeds from day to day it progressively reduces soil water content and Ψ_{soil}. Since there must be a gradient of decreasing potential through the water pathway from soil to atmosphere to provide the driving force for water flow, there is also a concomitant decline in Ψ_{plant} and plant water content, and consequently an internal water deficit develops and increases in magnitude.

At dawn on any day when internal gradients within the plant and soil have been more or less eliminated by the overnight equilibration period, it can be expected that $\Psi_{soil} \cong \Psi_{plant}$ even though the actual levels of Ψ_{soil} and Ψ_{plant} decline each day. If the overnight equilibration period has not been adequate to eliminate internal gradients it may be that Ψ_{soil} is higher than Ψ_{plant}, but the situation where Ψ_{soil} is lower than Ψ_{plant} cannot develop unless there is absorption of water by the aerial plant organs and the normal gradient is reversed (see Chapter 7).

In addition to this decline in Ψ_{plant}, which is an inevitable result of progressive drying of the soil profile, there is also a diurnal rhythm in Ψ_{plant}, superimposed on the general level, and caused by the relative rates of daily transpiration and absorption. At the beginning of each

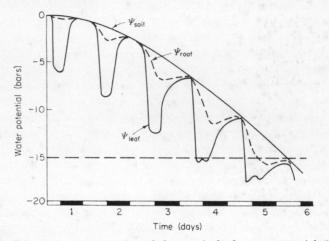

FIG. 9.1. Schematic representation of changes in leaf water potential (Ψ_{leaf}), root surface water potential (Ψ_{root}), and soil mass water potential (Ψ_{soil}), as transpiration proceeds from a plant rooted in initially wet ($\Psi_{root} \cong 0$) soil. The same evaporative conditions are considered to prevail each day. The horizontal dashed line indicates the value of Ψ_{leaf} at which wilting occurs.

day transpiration initially removes water from the leaves, and reduces Ψ_{leaf} below its dawn value, without compensating absorption. Although absorption commences as soon as potential gradients extend down to and across the soil-root interface, the quantitative lag of total absorption behind total transpiration, and hence the magnitude of the internal water deficit, continues to increase until the rate of absorption equals the rate of transpiration and is only reduced when it becomes more rapid.

The interrelationships between these two phenomena are illustrated schematically in Fig. 9.1 (see also Slatyer and Denmead, 1964; Cowan, 1965). Assuming that the same evaporation conditions prevail each day, the upper limiting curve shows the progressive decline in the water potential of the soil mass, Ψ_{soil}, as the soil dries from an initially wet condition ($\Psi_{soil} \cong 0$). The other curves show the water potential at the root surface, Ψ_{root}, and in the leaves, Ψ_{leaf}, assuming that transpiration proceeds for 12 hr and then ceases for 12 hr. It can be seen that when the soil is wet (days 1, 2) small differences in ($\Psi_{soil} - \Psi_{root}$) are needed to sustain flow. As soon as transpiration falls, there is a rapid recovery of plant water content to the stage $\Psi_{leaf} = \Psi_{root}$ and, by sunrise, $\Psi_{leaf} = \Psi_{root} = \Psi_{soil}$. As Ψ_{soil} continues to fall, however, (day 3, 4) the hydraulic conductivity of the soil declines rapidly so that larger values of ($\Psi_{soil} - \Psi_{root}$) are required to maintain flow at the desired level. By day 4 partial stomatal closure retards flow below potential levels, but even so recovery at night to the point $\Psi_{leaf} = \Psi_{root} = \Psi_{soil}$ becomes progressively slower.

As leaf water content and Ψ_{leaf} decline, leaf cell turgor also declines and, depending on the cell volume/turgor pressure relationship and the structural characteristics of the leaves concerned, they may gradually droop and take on a progressively more wilted appearance. There is good evidence that, in those leaves which do exhibit complete wilting, this occurs at the point of zero turgor pressure ($P_{leaf} = 0$), where from Eq. (5.3) $\Psi_{leaf} = -\pi_{leaf}$ (see Slatyer, 1957a, 1957b; Gardner and Nieman, 1964).

Slatyer (1957a) proposed that at the point where Ψ_{leaf} exhibited this value, and when $\Psi_{leaf} = \Psi_{root} = \Psi_{soil}$, the plant would be in a state of permanent wilting and the soil water content would be the permanent wilting percentage. This point is indicated on Fig. 9.1 by the line drawn horizontally across the diagram at an arbitrary level of -15 bars, but it should be recalled that π_{leaf} at this stage can range from 5–200 bars (Slatyer, 1957a).

On day 4 it can be seen that, during the day, Ψ_{leaf} falls below the line for several hours. This is assumed to be reflected in the diurnal wilting which may occur, even in water culture, under conditions of high evaporative demand. By day 5, however, Ψ_{soil} has also fallen below the line. Thus on the night of day 5 it is impossible for recovery of Ψ_{leaf} to lead to recovery of positive turgor in the plant, since at the point $\Psi_{leaf} = \Psi_{root} = \Psi_{soil}$, the value of Ψ_{leaf} would still be below the line. The plant would then be permanently wilted with recovery impossible without soil water recharge. In the diagram it is indicated that the overnight period is too short for recovery to the equilibrium point, but it is apparent that, regardless of the length of the recovery period, positive turgor could not be regained.

Although this is only schematic, it is of interest that actual data, obtained by Gardner and Nieman (1964) and presented in Fig. 9.2

provide very good experimental confirmation. In this case a single pepper plant rooted in a container of soil was grown under controlled conditions of evaporative demand, and Ψ_{leaf} and Ψ_{soil} measured

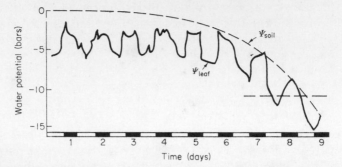

FIG. 9.2. Observed changes in leaf water potential (Ψ_{leaf}) and soil water potential (Ψ_{soil}) of a pepper plant rooted in clay loam soil, under controlled evaporative conditions (after Gardner and Nieman, 1964).

continuously. The value of Ψ_{leaf} ($= -\pi_{leaf}$) at zero turgor pressure ($P_{leaf} = 0$) was -14 bars. Not only was the expected general de-hydration pattern very closely followed but diurnal wilting was observed on day 8, when Ψ_{leaf} fell below the value at zero turgor pressure, and permanent wilting developed when Ψ_{soil} also reached this point. Further confirmatory data were obtained by Slatyer (1957b) and are presented in Table 9.I. They show the very close association between the values of Ψ_{soil}, Ψ_{leaf} and π_{leaf} when permanent wilting was observed in cotton, privet and tomato plants rooted in containers, even though the respective values of Ψ_{soil} at this point varied widely in the different species, being -38, -48 and -20 bars, respectively.

TABLE 9.I

Leaf osmotic pressure, (π_{leaf}), leaf water potential (Ψ_{leaf}), soil water potential (Ψ_{soil}), and soil water content at first† permanent wilting percentage‡

Plant	π_{leaf} (bars)	Ψ_{leaf} (bars)	Ψ_{soil} (bars)	Soil water content (g water g^{-1} soil)
Tomato	18	-19	-20	0·118
Privet	47	-45	-48	0·97
Cotton	38	-43	-38	0·102

† Using the terminology of Furr and Reeve (1945).

For comparative purposes the soil water content at $\Psi_{soil} = -15$ bars was 0·122 g water g^{-1} soil.

‡ Data of Slatyer (1957b).

Hence, it appears that the permanent wilting percentage, rather than being a soil characteristic, is determined by π_{leaf} and is the soil water content corresponding to Ψ'_{soil} when $\Psi'_{soil} = \Psi'_{leaf}$ and $\Psi_{leaf} = -\pi_{leaf}$, so that there is zero turgor pressure in the leaves of the plant being examined. Much more information as to the significance of the permanent wilting percentage, its probable variability and errors associated with its measurement, is given by Slatyer (1957a).

B. Diurnal Changes in Internal Water Deficits

The magnitude of the internal water deficit, expressed in terms of either Ψ'_{plant} or tissue water content, has been shown to be determined firstly by the effective value of Ψ'_{soil} (which establishes a base level of Ψ'_{plant}) and secondly by the degree to which absorption lags behind transpiration on any one day (which influences the absolute levels of Ψ'_{plant} at any point in the plant water system and the gradients of Ψ' which develop within it). The first phenomenon, a progressive and inevitable result of the removal of water from the soil, need not be specifically discussed here. The plant can only exercise control over the latter by stomatal closure.

Stomatal control of transpiration leads to an interesting pattern of diurnal changes in the internal water deficit. Under normal conditions of high soil water status, and in the absence of stomatal closure, the water potential gradients which develop through the plant can be expected to steepen as transpiration proceeds following an overnight period until rate of absorption equals rate of transpiration. Thus the absorption lag, and hence the internal water deficit, can be expected to be relatively small.

However, under more extreme conditions of either greater evaporative demand or reduced supply, the exent to which the water potential gradient through the plant would have to be steepened to satisfy the demand, may be associated with a reduction of Ψ'_{leaf} to a value which induces a degree of stomatal closure (see Fig. 7.9). It is under these conditions that the regulatory function of the stomata becomes apparent and serves to prevent further desiccation rather than to permit flow to be maintained at potential levels.

As a consequence, the slope of the water potential gradient, and hence the magnitude of the absorption lag, tends to be reduced compared with the value which would have developed without closure. Under conditions where water supply is severely impeded, for example when Ψ'_{soil} is lower than its value at the permanent wilting percentage, the stomata may remain closed all day. Consequently very small water potential gradients develop, so that although absorption is slow, transpiration is also slow and the absorption lag is frequently as small as under conditions of high water supply and high transpiration rates, even though the whole plant is under severe stress.

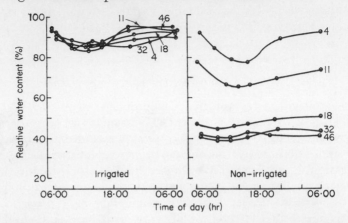

FIG. 9.3. Diurnal fluctuations in relative water content of leaves (phyllodes) of the desert shrub *Acacia aneura*, as a period of protracted dry weather followed soil water recharge. Numbers refer to days since rain (after Slatyer, 1962c).

These features are well illustrated in Fig. 9.3 after Slatyer (1962c) where diurnal fluctuations in relative leaf water content of a desert shrub *Acacia aneura* are shown for an artificially watered plant and for a naturally watered plant on five occasions over a rainless period of forty-six days. The maximum and minimum diurnal values for the artificially watered shrubs were approximately 95 and 85%, respectively, a range of only 10% but, as the soil commenced drying (day 4), the minimum values in the naturally watered shrubs dropped to 75%, giving a diurnal range of more than 15%. As the soil dried further, the absolute water deficit increased progressively but stomatal control reduced the diurnal fluctuations so that the values on day 11 were 75 and 65%, respectively. Finally, as extremely severe stress developed, comparable to water potential values of the order of $\Psi_{leaf} = -110$ bars (day 46), the diurnal fluctuation was only from 42 to 38%.

This was, of course, an extremely drought-tolerant shrub, but similar data have been obtained for normal crop plants (Slatyer, 1955) even though the magnitude of the water deficits was not nearly as great, nor was stomatal control as effective. It should again be noted from Fig. 9.3 that, although the diurnal water deficit fluctuated in the manner described, the progressive decline in relative water content and Ψ_{leaf}

over the whole period examined shows how the total deficit increased from day to day. (Water potential/relative water content relationships for this species are given in Fig. 5.5).

A phenomenon which can markedly influence the diurnal pattern of the absorption lag is the midday depression in transpiration, due to stomatal closure. This can reduce transpiration rate below the rate of absorption so that, instead of the maximum water deficit existing at the time of greatest evaporative demand, there is a tendency for turgor recovery at this time. This phenomenon, also referred to as "midday closure" of stomata, commonly occurs under field conditions of high evaporative demand and has even been observed with plants in water culture under extreme evaporative conditions.

In the absence of water stress, midday closure may occur as a result of high temperatures which cause an increase in intercellular CO_2 concentration (Heath and Orchard, 1957; Meidner and Heath, 1959). Internal CO_2 concentrations may also mediate the water deficit effect (Heath and Mansfield, 1962) to some degree, but midday closure usually develops when soil water supply first commences to limit absorption and water potential gradients through the plant become steep enough to reduce Ψ_{leaf} to the level associated with stomatal closure. As mentioned in Chapters 7 and 8, the stomatal reaction often tends to "overshoot" in the sense that complete closure may occur even though partial closure would be adequate to prevent extreme desiccation. When this occurs transpiration rate is sharply reduced and turgor recovery occurs. Under the conditions of overshoot, and before the stomata reopen, recovery may proceed to a stage where Ψ_{leaf} is higher than that appropriate to induce stomatal closure during the previous drying phase, causing a peak in the normal trough of the diurnal relative water content curve. This phenomenon, illustrated in Fig. 9.4, has been reported on a number of occasions (see Heath, 1959).

When soil water is freely available, recovery to the stage $\Psi_{soil} = \Psi_{leaf}$ appears to be complete within 2–3 hr after sunset, but there is some evidence that this phase is followed by a reduction in leaf water content during the rest of the night, presumably associated with internal redistribution of the water to other organs and tissues (see Kramer, 1949). However, these observations are not borne out by the data of Figs 9.3 and 9.4. Furthermore, in the references quoted by Kramer (1949), relative water content was not measured, water content being expressed, instead, on a dry weight basis. It is possible, therefore, that the observations are artifacts associated with changing dry weight of the tissue as a result of solute translocation.

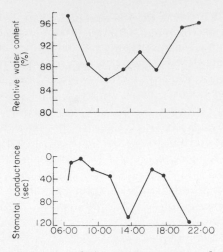

F$_{IG}$. 9.4. Diurnal changes in relative water content and stomatal conductance (measured with an air flow porometer) of upper leaves of a bulrush millet under high evaporative demand conditions (after Begg *et al.*, 1964).

C. *Relative Magnitude of Water Deficit in Different Tissues*

When $\Psi_{soil} \cong \Psi_{root} \cong \Psi_{leaf}$, such as frequently occurs at night, it can be reasonably assumed that the water deficit, expressed in terms of Ψ_{leaf} rather than relative water content, is the same throughout the plant. When transpiration is in progress however, gradients of potential develop through the plant system and quite different deficits can exist in different tissues and organs.

As might be expected, the general situation is one in which there tends to be a gradient of potential from the absorbing zones in the roots to the transpiring sites in the leaves, with those leaves from which most active transpiration is proceeding being those with lowest potential and highest deficit. Leaves not receiving direct sunlight, and fruit, which generally transpire at slower rates, generally have smaller deficits. Very few measurements of these phenomena have been made but those which are available provide reasonable confirmation, Bliss, Kramer and Wolf (1957) for example, observing differences in relative water content from 8–22% between the basal and tip leaves of tobacco growing under typical summer conditions.

Redistribution of water between organs is also of common occurrence. Although most fruits transpire very little, marked changes in fruit volume occur during the day in response to transpiration (Bartholomew, 1926; Hendrickson and Veihmeyer, 1941; Magness, Degman and Furr, 1935) and it has been demonstrated with cotton that there is less wilting

of leaves on plants with large numbers of developing bolls, due presumably to water movement from the bolls to the leaves (Anderson
and Kerr, 1943). As long as adequate soil water is available these
changes appear completely reversible, and the deficits are eliminated
each night. Redistribution of water between branches has similarly
been demonstrated (Dixon, 1914; Brierley, 1934, 1936).

An interesting phenomenon, which leads to redistribution between
plants, is that of root grafting, reported on particularly by la Rue
(1952) and Bormann (1957, 1961). It has been shown by these workers
that anatomical connections develop between interwoven roots of
adjacent plants and that entire groups of plants may share anatomical
unions through which nutrients and water can move. In an experiment
with split root systems of tomatoes, Bormann (1957) demonstrated
that water could be transferred between plants, the roots of which
were associated in a common soil container, and even through the roots
of an intervening plant which itself had access only to water transferred
from the roots of the other plants. In both cases water was transferred
in quantities sufficient to delay, and in some cases prevent, the onset of
wilting. These observations are not only of particular interest in themselves but indicate the care which must be taken in making field
measurements on plants in communities if the soil water status under
adjacent plants is markedly different from those being examined.

D. Differences in Water Deficit Within Tissues

Just as water deficits tend to be greatest in those leaves which are
transpiring most rapidly, gradients of water potential tend to develop
within any one leaf, due to leaf shape, the influence of partial shading by
another leaf, or similar phenomena which result in a different exposure
and hence different local transpiration rates at different points on the
leaf surface. In addition, gradients tend to be developed along the leaf,
particularly in long leaves, and can become pronounced if the internal
water distribution system is barely adequate to supply the terminal
cells.

These phenomena have been investigated by, amongst others,
Slavik (1963b) who has reported that the basal sections of a transpiring
tobacco leaf have consistently higher relative water content values
than the apical sections. Previously unpublished data of the author,
obtained in conjunction with a detailed study of diurnal water and
energy exchanges in a crop of bulrush millet (see also Fig. 9.4), are given
in Fig. 9.5 and provide an example of this tendency and also of the
effect of leaf orientation. The leaves examined were approximately

70 cm long. They emerged from the leaf sheath at a steep upward angle from the stem but then drooped so that the central part of the leaf was more or less horizontal, but the terminal 20 cm was oriented almost vertically downward. The observations show that early in the morning the gradients through the leaf were negligible; three hr after sunrise the greatest deficit existed in the central most exposed section of the leaf, but two hr later continued transpiration from the tip, possibly combined with an enhanced lag in water flow to the tip,

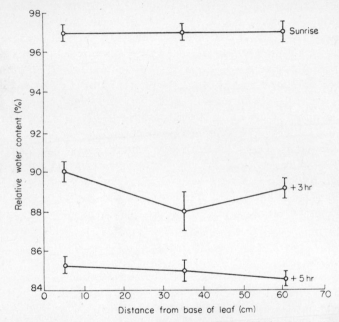

Fig. 9.5. Changes in relative water content along the leaf blade of actively transpiring bulrush millet leaves (cf. Fig. 9.4) under conditions favouring rapid transpiration. Each point is the mean of six determinations; vertical bars indicate standard error of each point.

caused a progressive gradient to develop through the leaf from base to tip. Even so, the overall gradient can be seen to be small, corresponding to a water potential difference of only 2–3 bars.

It has generally been assumed that there are small differences in water potential across cell membranes and that, because water permeability is relatively high, these will tend to be rapidly eliminated by water flow in the direction of the gradient so established (see Chapter 6).

Although there is evidence of transient lags in equilibration of labelled water with cell water in transpiring plants, compared with non-transpiring plants (Raney and Vaadia, 1965b), the water potential differences associated with these lags are likely to be very small.

2. SIGNIFICANCE OF WATER DEFICITS TO PHYSIOLOGICAL PROCESSES

That water deficits interfere with plant growth, and, if severe, cause the death of plants, is undoubtedly one of the most common and self-evident observations which can be made. Because of this it is somewhat surprising that, although the influence of water deficits on such processes as transpiration, photosynthesis, and growth has been investigated for many years and is well documented, the specific sequence of metabolic events associated with the imposition of water deficits has not been studied to the same extent and is still only understood in general terms. This is largely due to the fact that plant growth is the end product of a number of separate processes, among them, cell division, cell enlargement and photosynthesis, and while closely integrated, each has, to a considerable extent, its own controls.

Many of the effects of water deficits on uptake, permeability and transport, and on anatomical and physiological characteristics of roots and shoots, have already been discussed (see Chapters 5–8). For the present, therefore, attention is devoted to the influence of water deficits on various physiological processes which have not been treated previously. Reviews by Stocker (1960), Kessler (1961), Vaadia, Raney and Hagan (1961) and Gates (1964) provide valuable background information.

A. Effects of Hydration on Protoplasmic Properties

Recent research on protein hydration (Klotz, 1958; Privalov, 1958; Singer, 1962) has demonstrated the close interaction between the nature of the hydration shell surrounding protein molecules and the physico-chemical properties of the proteins themselves. The hydration shell consists of an ice-like sheath of several layers of water molecules surrounding and linking the protein molecules. To an increasing degree, the maintenance of this structure is considered to be essential for the maintenance of the properties and functions of the proteins themselves, denaturation following disruption of the hydration lattice, and renaturation following the removal of the denaturing agent, and leading to reconstruction of the original hydration structure (Klotz, 1958).

The effect of heat is therefore thought to reduce the extent of the ice-like structure and, if the increase in temperature is not sufficient to disturb the polypeptide configuration, re-establishment of the original structure follows on cooling. The effect of decreasing pH, may, in some cases be to increase the extent of the ice-cage (Klotz and Ayers, 1957). The effect of electrolytes and non-electrolytes will probably depend on the degree to which they cause re-orientation of the ice-cage. Urea, for example, because of its strong hydrogen bonding charactersitics, may break down the frozen structure of the water envelope. It is probable that some ions may help to stabilize the structure, and protect it against denaturation caused by other agencies (Boyer *et al.*, 1946). There is little evidence as to the effect of reduced hydration, *per se*, on the cage structure, but the observed effects of water stress on protoplasmic viscosity suggest that, in the range of water potentials normally encountered by wilting plants, say, $\Psi = -50$ to -10 bars (corresponding, in general, to water activities between $0 \cdot 96$ and $0 \cdot 99$), marked changes in structure may occur.

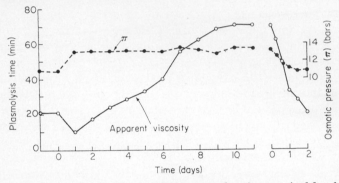

FIG. 9.6. Changes in protoplasmic viscosity (measured as time required for plasmolysis) and vacuolar osmotic pressure of *Lamium maculatum* during imposition, continuance and removal of soil water stress (after Stocker and Ross, 1956).

Stocker (1960) considers the response of protoplasm to desiccation to follow a two stage process, leading to an altered, but stable, new structure. In the first reaction stage, he envisages a partial melting of the ice-cage, with associated rapid decrease in protoplasmic viscosity. Subsequently, during the second resitution phase, a new protein-water structure develops with a modified and possibly enlarged hydration structure. As long as the stress is imposed and removed gradually, and is not too severe, he considers this process to be completely reversible following rehydration. An example of the changes in protoplasmic viscosity which accompany the onset, continuance, and removal of stress, is given in Fig. 9.6.

The figure shows the changes in protoplasmic viscosity and vacuolar osmotic pressure of leaves of *Lamium maculatum* when the moist soil

($\Psi_{soil} \cong 0$) in which it had been growing was rapidly dried to a value approaching the permanent wilting percentage, and then maintained at about this value (by markedly reducing evaporative conditions) for 11 days (Stocker and Ross, 1956). An initial reduction in viscosity occurred as stress was imposed, followed by a progressive rise to values much higher than those originally prevailing during the subsequent stress period. Upon rewatering, the viscosity dropped to the original values within 2 days. The absence of progressive changes in internal osmotic pressure during the restitution phase suggests that solute effects on the hydration characteristics were not important.

Thus it seems that any changes in structure, which lead to denaturation, can be expected to have a marked effect on protein metabolism, and reported changes in enzyme activity by heat and other factors may frequently be due to this phenomenon. Evidence of this type has recently been quoted by Chen, Kessler and Monselise (1964), quite low sucrose concentrations causing marked changes in protein conformation and reductions in enzyme activity. When stress is prolonged, as in the example of Fig. 9.6, it is probable that a marked degree of inactivation of many enzyme systems occurs. This is borne out by Stocker and Ross' (1956) evidence that respiration rates, which rose as stress was imposed, progressively declined through the restitution phase.

Overall, it is probable that the integrity of specific protein-water structures, and of the entire cytoplasm, is essential for the continuance of most physiological processes at maximum rates. Most processes are probably not unduly suppressed by the degree of stress which exists diurnally in well watered plants, but as soil water stress increases key processes will become progressively inactivated. Almost certainly the reactions of different species or varieties to water stress are due largely to different degrees of tolerance of specific metabolic systems, as suggested by Langridge (1963) for heat tolerance.

B. Effect of Water Deficits on Cell Enlargement

Although any factor which affects cell metabolism must affect cell enlargement and plant growth, some effects of water deficits on plants appear to be more directly mediated by turgor pressure, P.

Direct effects on cell turgor can be recognized in two important phenomena: stomatal closure and cell enlargement. As pointed out in Chapter 8, the state of guard cell turgor can be rather complicated, but it still directly regulates stomatal aperture and, as a result, ultimately influences both transpiration and photosynthesis. Complete, or almost complete, closure can therefore markedly reduce both processes and so ultimately reduces growth. For this reason it has sometimes been suggested that many of the detrimental effects of water deficits, at moderate stress levels, can be attributed to stomatal closure, particularly since the energy balance of the leaf is also affected and leaf temperature may rise to damaging levels.

The effect of turgor pressure on cell enlargement also has many indirect implications, since reduced cell size results ultimately in reduction in leaf area and hence in effective photosynthetic surface. The relationship has been investigated by a number of workers (see, for example, Broyer, 1950; Burström, 1953b; Ordin, Applewhite and Bonner, 1956; Ordin, 1958, 1960; Probine and Preston, 1962; Brouwer, 1963) and in general a close relationship between P and enlargement (or elongation) has been demonstrated.

Ordin (1958, 1960) investigated the mode of action of water stress on the relation between cell enlargement and turgor pressure in some detail, attempting to determine whether it depended on the direct influence of reduced P or on the concomitant increase in π. By inducing water stress with relatively non-permeating (mannitol) and relatively permeating (sodium chloride) solutes he was able to impose the same water deficit, and Ψ, with different internal values of π and P. It was argued that, if reduction in P was the main factor affecting enlargement, the mannitol treatment should give the greatest suppression. The experimental results, obtained with *Avena* coleoptiles (Ordin, 1960), supported this contention. The data also indicate that increased π did not suppress the metabolism of non-cellulosic poly-saccharides measured by incorporation of C^{14} from applied glucose C^{14} although cellulose formation was suppressed. Reduced P, however, affected both cell wall metabolism and elongation, and because of the close association between these factors, Ordin proposed that some aspect of cellulose synthesis may have been involved in the elongation response to reduced turgor. Subsequently Plaut and Ordin (1961) obtained similar results with entire leaves of sunflower and almond when stress was imposed either by desiccation or osmotic incubation.

Data of other workers, such as those of Wadleigh and Gauch (1948), also show a close relationship between P and leaf enlargement (see Fig. 9.7). With cotton leaves, these workers found a progressive decline in rate of cotton leaf elongation, with increasing soil moisture stress, until elongation ceased at approximately zero turgor pressure.

In non-vacuolated meristimatic tissue, such as in rapidly dividing root and shoot tips and developing fruiting bodies, it is apparent that turgor pressure does not control the rate of enlargement and, in such cases, direct hydration effects are presumably operative, probably acting not only on enlargement *per se* but also on cell division. Because this situation is more complicated experimental data tend to be more conflicting although studies with fruit, already reported (Bartholomew, 1926; Hendrickson and Veihmeyer, 1941) indicate reproduceable

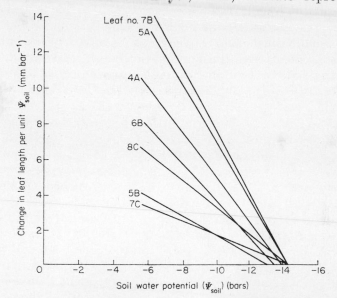

FIG. 9.7. Rate of change of length of cotton leaves in response to soil water stress. The intercept, which indicates the value of Ψ_{soil} at which growth ceased, is close to the value at zero turgor pressure (after Wadleigh and Gauch, 1948).

diurnal rhythms in expansion very closely related to changes in leaf turgor. However, the Ψ values first associated with reductions in expansion rates of non-vacuolated or partly vacuolated tissues vary considerably, as do the values associated with cessation of elongation. Thus Balls (1908) with cotton and Loomis (1934) with corn have shown the extreme sensitivity of shoot elongation to water deficits, complete inhibition occurring during periods of active transpiration, whereas with tomato, Wilson (1948) and Slatyer (1957b) have shown elongation continuing even when the leaf tissue was completely wilted and very low values of Ψ_{leaf} prevailed.

C. Effects of Water Deficits on Cell Division and Nitrogen Metabolism

In general, cell division appears less influenced by water deficits than cell elongation, but the data on this point also tend to be rather conflicting. Reasonable evidence for this view is given by the observation that cell number is frequently of the same general order in plants exposed to water stress compared with controls, although cell size is greater in the latter (Maximov, 1929) and by the phenomenon of more rapid relative growth rates in stressed plants, following re-watering, compared with controls (Petrie and Arthur, 1943; Gates, 1955a, 1955b). This could result from a requirement for expansion alone, rather than division and expansion, after removal of stress.

Gardner and Nieman (1964) have recently shown that rate of increase in DNA content of cotyledonary leaves of radish is reduced by about 60% by a reduction in leaf water potential from $\Psi = 0$ to $\Psi = -2$ bars and, at the stage of $P = 0$ ($\Psi = -8$ bars), the rate is only about 20%

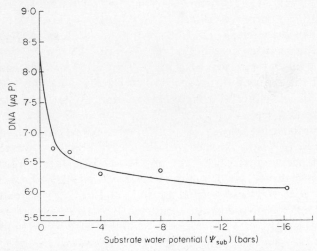

FIG. 9.8. DNA content of radish leaves (expressed as μg of phosphorus per 20 cotyledonary leaves) after 28 hr incubation in a range of mannitol solutions of different water potential (Ψ_{sub}). The dashed line indicates DNA content prior to incubation (after Gardner and Nieman, 1964).

of the control (see Fig. 9.8). However as desiccation proceeded to $\Psi = -16$ bars, the rate of increase in DNA was not further affected to any marked extent. Since cell number is linearly related to DNA content (Nieman and Poulson, 1962) this suggests that the initial reduction in rate of DNA replication is associated with a proportional decrease in rate of cell division, and it is apparent that this was much greater, relatively, than the corresponding decline in turgor pressure ($\cong 25\%$). Subsequently, however, cell division continued although at a slow rate while $P \to 0$ and, even at Ψ values substantially lower than those at zero turgor, the rate was still about 20% of the control.

While these data are not necessarily inconsistent with the general view given above, it can be appreciated that, depending on the degree of stress, the relative effects of water deficits on cell division can be greater or less than on turgor pressure and cell enlargement.

The effect of water stress on nucleic acid synthesis also has been investigated by Gates and Bonner (1959). With whole tomato plants exposed to a period of soil water stress which reduced Ψ_{soil} almost to the permanent wilting percentage, they observed a reduced rate of increase in RNA and then a decline in total RNA, as water stress increased towards wilting. At the end of the 7-day period both total RNA and DNA levels were at levels close to the original ones. When P^{32} was applied to the leaves of the wilted and control plants, however, it was shown that RNA became labelled in both treatments to about the same extent, so that the failure of the stressed plants to accumulate RNA appeared to result not from failure to synthesize new material but due to accelerated destruction of existing RNA. Again, these experiments may be interpreted to suggest continued cell division during stress, and the reduced apparent rate of DNA replication observed by Gardner and Nieman (1964) may also be due in part to destruction of pre-existing material.

Another relevant study dealing with nitrogen metabolism and water stress is that of Chen, Kessler and Monselise (1964) in which changes in total nitrogen, protein nitrogen and amino acid levels were followed in drought-tolerant and drought-sensitive species of citrus seedlings. The plant material was rooted in soil, and soil water content was reduced from $\Psi \cong 0$ to a stage significantly drier than the permanent wilting percentage over a period of about 10 days. These workers considered the observed changes followed a three stage pattern, which overlapped with Stocker's (1960) reaction and restitution phases. In their first activation stage, which occurred in the soil water range above the permanent wilting percentage, there was a rapid drop in water content, but a slight increase in protein level, perhaps associated with continued RNA synthesis as observed by Gates and Bonner (1959) and Kessler (1961). In the second phase, which occurred in the wilting range, it was thought that significant protein hydrolysis took place, as observed by Petrie and Wood (1938a, 1938b), Maximov (1941), Chibnall (1954) and Mothes (1956). Finally, an apparent increase in protein levels occurred under extreme drought conditions but it was thought this may have reflected an increase of peptides rather than true protein synthesis.

It seems probable that, if stress is gradually imposed over a period of days, slow changes in protein and protoplasmic structure and function result in Stocker's (1960) reaction and restitution phases merging into a general response to stress which is associated with a gradual inactivation of nucleic acid synthesis and an accelerated breakdown of metabolites in less active cells and tissues. As stress becomes more severe, synthesis is probably completely inhibited, and protein breakdown leads to accelerated migration of nitrogen and phosphorus compounds from laminae to stems, as observed by Gates (1957).

D. Effect of Water Deficits on Photosynthesis

Photosynthesis is dependent on three main groups of processes: diffusive processes associated with the supply of CO_2 to the photosynthetic sites, photochemical processes associated with the utilization of light energy for photosynthetic purposes and "dark" chemical processes associated with the chemical reduction of CO_2. A fourth group, associated with the transport of photosynthate away from the active sites may also be included.

It is apparent that not only can any one of these processes limit photosynthesis, but all can be affected by water stress. Furthermore, the net rate of assimilation normally identified with plant growth is determined not by photosynthesis alone but also by respiration which may be influenced differently by water stress. Consequently, it is not surprising that the effects of water deficits on photosynthesis are still not fully understood even though a number of experiments have been conducted in this general subject area (see for example, Verduin and Loomis, 1944; Thomas and Hill, 1949; Ashton, 1956; Brix, 1962; Slavik, 1963a; El-Sharkaway and Hesketh, 1964).

In general, reductions in apparent photosynthesis commence at water potentials close to zero ($\Psi = -1$ to -3 bars) and decline more or less linearly with turgor pressure to a value of zero at approximately $P = 0$, thereafter becoming negative as respiration exceeds photosynthesis (see Fig. 9.9) (Schneider and Childers, 1941; Loustalot, 1945; Bordeau, 1954; Brix, 1962). Occasionally a slight increase in apparent photosynthesis is observed between $\Psi = 0$ and the point at which the first detectable decline is observed (Bordeau, 1954; Brix, 1962) but the reasons advanced for this, which include enhanced CO_2 transfer following a small increase in stomatal aperture, are not fully satisfactory.

Two main modes of action of water deficits on photosynthesis can be recognized. In the first place, stomatal closure and reduced rates of CO_2 exchange can influence the supply of CO_2. Secondly, a direct effect of the water deficit on the biochemical processes involved in photosynthesis can be expected.

In the past, much emphasis has been placed on the former phenomenon, since a close parallelism is frequently observed between transpiration and photosynthesis, as water stress is imposed (see for example, Kramer, 1949) and it has been assumed that stomatal closure should affect both processes to more or less the same degree. Typical

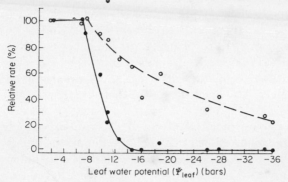

FIG. 9.9. Effect of decreasing leaf water potential on relative rates of photosynthesis (●) and respiration (○) of tomato plants (after Brix, 1962).

results of this type are given in Fig. 9.10 after Brix (1962) using the same data as in Fig. 9.9. However, although photosynthesis involves CO_2 transport across the same external and internal diffusive resistances as water vapour in the transpiration pathway (corrected for differences in the diffusion coefficient for CO_2 in air, rather than for water vapour) it also involves an additional resistance in series, associated with liquid phase diffusion of CO_2 from the surfaces of the mesophyll cell walls to the photosynthesizing sites in the chloroplasts (Penman and Schofield, 1951; Gaastra, 1959; Bierhuizen and Slatyer, 1964b).

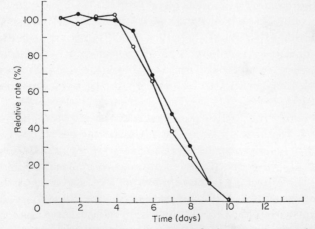

FIG. 9.10. Decline in relative rates of photosynthesis (●) and transpiration (○) of tomato plants, with decreasing soil water, over a period of 10 days (after Brix, 1962).

This resistance generally constitutes a substantial proportion of the total diffusive resistance for CO_2 when the stomata are not closed. Consequently, even if CO_2 supply is limiting photosynthesis, a specified degree of stomatal closure can be expected to influence water vapour transport to a greater extent than CO_2 transport and, if CO_2 supply is not limiting, no effect will be observed until limitations commence. Therefore the parallelism, rather than indicating a limitation of CO_2 supply, may indicate a relationship between the turgor operated stomatal mechanism in its effect on transpiration and a turgor, or hydration effect on biochemical rather than physical aspects of photosynthesis.

Under conditions of moderate-high light intensity, CO_2 frequently appears to limit photosynthesis, when plants are not under water stress. This is shown by experiments which indicate a linear increase in photosynthesis with increasing bulk CO_2 levels (Gaastra, 1959; Brix, 1962; Bierhuizen and Slatyer, 1964b). As stress is imposed, however, other factors become progressively more important. Some of the most convincing demonstrations of direct hydration effects are those where photosynthesis has been measured with aquatic species in water culture to which various amounts of osmotic solutes have been added, when a close relationship between water potential and photosynthesis has been demonstrated, although photosynthesis may not cease until the water potential reaches a level significantly lower than that prevailing at zero turgor pressure (Walter, 1929; Greenfield, 1942).

Other experiments with lower plants, not containing stomata, show similar effects (Stocker and Holdheide, 1937; Einsgraber, 1954), although the assumption of a direct effect of hydration on photosynthesis requires that epidermal permeability to CO_2 is unaffected by water content. With higher plants, Slavik (1963a) concluded from measurements on tobacco leaves that greater rates of photosynthesis in the leaf base compared with the apex are caused primarily by increased hydration, expressed via lower osmotic pressures in the basal cells.

The effects of water deficits on respiration rates can result in significant changes in rates of apparent photosynthesis, since the relative significance of even a constant respiration rate increases rapidly as photosynthesis is reduced by water stress. Furthermore, recent evidence of significantly higher respiration rates under light than under dark conditions (Moss, 1966) suggests that the ratio of respiration to photosynthesis is likely to be much higher than the values of 0·1–0·2 quoted by Thomas (1955), at least for some species.

As water stress is imposed, an increase in respiration rate may first be observed, followed by a reduction in rate as the plant adapts to stress. These two phases may correspond to Stocker's (1960) reaction and restitution phases. If stress is imposed gradually, the first phase may not become apparent, a progressive decline in respiration rate with increasing stress being observed. If stress is imposed rapidly, however, the initial increase can be quite marked, as in the case of diurnal wilting which is associated with a rapid increase in intercellular CO_2 concentration (Heath and Meidner, 1961). The respiration responses of different species as stress is imposed (Brix, 1962) are probably due in part to differences in rate of stress imposition.

E. Sequential Effects of Water Deficits on Plant Growth

From the foregoing sections dealing with more specific effects of water deficits, it is now appropriate to indicate a possible sequence of events associated with stress imposition as it affects plant growth. This account is almost certainly oversimplified; it is presented with the objective of stimulating interest in this important area of research.

In a typical plant, growing with adequate water and nutrient supply and normal transpiration rates, only small diurnal water deficits will arise. Under these conditions, it seems probable that the only observable effects of water stress will be a reduced rate of cell division in those cells where the maximum deficit is experienced (Gardner and Nieman, 1964) and a reduced rate of elongation of the shoot apices (Balls, 1912; Loomis, 1934). Growth of the whole plant should proceed virtually unimpeded, since both the degree and duration of the water deficit will be restricted to a few hours.

As the soil dries, however, and a base level of stress is imposed by Ψ_{soil} in addition to the superimposed diurnal lag of absorption behind transpiration, there will be an accelerated breakdown of RNA and probably DNA, even though RNA synthesis may continue and cell division itself may only decline at a relatively slow rate (Gates and Bonner, 1959). Initially, there will be a tendency for suppression of metabolism to occur only during the diurnal period of maximum water deficit, but this period will become longer each day. Stomatal closure at this time will retard transpiration and increase leaf temperature, and may cause reduction in photosynthesis through its effect on CO_2 exchange. Apparent photosynthesis may also be influenced by the reduction in turgor and by the increased respiration which can be expected during the reaction phase of stress imposition (Stocker, 1960; Brix, 1962). Reductions in rate of cell enlargement, and hence in the expansion of leaf area, will progressively increase as cell turgor falls. All these factors will cause a gradual falling off in growth rates.

As stress becomes more severe, and Ψ_{soil} approaches the permanent wilting percentage, the superimposed diurnal water deficit will become less important, Ψ_{plant} will be dominated by Ψ_{soil} and turgor pressure will approach zero (Slatyer, 1957a). The rate of cell division will probably be markedly reduced and cell enlargement will virtually cease (Gardner and Nieman, 1964). The stomata will be closed for most of the day with transpiration reduced to about the rate for cuticular transpiration. There will be an associated and marked increase in leaf temperature. As most metabolic processes slow down, respiration will gradually decrease. Despite this decrease, apparent photosynthesis will also fall to values close to zero, because of impeded CO_2 transport and the direct effects of hydration on the photosynthetic reactions themselves. Overall plant growth rates (expressed as rate of change in dry matter) will approach zero.

Associated with the carbohydrate and protein breakdown involved in the disruption of normal cell metabolism will be an accelerated migration of soluble leaf phosphorus and nitrogen compounds, particularly from the older leaves, to the stem (Gates, 1957). There may also be an increase in the concentration of sugars and other osmotically active carbohydrate breakdown products, but this will depend to some degree on respiration rates and the total amounts of carbohydrate available (Spoehr and Milner, 1939).

As desiccation continues, cell division and elongation also cease and there will be a progressive loss of dry weight through continued respiration, overall growth rates becoming negative. Finally, the degree of protoplasmic dehydration reaches critical levels and individual cells and tissues die. Generally speaking, if stress is imposed slowly death occurs first in the older leaves but if it is rapid the youngest leaves with lowest Ψ values may die first. Root hairs also appear to die at relatively low stress levels and marked root suberization develops as desiccation proceeds (see Kramer and Kozlowski, 1961). In some species and situations the tops die out completely before the roots but in other cases there is evidence that most of the roots die before the shoots.

If the water supply is renewed before death occurs, the recovery to normal metabolic behaviour appears to take several days, even in situations where only a single brief period of desiccation to the permanent wilting percentage has occurred (Gates, 1955a, 1955b). In many cases recovery never appears to be complete, as is discussed further below. There is evidence that recovery is delayed firstly by the marked reduction in rates of water absorption caused by death of root hairs, or roots, and increased suberization of the root system, which reduce the permeability of the root to water and nutrients. Although these features have been considered in Chapter 7, the data of Brix (1962) showing that time for recovery of photosynthesis in a wilted Loblolly pine seedling could be reduced from 50 to about 10 hr by cutting off the root system and placing the cut stem in water, reflects this phenomenon particularly well. However, Brix also observed that in wilted tomato plants full recovery did not occur, even though turgor was regained.

It is not surprising that, even after turgor recovery has occurred and leaf expansion has been resumed, normal metabolism, including cell division and photosynthesis, takes a considerable time to become re-established, because water stress has caused complete dislocation of the growth system as indicated by migration of phosphorus and nitrogen out of the leaves. Stomatal function, also, may take some time to recover and, in many cases, it is probable that the increased rate of senescence in the previously active leaves is associated with a partial loss of stomatal function (Slatyer and Bierhuizen, 1964b).

It is thought that observations of more rapid relative growth rates after a period of water stress than before (Gates, 1955a; Owen, 1958) (see Fig. 9.11) can be reconciled with this view, since phosphorus and nitrogen migration is least pronounced in the most active and developing leaves and in meristematic tissues. On removal of stress, these tissues tend to expand rapidly and further cell division may be stimulated by the increased availability of nitrogen and phosphorus compounds in the vascular system, even though some of the older leaves may not recover and some of the previously active leaves may have senesced more rapidly than under normal conditions.

The stage of growth at which water stress occurs can exert important influences on the final yield of some crop plants, particularly in determinate species such as annual cereals where water stress at the time of fruit set may result in an inevitable yield decrement. With corn, for example, Robins and Domingo (1953) found that maximum yield was reduced by water stress at the tasseling stage and Denmead and Shaw (1960) found a reduction of about 50% was caused by water stress at the silking stage. For wheat, somewhat similar tendencies occur but Watson (1952), Asana and Saini (1958) and others have shown that continued active photosynthesis after fruit set is also a most important determinant of final yield. The effect of water stress, or of sudden changes in internal water status, on boll set in cotton is a widely recognized phenomenon (Eaton, 1955) and most other crops have periods in their life cycle when they exhibit pronounced sensitivity to stress.

In summary, water stress appears to cause significant and progressive decrements in most processes concerned with plant growth as soon as

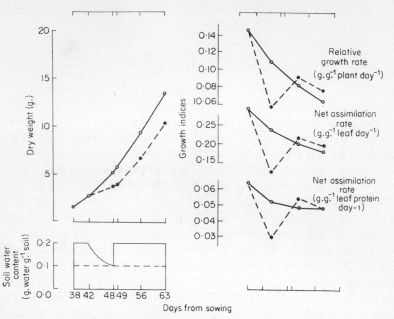

FIG. 9.11. Changes in dry weight, soil water content, relative growth rate, net assimilation rate on a lamina dry weight basis, and net assimilation rate on a lamina protein basis, of control tomato plants (O) and those which were briefly stressed to the permanent wilting percentage (●) (after Gates, 1955a).

water potentials drop significantly below fully turgid values. Many processes appear to decline more or less proportionally with decreases in turgor pressure, Even though turgor pressure itself may only directly affect rate of cell enlargement, the stage of zero turgor pressure, where water in the cell passes from a state of positive pressure to one of tension, may be of considerable significance. Also, indirect effects of decreasing turgor pressure associated with stomatal closure, reduced transpiration rate, and impeded CO_2 uptake are of importance, particularly since almost complete stomatal closure is often associated with zero turgor. It is not surprising therefore that most processes concerned with plant growth cease at about the Ψ_{leaf} value associated with zero turgor pressure in the cells of the active leaf tissue. This again highlights the importance of permanent wilting as a phenomenon of considerable importance in plant growth and production, and demonstrates the significance of the permanent wilting as a lower limit of soil water availability for growth in any one plant, even though the water potential at which it occurs may vary widely from plant to plant. Again it should be emphasized that transpiration continues, although at a slow rate, beyond this stage, continuing to desiccate the plant until death occurs.

F. *Availability of Soil Water for Plant Growth*

The foregoing sections of this chapter have shown clearly that even small internal water deficits can be expected to reduce plant growth and that, as they increase, progressive reductions in growth can be expected until positive growth ceases at a value of Ψ_{leaf} approximately equivalent to the state of zero turgor pressure in the leaf tissue. However, it has also been emphasized (see also Chapters 4 and 7) that the degree to which internal water deficits develop in the leaves depends on the evaporative demand, on the soil water potential at the root surfaces, Ψ_{root}, and on the gradients of water potential within the plant. In turn these gradients depend on the degree of stomatal closure, and Ψ_{root} depends on the volume of soil per unit length of root, the bulk value of Ψ_{soil}, and the hydraulic conductivity of the soil.

For crops with high root density and large root zone, small values of $\Psi_{soil} - \Psi_{leaf}$ are adequate to sustain flow, particularly under conditions of low evaporative demand, and under these conditions transpiration may continue at potential rates almost until the permanent wilting percentage is reached (see Figs 7.11, 7.12). For plant growth, however, more rapid growth is generally observed with increasing frequency of irrigation (Salter, 1954, 1958) even under the conditions just described. Of course, maximum growth rates may not necessarily be associated with maximum yield of the economic crop product, or with maximum economic return to the grower, but from the point of view of total physiological yield the foregoing statement is well supported by theory and experiment (Richards and Wadleigh, 1952; Slatyer, 1957a; Hagan and Vaadia, 1960).

Despite this evidence for a progressive decline in soil water availability for plant growth, as a soil dries following water recharge, a considerable amount of controversy existed on this subject until quite recently, with a group of researchers arguing that soil water was equally available for plant growth between field capacity and the permanent wilting percentage. The evidence for "equal availability" was based primarily on field data dealing with deep-rooted tree crops (see for example, Veihmeyer and Hendrickson, 1950; Veihmeyer, 1956) and was confounded with evidence of "equal availability" for transpiration. As has been brought out in this discussion, transpiration may proceed virtually unimpeded until a significant degree and duration of diurnal stomatal closure occurs, and it is not difficult to understand how, particularly under tree crop conditions, this stage may not be reached until the permanent wilting percentage is approached.

Evidence of equal availability for plant growth, however, is lacking, even though it is readily conceded that irrigation at a frequency below that required for maximum growth may be desirable for maximum economic yield. An excellent summary of the experiments on which the hypothesis of "equal availability" for growth was based is given by Richards and Wadleigh (1952). In every case they were able to show clearly that, while transpiration may have been relatively unaffected by significant reductions in Ψ_{soil}, plant growth was progressively and severely inhibited as the permanent wilting percentage was approached.

It is thought that two factors, in particular, may have helped to provide evidence of "equal availability". In the first place, because of the shape of the $\Psi(\theta)$ curve for soil water retention (see Fig. 4.5) much of the water retained at water contents above the permenent wilting percentage (which is in the range $\Psi_{soil} = -10$ to -20 bars for most crop plants) is retained at water contents above $\Psi_{soil} = -2$ bars. As can be appreciated from Chapters 4 and 7, water at these values of Ψ_{soil} flows to the plant under relatively small water potential gradients and only small growth restrictions may occur until Ψ_{soil} falls below this level. Availability therefore decreases most rapidly over the small water content range corresponding to, say, $\Psi_{soil} = -2$ to -10 bars and, with the probable magnitude of errors in soil water content determination (Aitchison, Butler and Gurr, 1951), it can easily be appreciated how evidence of equal availability down to "about" the permanent wilting percentage could be obtained.

Secondly, in any one root zone, there tends to be most rapid extraction of soil water following recharge from the zones of maximum root density nearest to the soil surface. These zones may be dried to values approaching the permanent wilting percentage while, in deeper layers, the soil may still be close to field capacity ($\Psi_{soil} \cong 0$). The zone of maximum absorption then gradually moves downward and outward so that water will continue to be absorbed mainly from moist soil until most of the root zone is depleted (Davis, 1940; Russell and Danielson, 1956). Since water conductivity in the vascular elements is high, the integrated value of Ψ_{xylem} may therefore be very little lower than the values of Ψ_{soil} in the zones of maximum absorption even though much of the root zone may be at soil water contents close to the permanent wilting percentage. In this way a crop with high root density and deep root zone may only be subjected to small internal water deficits as long as part of the root zone is at values of $\Psi_{soil} \cong 0$. Begg, *et al.* (1964) found evidence of this phenomenon in a deeply rooted bulrush millet crop when the surface 2 m of soil was dried to values close to the permanent wilting percentage.

3. EFFECTS OF SALINITY ON INTERNAL WATER DEFICITS AND PLANT GROWTH

It is a matter of general observation that the presence, in the soil or root medium, of salts in concentrations much in excess of those required for normal plant functions, frequently results in some decrement in growth and yield and, when the total concentration of soluble salts reaches osmotically significant levels, this is almost invariably true. In consequence, soil salinity is an important problem in agriculture, although it is generally confined to arid and semi-arid regions where natural leaching by percolating rainfall has not removed excess soluble salts from the root zone. It is of particular significance in irrigation areas where, in the absence of satisfactory water application and drainage procedures, salt accumulation can result from either too little or too much water application. Inadequate rates of application to leach the salts through the profile, combined with the continued evaporation of pure water can result in accumulation even when the quality of the applied water is very high, while excess water can raise the water table until it reaches the root zone.

Detailed accounts of soil salinity, as both a physiological and agronomic problem, can be found in many specific reviews and monographs, including those of Richards (1954), Bernstein and Hayward (1958), Hayward and Bernstein (1958) and Bernstein (1962). In the present instance attention will be confined to the mode of action of soil or substrate salinity in its effect on plant responses, particularly those mediated by internal water deficits.

A. Effects of Salinity on Plant Growth

The general pattern of plant' response to the imposition of saline substrates of sufficient concentration to significantly reduce Ψ_{sub} is one of growth suppression more or less in proportion to the solute concentration, but the degree of suppression varies with different species and with different commercial strains within a species. A good example of this phenomenon is shown in Fig. 9.12 from an experiment of Greenway (1962) in which a relatively salt-tolerant and relatively sensitive variety of barley were compared. NaCl was mixed into sand cultures at concentrations of 125 and 250 milliequivalents per litre of culture solution (approximately equal to 6 to 12 bars osmotic pressure). The plants were irrigated with the culture solution at frequent intervals each day. The results shown in Fig. 9.12 indicate that not only was total dry weight severely depressed in each treatment, and differentially depressed in the two varieties, but that all the components of grain yield were also depressed—ears per plant, grains per ear, and weight per grain.

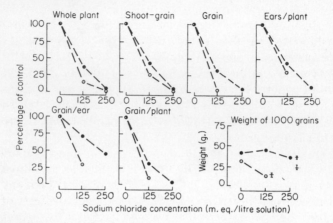

Fig. 9.12. Effects of two levels of substrate salinity (125 and 250 m equiv NaCl per litre) on total growth and yield components of a relatively salt-tolerant (●) (var. Bolivia) and a relatively salt-sensitive (○) (var. Chevron) variety of Barley (after Greenway, 1962).

Similar results have been obtained by other workers (see for example, Bernstein, 1962) even when halophytes have been used (Black 1956, 1960). In general, the degree of growth suppression has been more pronounced when non-electrolyte substrates containing slowly permeating solutes have been used, than with rapidly permeating electrolyte solutes (Slatyer, 1961; Jarvis and Jarvis, 1963a).

The effect of salinity on the yield of marketable product is not necessarily influenced to the same degree as total physiological yield although, as a first approximation, simple proportionality may often exist. With vegetable crops, since succulence (fresh weight/dry weight) may be increased by salinity, moderate reductions in yield of dry matter may be compensated for, to a considerable extent, when the economic yield is measured as fresh weight. This can be expected to be particularly true of species such as beet which are relatively salt-tolerant (Magistad *et al.*, 1943). An anomaly also exists in the case of guayule, in which rubber accumulation is enhanced by moderate degrees of water stress, although it is depressed as stress becomes severe (Wadleigh, Gauch and Magistad, 1946). Much information on these phenomena is given by Hayward and Bernstein (1958).

The recorded effects of salinity on transpiration are much more variable, different investigators frequently obtaining conflicting evidence (Eaton, 1941; Slatyer, 1961; Jarvis and Jarvis, 1963b; Brouwer, 1963). This is almost certainly caused by differences in stomatal behaviour in different experiments and treatments. If internal osmotic adjustment, following treatment imposition, leads to turgor recovery in the experimental plants, the stomata are likely to behave in generally similar fashion to control plants (Boyer, 1965), whereas if there is incomplete osmotic adjustment or if secondary effects on, say, root permeability occur, reduced turgor may well result in different stomatal behaviour. In general, apart from an initial reduction in transpiration rate when a saline substrate is first imposed (Slatyer, 1961), the effects of similar substrate concentrations on transpiration are much less pronounced than on growth.

Various modes of action of salinity in affecting plant responses have been advanced from time to time. These include a reduced water availability in the substrate, that is a "physiological drought" in the sense that effects of salinity on Ψ_{sub} may be similar to effects of low soil water content, and an excessive ion accumulation in the plant tissues, possibly combined with a reduced uptake of essential mineral elements.

The effects of salinity on internal water deficits will now be considered, followed by a brief discussion on the effects of salinity on physiological processes.

B. *Effects of Salinity on Internal Water Deficits*

In evaluating the effects of salinity on plant response the view has been developed, particularly by Wadleigh and Ayers (1945), Wadleigh (1946) and Richards and Wadleigh (1952) that the effect of the osmotic pressure of the soil solution, π_{soil}, is identical with that caused by an equivalent value of matric pressure, τ_{soil}, due to reduced soil water content, and also that the two influences are additive in their effect on the availability of soil water for plants. This is to be expected from the thermodynamic considerations outlined in Chapter 3, since both the osmotic solutes and the soil matrix reduce and, in the absence of other component potentials, determine the soil water potential, Ψ_{soil}.

The concept of a total soil water stress, containing both matric and osmotic forces, has been supported by experimental evidence that the degree of growth inhibition caused by a specified level of Ψ'_{soil} is approximately the same regardless of whether Ψ'_{soil} is composed mainly π_{soil} or mainly of τ_{soil} components (Magistad, Ayers, Wadleigh and Gauch, 1943; Wadleigh and Ayers, 1945; Wadleigh, Gauch and Magistad, 1946). This is illustrated in Fig. 9.13 from Wadleigh and

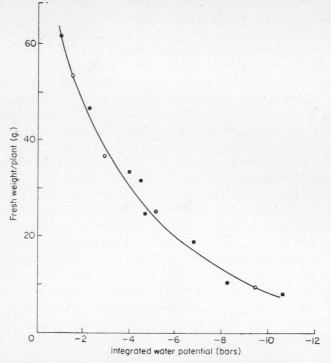

FIG. 9.13. Effects of soil water potential on growth of bean plants. The low (●), medium (○), and high (■) stress treatments involved re-watering at different degrees of soil water depletion. Each treatment was divided into four sub-treatments receiving nil, 0·1, 0·2 and 0·3 g NaCl per 100 g soil (after Wadleigh and Ayers, 1945).

Ayers (1945) in which growth of bean plants was followed for a range of combinations of τ_{soil} and π_{soil} treatments.

However, it is apparent that if there is a significant uptake of the soil solutes into the plant, and a concomitant increase in π_{plant}, the osmotic relations within the plant will be considerably altered. Thus, while the effects of π_{soil} and τ_{soil} may appear similar in their general effects on plant growth, their effects on internal plant water relations can only be expected to be similar if the plant behaves as an ideal osmometer and there is no solute entry. Furthermore, while those experiments just cited appear convincing, it is not possible to maintain a specified level of τ_{soil} (at the levels used in these experiments) and the experimental procedure adopted in these studies was to re-water the plants when τ_{soil} reached the specified limiting values. Thus the average value of Ψ_{soil} in these treatments was always higher than in the experiments where a τ_{soil} component of Ψ_{soil} existed. From this point of view the results of the experiments could be interpreted as showing that the same degree of growth suppression was caused by higher values of π_{soil} than τ_{soil}.

Because of this point, and because it is apparent that there is not only entry of soil solutes as part of normal nutrient uptake but also substantial entry of the excess salts (Eaton, 1927, 1942; Bernstein, 1961, 1963; Slatyer, 1961) it is clear that a revised explanation of the internal osmotic effects of saline soils and substrates is required. As one extreme attitude, Walter (1955) has stated that, since the soil solutes are freely diffusible into the plant system, π_{soil} must be balanced by a compensating increase in π_{plant} so that the turgor pressure, P_{plant}, remains unaffected. Thus he considers that no effect of π_{soil} can exist, and the observable effects on plant response must be due to disturbed nutrient uptake patterns or specific ion toxicities. However, this view cannot be readily reconciled with the observation that iso-osmotic concentrations of various salts and non-electrolyte solutes frequently induce similar effects (Eaton, 1941; Long, 1943; Hayward and Spurr, 1944; Brouwer, 1963).

Another view was advanced by Philip (1958a) and Bonner (1959) who suggested that a vapour gap may develop between root and soil, as both shrink during soil water depletion. It was proposed that initially the Walter (1955) view would hold and the effective soil water stress would be determined by the matric component alone, but that when a vapour gap developed the total soil water stress view would hold, since the vapour zone would constitute an ideal differentially permeable membrane. It has already been mentioned in Chapter 4 that Bernstein, Gardner and Richards (1959) objected to this hypothesis on the basis that even if a vapour gap developed (which they doubted over the normal range of soil water contents), the rates of water flow would be too small to account for observed rates of transpiration. A more serious criticism, however, is that salinity effects are apparent in culture solutions as well as in soils and in culture solutions there is no possibility of vapour gap development.

A more satisfactory explanation appears to have been developed recently following the independent confirmation by Bernstein (1961) and Slatyer (1961) of Eaton's (1927) view that there is effective osmotic adjustment in both roots and shoots to imposed osmotic substrates; that is, that the internal osmotic pressure, π_{plant}, increases by an amount approximately equal to the osmotic pressure of the externally applied solutes, π_{sub}. Slatyer (1961) has shown that, in partial confirmation of Walter's view, the turgor pressure P also returns to its original value but, in opposition to Walter's view, it can be appreciated that the plant is then not fully turgid ($P = \pi_{plant}$), since Ψ_{sub} (or Ψ_{soil}) and hence Ψ_{plant} do not return to their original values (which can be assumed to have been zero in wet soil or water culture) but are reduced by an amount equal to π_{sub}.

Thus some reduction of metabolic function may be attributable to direct osmotic affects on internal water deficits, but it should not be as pronounced as that caused by a similar reduction in Ψ_{sub} caused by a reduction in soil water content. By comparison, if non-permeating solutes are responsible for the decrease in Ψ_{sub} (essentially a laboratory, rather than a field, phenomenon) the effect induced can be expected to be similar to that caused by reductions in soil water content. Relative differences of this type, between iso-osmotic concentrations of permeating and non-permeating solutes have been observed by Slatyer (1961) and Jarvis and Jarvis (1963a).

While this explanation supports the re-interpretation of data such as those of Wadleigh and Ayers (1945) as to matric versus osmotic effects in soils, it describes a short-term response to imposed salinity rather than a mechanism for long-term tolerance. For many plants growing in saline habitats, there is evidence of effective osmotic adjustment in both crop plants (Eaton, 1927, 1942) and halophytes (Black, 1956, 1960) and, under such conditions, it can be expected that the main effects of salinity are associated with ion accumulation in the plant rather than with reduced water availablity in the substrate. Where there is incomplete adjustment however, (Scholander *et al.*, 1962, Greenway and Thomas, 1965) at least some of the effects may be attributable to reduced water availability.

C. Some Effects of Salinity on Metabolism

The presence of electrolytes in the root medium, in what might be termed "osmotic" concentrations, generally results in much enhanced concentrations of electrolytes in the plant, with adverse effects on the ion balance within cells and tissues. The composition of the electrolytes absorbed may broadly reflect that of the substrate, but tolerance to salinity appears to require a high degree of selectivity in ion uptake. Thus van den Berg (1952) concluded that salt tolerance of various agricultural crops was related to regulation of ion uptake and, with special reference to chloride and sodium ions, a number of studies have indicated that tolerance is associated with relatively low rates of absorption (Hayward and Wadleigh, 1949; Ehlig, 1960; Greenway, 1962). As with normal salt uptake, the primary discriminatory barrier appears to be in the root, so that xylem sap concentration of sodium and chloride ions may be very low, even though accumulation may proceed in individual cells in the shoot (Scholander *et al.*, 1962).

The presence of enhanced quantities of electrolytes in plant cells is likely to have direct effects on protein hydration, because of the effect of the ions on the character of the hydration shell surrounding the protein molecules (Klotz, 1958). This in turn is likely to affect protoplasmic viscosity and volume relationships, perhaps leading, in some cases, to the enhanced succulence which may be observed under salinity conditions (Nieman, 1962). Increases in cytoplasmic and vacuolar osmotic pressure, an inevitable result of ion accumulation and osmotic adjustment, may also affect protein hydration. Specific effects on enzyme inactivity may be associated with these phenomena (Laties, 1954; Hackett, 1961).

It is of interest that Kessler *et al.* (1964) reported changes in protein conformation and enzyme activities at low concentrations of sucrose or NaCl, suggesting an explanation for the similar effects sometimes induced by iso-osmotic substrates of varied composition (Gauch and Wadleigh, 1944). However, any such effect can also be expected to be mediated at the molecular level, by intermolecular binding forces, so that different effects should be caused by, say, K^+ ions which may not disturb the ice-like structure of the hydration water, and Na^+ ions, which probably do (Kavanau, 1964).

In consequence, it can be expected that an excess accumulation of electrolytes in plant cells, particularly of ions such as Na^+ or Cl^-, will result in progressive changes in protein hydration and conformation and in enzyme activity, resulting in a gradual dislocation of metabolism. Kessler *et al.* (1964) also reported that salinity strongly suppressed RNA and DNA accumulation, particularly when salinity treatments were first imposed. They observed differential effects on RNA and DNA, since the decrease in RNA was attributed to intensified activity of cytoplasmic RNase, whereas DNA was reduced due to impaired synthesis. Not all enzyme systems show suppressed activity as salinity increases, as indicated by this evidence. Nieman (1962) also considered there may be some increases in enzymatic activity and noted that respiration increased in a wide variety of species as salinity levels were increased.

Increase in respiration rate provides another direct mechanism for growth suppression, affecting net assimilation rate. Increases in respiration can be expected as a result of the energy requirements for selectivity in ion absorption in the presence of high external substrate concentrations, and less tolerant species, with less efficient discriminating mechanisms, can be expected to expend more energy in this function. This pattern was observed by Nieman (1962) in a study of respiration and photosynthesis of a number of species of varying tolerance to salinity. However, his measurements were made on leaf discs in Warburg equipment where the limitations of, for example, stomatal closure on photosynthesis would not be apparent. By comparison, Boyer (1965) obtained evidence of a slow decline in net photosynthesis per unit leaf area, to a value about 30% below that of the controls, in whole cotton plants grown in salinized culture solutions with concentrations ranging up to $\pi_{sub} = 8\cdot 5$ bars. The stomata were reported to be open and the effect was attributed directly to salinity. Although dark respiration did not appear to change significantly, cotton is relatively salt-tolerant and this result is not necessarily at variance with that of Nieman, who did not include cotton among the species he examined.

In summary, the progressive reductions in growth rate caused by increasing salinity appear to be caused primarily by the effect of excess ion accumulation in the affected plants. Direct osmotic effects, acting through reduced water availability to the plants, appear to be of secondary importance, except as initial responses to the imposition of saline substrates. These differences, between soil water stress mediated by water availability and internal water deficits, and salinity stress mediated by excess ion accumulation, are illustrated by experiments dealing with plant response to stress removal. On removal of soil water stress relative growth rates more rapid than those in control plants are frequently observed (Gates, 1955a, 1955b), whereas relative growth rates following salinity treatments are less rapid than in the control (Greenway, 1962). This is probably not only a result of different forms of metabolic inhibition during stress, but to the length of time required, following salinity treatments, for the excess ion accumulation within the plant to be diluted by new growth.

Bibliography

Addoms, R. M. (1946). Entrance of water into suberized roots. *Pl. Physiol.*, *Lancaster* **21**, 109–111.

Ahrens, K. (1939). Bestimmung des Turgordruckes an einer Einzelzelle mit der Manometer. *Planta* **30**, 113–117.

Aitchison, G. D., Butler, P. F. and Gurr, C. G. (1951). Techniques associated with the use of gypsum block soil meters. *Aust. J. appl. Sci.* **2**, 56–75.

Allen, L. H., Yocum, C. S. and Lemon, E. R. (1964). Photosynthesis under field conditions. VII. Radiant energy exchanges within a corn crop canopy and implications in water use efficiency. *Agron. J.* **56**, 253–259.

Allerup, S. (1961). Stem cutting and water movement in young barley plants. *Physiologia Pl.* **14**, 632–637.

Allyn, R. B. (1942). A calibrated soil probe for measuring field soil moisture. *Soil Sci.* **53**, 273–285.

de T. Alvim, P. (1965). A new type of porometer for measuring stomatal opening and its use in irrigation studies. *UNESCO Arid Zone Res.* **25**, 325–329.

van Andel, O. M. (1953). The influence of salts on the exudation of tomato plants. *Acta bot. neerl.* **2**, 445–521.

Anderson, A. B. C. (1943). A method of determining the soil moisture content based on the variation of the electrical capacitance of the soil at low frequency, with moisture content. *Soil Sci.* **56**, 29–41.

Anderson, A. B. C. and Edlefson, N. E. (1942). The electrical capacity of the 2-electrode plaster of paris block as an indicator of soil moisture. *Soil Sci.* **54**, 35–46.

Anderson, D. B. and Kerr, T. (1943). A note on the growth behaviour of cotton bolls. *Pl. Physiol.*, *Lancaster* **18**, 261–269.

Andersson, N. E., Hertz, D. H. and Rufelt, H. (1954). A new fast recording hygrometer for plant transpiration measurements. *Physiologia Pl.* **7**, 753–767.

Angus, D. E. and Bielorai, H. (1965). Transpiration reduction by surface films. *Aust. J. agric. Res.* **16**, 107–112.

Arisz, W. H., Helder, R. J. and van Nie, R. (1951). Analysis of the exudation process in tomato plants. *J. exp. Bot.* **2**, 257–297.

Army, T. J. and Ostle, B. (1957). The association between free-water evaporation and evapotranspiration of spring wheat under the prevailing climatic conditions of the plains area of Montana. *Proc. Soil. Sci. Soc. Am.* **21**, 469–472.

Arvidsson, I. (1951). Austrocknungs- und Dürrersistenzverhaltnisse eineger Wasserabsorption durch oberirdische Organe. *Oikos* Supplement I.

Asana, R. D. and Saini, A. D. (1958). Studies in physiological analysis of Yield IV. The influence of soil drought on grain development, photosynthesis surface and water content of wheat. *Physiologia Pl.* **11**, 666–674.

Ashton, F. M. (1956). Effects of a series of cycles of alternating low and high soil water contents on the rate of apparent photosynthesis in sugar cane. *Pl. Physiol.*, *Lancaster* **31**, 266–274.

Aslyng, H. C. (1963). Soil physics terminology. *Int. Soc. Soil Sci.*, Bull. **23**, 1–4.

Ayers, H. D. and Wikramanayake, V. E. A. (1958). The effect of the water storage capacity of the soil on mass infiltration. *Can. J. Soil Sci.* **38**, 44–48.

Aykin, S. (1946). The relations between water permeability and suction potential in living and non-living osmotic systems. *Istamb. Üniv. Fen. Fak. Mecm.* **11**; 271–295.

Babcock, K. L. (1963). Theory of the chemical properties of soil colloidal systems at equilibrium. *Hilgardia* **34**, 417–542.

Babcock, K. L. and Overstreet, R. (1955). The thermodynamics of soil moisture: A new approach. *Soil Sci.* **80**, 257–263.

Babcock, K. L. and Overstreet, R. (1957a). The extra-thermodynamics of soil moisture. *Soil Sci.* **83**, 455–464.

Babcock, K. L. and Overstreet, R. (1957b). A note on the "Buckingham" equation. *Soil Sci.* **84**, 341–343.

Bahrani, B. and Taylor, S. A. (1961). The influence of soil moisture potential and evaporative demand on actual evapotranspiration from an alfalfa field. *Agron. J.* **53**, 233–237.

Balls, W. L. (1908). "The Cotton Plant in Egypt". Macmillan, London.

Bange, G. G. J. (1953). On the quantitative explanation of stomatal transpiration. *Acta bot. neerl.* **2**, 255–297.

Barrs, H. D. (1964). Heat of respiration as a possible cause of error in the estimation by psychrometric methods of water potential in plant tissue. *Nature, Lond.* **203**, 1136–1137.

Barrs, H. D. (1965a). Psychrometric measurement of leaf water potential: lack of error attributable to leaf permeability. *Science, N.Y.* **149**, 63–65.

Barrs, H. D. (1965b). Comparison of water potentials in leaves as measured by two types of thermocouple psychrometer. *Aust. J. biol. Sci.* **18**, 36–52.

Barrs, H. D. and Weatherley, P. E. (1962). A re-examination of the relative turgidity technique for estimating water deficits in leaves. *Aust. J. biol. Sci.* **15**, 413–428.

Bartholomew, E. T. (1926). Internal decline of lemons. III. Water deficits in lemon fruits caused by excessive leaf evaporation. *Am. J. Bot.* **13**, 102–117.

Basler, E., Todd, G. W. and Meyer, R. E. (1961). Effects of moisture stress on absorption, translocation and distribution of 2·4-Dichlorophenoxyacetic acid in bean plants. *Pl. Physiol., Lancaster* **36**, 573–576.

Batchelor, L. D. and Reed, H. S. (1923). The seasonal variation of the soil moisture in a walnut grove in relation to the hygroscopic coefficient. *Calif. Agric. Exp. Sta.* Tech. Paper 10.

van Bavel, C. H. M. (1952). Gaseous diffusion and porosity in porous media. *Soil Sci.* **73**, 91–104.

van Bavel, C. H. M. and Verlinden, F. J. (1956). Agricultural drought in North Carolina. *North Carolina Agric. Exp. Sta.* Tech. Bull. 122.

van Bavel, C. H. M., Underwood, N. and Ragar, S. R. (1957). Transmission of gamma rays by soil and soil densitometry. *Proc. Soil Sci. Soc. Am.* **21**, 588–599.

van Bavel, C. H. M., Fritschen, L. J. and Lewis, W. E. (1963). Transpiration by Sudangrass as an externally controlled process. *Science, N.Y.* **141**, 269–270.

van Bavel, C. H. M., Nakayama, F. S. and Ehrler, W. L. (1965). Measuring transpiration resistance of leaves. *Pl. Physiol., Lancaster* **40**, 535–540.

Beament, J. W. L. (1961). The water relations of insect cuticle. *Biol. Rev.* **36**, 281–320.

Beament, J. W. L. (1965). The active transport of water: evidence, models and mechanisms. *Symp. Soc. exp. Biol.* **19**, 273–298.

Beck, W. A. (1928). Osmotic pressure, osmotic value, and suction tension. *Pl. Physiol., Lancaster* **3**, 413–440.

Begg, J. E. (1965). High photosynthetic efficiency in a low-latitude environment. *Nature, Lond.* **205**, 1025–1026.

Begg, J. E., Bierhuizen, J. F., Lemon, E. R., Misra, D. K., Slatyer, R. O. and Stern, W. R. (1964). Diurnal energy and water exchanges in bulrush millet in an area of high solar radiation. *Agric. Met.* **1**, 294–312.

Belcher, D. J., Cuykendall, T. R., and Sack, H. S. (1950). The measurement of soil moisture and density by neutron and gamma-ray scattering. Tech. Rept. 127, Civil Aeronautics Admin., U.S.A.

Belsky, M. M., Siegenthaler, P. A. and Packer, L. (1965). Evidence for conformational changes in *Euglena* chloroplasts. *Pl. Physiol., Lancaster* **40**, 290–293.

Bennet-Clark, T. A. (1959). Water relations of cells. *In*, "Plant Physiology—A treatise" (F. C. Steward, ed.) Vol. 1, pp. 105–191. Academic Press, New York and London.

Bennet-Clark, T. A. and Bexon, D. (1940). Water relations of plant cells. *New Phytol.* **39**, 337–361.

Bennet-Clark, T. A., Greenwood, A. D., and Barker, J. W. (1936). Water relations and osmotic pressure of plant cells. *New Phytol.* **35**, 277–291.

van den Berg, C. (1952). The influence of absorbed salts on growth and yield of agricultural crops of salty soils. *Versl. Landb., Ned.* **58**, 1–118.

Bernal, J. D. (1965). The structure of water and its biological implications. *Symp. Soc. exp. Biol.* **19**, 17–31.

Bernstein, L. (1961). Osmotic adjustment of plants to saline media. I. Steady state. *Am. J. Bot.* **48**, 909–918.

Bernstein, L. (1962). Salt-affected soils and plants. *UNESCO Arid Zone Res.* **18**, 139–174.

Bernstein, L. (1963). Osmotic adjustment of plants to saline media. II. Dynamic phase. *Am. J. Bot.* **50**, 360–370.

Bernstein, L. and Hayward, H. E. (1958). Physiology of salt tolerance. *A. Rev. Pl. Physiol.* **9**, 25–46.

Bernstein, L., Gardner, W. R. and Richards, L. A. (1959). Is there a vapor gap around plant roots? *Science, N.Y.* **129**, 1750–1753.

Biddulph, O., Nakayama, F. S. and Cory, R. (1961). Transpiration and ascension of calcium. *Pl. Physiol., Lancaster* **36**, 429–436.

Bierhuizen, J. F. and Slatyer, R. O. (1964a). An apparatus for the continuous and simultaneous measurement of photosynthesis and transpiration under controlled environmental conditions. *CSIRO Aust. Div. Land Res.* Tech. Paper No. 24.

Bierhuizen, J. F. and Slatyer, R. O. (1964b). Photosynthesis of cotton leaves under a range of environmental conditions in relation to internal and external diffusive resistances. *Aust. J. biol. Sci.* **17**, 348–359.

Bierhuizen, J. F., Slatyer, R. O. and Rose, C. W. (1965). A porometer for laboratory and field operations. *J. exp. Bot.* **16**, 182–191.

Billings, W. D. and Morris, R. J. (1951). Reflection of visible and infrared radiation from leaves of different ecological groups. *Am. J. Bot.* **38**, 327–331.

Bircumshaw, L. L. and Riddiford, A. C. (1952). Transport control in heterogeneous reactions. *Q. Rev. chem. Soc.* **6**, 157–185.

Björkman, O. and Holmgren, P. (1963). Adaptability of the photosynthetic apparatus to light intensity in ecotypes from exposed and shaded habitats. *Physiologia Pl.* **16**, 889–914.

Black, R. F. (1956). Effect of NaCl in water culture on the ion uptake and growth of *Atriplex hastata* L. *Aust. J. biol. Sci.* **9**, 67–80.

Black, R. F. (1960). Effects of NaCl on the ion uptake and growth of *Atriplex vesicaria* Heward. *Aust. J. biol. Sci.* **13**, 249–266.

Black, J. N. and Watson, D. J. (1960). Photosynthesis and the theory of obtaining high crop yields, by A. A. Ničiporovič. An abstract with commentary. *Fld. Crop Abstr.* **13**, 169–175.

Blaney, H. F. and Criddle, W. D. (1950). Determining water requirements in irrigated areas from climatological and irrigation data. *U.S. Dept. Agric. Soil. Cons. Serv.* Tech. Paper 96.

Bliss, L. C., Kramer, P. J. and Wolf, F. A. (1957). Drought resistance in tobacco. *Tobacco Sci.* **1**, 120–123.

Boaler, S. B. and Hodge, C. A. H. (1962). Vegetation stripes in Somaliland. *J. Ecol.* **50**, 465–474.

Boaler, S. B. and Hodge C. A. H. (1964). Observations on vegetation arcs in the northern region, Somali Republic. *J. Ecol.* **52**, 511–544.

Bodman, G. B. and Colman, E. A. (1944). Moisture and energy conditions during downward entry of water into soils. *Proc. Soil Sci. Soc. Am.* **8**, 116–122.

Bogen, H. J. (1940). Ionenwirkung auf die Permeabilität von *Rhoco discolor*. *Z. Bot.* **36**, 65–106.

Bogen, H. J. (1941). Ionenwirkung auf die Permeabilität von *Gentiana cruciata*. *Planta* **32**, 150–175.

Bolt, G. H. and Frissel, M. J. (1960). Thermodynamics of soil moisture. *Neth. J. agric. Sci.* **8**, 57–78.

Bolt, G. H. and Miller, R. D. (1958). Calculation of total and component potentials of water in soil. *Trans. Am. geophys. Un.* **39**, 917–928.

Bonner, J. (1959). Water transport. *Science, N.Y.* **129**, 447–450.

Bonner, J. (1962). The upper limit of crop yield. *Science, N.Y.* **137**, 11–15.

Bonner, W. and Bonner, J. (1948). The role of carbon dioxide in acid formation by succulent plants. *Am. J. Bot.* **35**, 113–117.

Bonner, J., Bandurski, R. S. and Millerd, A. (1953). Linkage of respiration to auxin-induced water intake. *Physiologia Pl.* **6**, 511–522.

Boon-Long, T. S. (1941). Transpiration as influenced by osmotic concentration and cell permeability. *Am. J. Bot.* **28**, 333–343.

Bordeau, P. F. (1954). Oak seedling ecology determining segregation of species in Piedmont Oak–Hickory Forests. *Ecol. Monogr.* **24**, 297–320.

Bormann, F. H. (1957). Moisture transfer between plants through intertwined root systems. *Pl. Physiol., Lancaster* **32**, 48–55.

Bormann, F. H. (1961), Intraspecific root grafting and the survival of eastern white pine stumps. *Forest Sci.* **7**, 247–256.

Bouyoucos, G. J. (1949). Nylon electrical resistance unit for continuous measurement of soil moisture in the field. *Soil Sci.* **67**, 319–330.

Bouyoucos, G. J. (1952). Improvements in the nylon method of measuring soil moisture in the field. *Agron. J.* **44**, 311–314.

Bouyoucos, G. J.(1954). Electrical resistance methods as finally perfected for making continuous measurement of soil moisture content under field conditions. *Mich. Agric. Exp. Sta.* Quart. Bull. **37**, 132–149.

Bouyoucos, G. J. and Mick, A. H. (1940). An electrical resistance method for the continuous measurement of soil moisture under field conditions. *Mich. Agric. Exp. Sta. Tech. Bull.* **172**.

Boyer, J. S. (1965). Effects of osmotic water stress on metabolic rates of cotton plants with open stomata. *Pl. Physiol., Lancaster* **40**, 229–234.

Boyer, P. D., Lum, F. G., Ballou, G. A., Luck, J. M. and Rice, R. G. (1946). The combination of fatty acids and related compounds with serum albumin I. *J. biol. Chem.* **162**, 181–198.

Brauner, L. (1930). Ueber polare Permeabilität. *Ber. dt. bot. Ges.* **49**, 109–118.

Brauner, L. and Hasman, M. (1952). Weitere Untersuchungen über den Wirkungsmechanismus des Heteroauxins bei der Wasseraufnahme von Pflanzenparenchymen. *Protoplasma* **41**, 302–326.

Breazeale, E. L. and McGeorge, W. T. (1953a). Exudation pressure in roots of tomato plants under humid conditions. *Soil Sci.* **75**, 293–298.

Breazeale, E. L. and McGeorge, W. T. (1953b). Influence of atmospheric humidity on root growth. *Soil Sci.* **76**, 361–365.

Breazeale, E. L., McGeorge, W. T. and Breazeale, J. F. (1950). Moisture absorption by plants from an atmosphere of high humidity. *Pl. Physiol., Lancaster* **25**, 413–419.

Breazeale, E. L., McGeorge, W. T. and Breazeale, J. F. (1951). Water absorption and transpiration by leaves. *Soil Sci.* **72**, 239–244.

Brewig, A. (1936). Regulationserscheinungen bei der Wasseraufnahme und die Wasserleitgeschwindigkeit in *Vicia faba*. *Jb. wiss. Bot.* **82**, 803–828.

Brierley, W. G. (1934). Absorption of water by the foliage of some common fruit species. *Proc. Am. Soc. hort. Sci.* **32**, 277–283.

Brierley, W. G. (1936). Further studies of the absorption of water by red raspberry foliage, and some evidence relative to the movement of water within the plant. *Proc. Am. Soc. hort. Sci.* **34**, 385–388.

Briggs, G. E. (1957). Some aspects of free space in plant tissues. *New Phytol.* **56**, 305–324.

Briggs, G. E., Hope, A. B. and Robertson, R. N. (1961). "Electrolytes and Plant Cells". Blackwell Scientific Publications, Oxford.

Briggs, L. J. and Shantz, H. L. (1912a). The wilting coefficient for different plants and its indirect determination. *U.S. Dept. Agric. Bur. Plant. Ind. Bull.* **230**.

Briggs, L. J. and Shantz, H. L. (1912b). The relative wilting coefficient for different plants. *Bot. Gaz.* **53**, 229–235.

Brix, H. (1962). The effect of water stress on the rates of photosynthesis and respiration in tomato plant and loblolly pine seedlings. *Physiologia Pl.* **15**, 10–20.

Brouwer, R. (1953). Water absorption by the roots of *Vicia faba* at various transpiration strengths. I. Analysis of the uptake and the factors determining it. *Proc. K. ned. Akad. Wet.* **C56**, 106–115.

Brouwer, R. (1954). Water absorption by the roots of *Vicia faba* at various transpiration strengths. III. Changes in water conductivity artificially obtained. *Proc. K. ned. Akad. Wet.* **C57**, 68–80.

Brouwer, R. (1963). The influence of the suction tension of the nutrient solutions on growth, transpiration and diffusion pressure deficit of bean leaves (*Phaseolus vulgaris*). *Acta bot. neerl.* **12**, 248–261.

Brouwer, R. (1965). Water movement across the root. *Symp. Soc. exp. Biol.* **19**, 131–149.

Brown, E. M. (1939). Some effects of temperature on the growth and chemical composition of certain pasture grasses. *Missouri Agric. Exp. Sta. Res. Bull.* **299**.

Brown, H. T. and Escombe, F. (1900). Static diffusion of gases and liquids in relation to the assimilation of carbon and translocation in plants. *Phil. Trans. R. Soc.* B, **193**, 223–291.

Broyer, T. C. (1947). The movement of materials into plants. Part I. Osmosis and the movement of water into plants. *Bot. Rev.* **13**, 1–58.

Broyer, T. C. (1950). Some gross correlations between growth enlargement and the solute and water relations of plants with special emphasis on the relation of turgor pressure to distension of cells. *Plant Physiol., Lancaster* **25**, 420–432.

Broyer, T. C. (1951). Experiments on imbibition and other factors concerned in the water relations of plant tissues. *Am. J. Bot.* **38**, 485–495.

Broyer, T. C. (1952). On volume enlargement and work expenditure by an osmotic system in plants. *Physiol. Plant., Lancaster* **5**, 459–469.

Broyer, T. C. and Hoagland, D. R. (1943). Metabolic activities of roots and their bearing on the relation of upward movement of salts and water in plants. *Am. J. Bot.* **30**, 261–273.

Bruinsma, J. (1958). Studies on the Crassulacean acid metabolism. *Acta bot. neerl.* **7**, 531–590.

Brun, W. A. (1961). Photosynthesis and transpiration from upper and lower surfaces of intact banana leaves. *Plant Physiol., Lancaster* **36**, 399–405.

Buckingham, E. A. (1907). Studies of the movement of soil moisture. *U.S. Dept. Agric.* Bull. **38**.

Budyko, M. (1956). "The Heat Balance of the Earth's Surface". U.S. Dept. Commerce, Office Tech. Services, Washington, D.C.

Buffel, K. (1952). New techniques for comparative permeability studies on the oat coleoptile with reference to the mechanism of auxin action. *Meded. Kn. vlaam. Acad.* **14**, No. 7.

Burbidge, N. T. (1946). Morphology and anatomy of the Western Australian species of Triodia R.Br. to internal anatomy of leaves. *Trans. R. Soc. S. Aust.* **70**, 221–229.

Burström, H. (1953a). Growth and water absorption of *Helianthus* tuber tissue. *Physiologia Pl.* **6**, 685–691.

Burström, H. (1953b). Studies on growth and metabolism of roots. IX. Cell elongation and water absorption. *Physiologia Pl.* **6**, 260–274.

Buswell, A. M. and Rodebush, W. H. (1956). Water. *Scient. Am.* **202**, 1–10.

Butler, P. F. and Prescott, J. A. (1955). Evapotranspiration from wheat and pasture in relation to available moisture. *Aust. J. agric. Res.* **6**, 52–61.

Butt, V. S. and Peel, M. (1963). The participation of glycollate oxidase in glucose uptake by illuminated *Chlorella* suspensions. *Biochem. J.* **88**, (Supplement) 31P.

Campbell, R. B., Bower, C. A. and Richards, L. A. (1949). Change of electrical conductivity with temperature and the relation of osmotic pressure to electrical conductivity and ion concentration for soil extracts. *Proc. Soil Sci. Soc. Am.* **13**, 66–69.

Carr, D. J. and Gaff, D. F. (1962). The role of the cell-wall water in the water relations of leaves. *UNESCO Arid Zone Res.* **16**, 117–125.

Cary, J. W. and Taylor, S. A. (1962). The interaction of the simultaneous diffusions of heat and water vapor. *Proc. Soil Sci. Soc. Am.* **26**, 413–416.

Catsky, J. (1960). Determination of water deficit in disks cut out from leaf blades. *Biologia Pl.* **2**, 76–78.

Catsky, J. (1962). Water saturation deficit in the wilting plant. The preference of young leaves and the translocation of water from old into young leaves. *Biologia Pl.* **4**, 306–314.

Catsky, J. (1963). Water saturation deficit and its development in young and old leaves. *In*, "The Water Relations of Plants" (A. J. Rutter and F. H. Whitehead eds.), pp. 101–112. Blackwell, London.

Chang, H. T. and Loomis, W. E. (1945). The effect of carbon dioxide on absorption of water and nutrients by roots. *Pl. Physiol.*, *Lancaster* **20**, 221–232.

Chapman, H. D. and Parker, E. R. (1942). Weekly absorption of nitrate by young bearing orange trees growing out of doors in solution cultures. *Pl. Physiol.*, *Lancaster* **17**, 366–376.

Chatfield, C. and Adams, G. (1940). Proximate composition of American food materials. *U.S. Dept. Agric. Circ.* **549**.

Chen, D., Kessler, B. and Monselise, S. P. (1964). Studies on water regime and nitrogen metabolism of citrus seedlings grown under water stress. *Pl. Physiol.*, *Lancaster*, **39**, 379–386.

Chibnall, A. C. (1923). A new method for the separate extraction of vacuolar and protoplasmic material from leaf cells. *J. Biol. Chem.* **55**, 333–342.

Chibnall, A. C. (1954). Protein metabolism in rooted runner bean leaves. *New Phytol.* **53**, 31–38.

Childs, E. C. (1957). The physics of land drainage. *In*, "Drainage of Agricultural Lands" (J. N. Luthin, ed.), pp. 1–78. Academic Press, New York and London.

Childs, E. C. and Collis-George, N. (1950). The permeability of porous materials. *Proc. R. Soc.* A. **201**, 392–405.

Chinard, F. P. (1952). Derivation of an expression for the rate of formation of glomerular fluid. *Am. J. Physiol.* **171**, 578–586.

Chow, V. T. (1964). "Handbook of Applied Hydrology". McGraw-Hill, New York.

Christensen, G. N. and Kelsey, K. (1958). The sorption of water vapor by the constituents of wood: determination of sorption isotherms. *Aust. J. appl. Sci.* **9**, 265–282.

Christiansen, J. E. (1944). Effect of entrapped air upon permeability of soils. *Soil Sci.* **58**, 355–365.

Coile, T. S. (1936). Soil samplers. *Soil Sci.* **42**, 139–142.

Collander, R. (1949). Permeability of plant protoplasts to small molecules. *Physiologia Pl.* **2**, 300–311.

Collander, R. (1950). The permeability of *Nitella* cells to rapidly penetrating non-electrolytes. *Physiologia Pl.* **3**, 45–57.

Collis-George, N. (1961). Free energy considerations in the moisture profile at equilibrium and effect of external pressure. *Soil Sci.* **91**, 306–311.

Colman, E. A. (1944). The dependence of field capacity on depth of wetting of field soils. *Soil Sci.* **58**, 43–50.

Colman, E. A. and Bodman, G. B. (1945). Moisture and energy conditions during downward entry of water into moist and layered soils. *Proc. Soil Sci. Soc. Am.* **9**, 3–11.

Colman, E. A. and Hendrix, T. M. (1949). The fiberglas electrical soil-moisture instrument. *Soil Sci.* **67**, 427–438.

Commoner, B., Fogel, S. and Muller, W. H. (1943). The mechanism of auxin action. The effect of auxin on water absorption by potato tuber tissue. *Am. J. Bot.* **30**, 23–28.

Cook, J. A. and Boynton, D. (1952). Some factors affecting the absorption of urea by McIntosh apple leaves. *Proc. Am. Soc. hort. Sci.* **59**, 82–90.

Cowan, I. R. (1965). Transport of water in the soil-plant-atmosphere system. *J. Appl. Ecol.* **2**, 221–239.

Crafts, A. S., Currier, H. B. and Stocking, C. R. (1949). "Water in the Physiology of Plants". Chronica Botanica Company, Waltham, Mass.

Currier, H. B. (1944). Water relations of root cells of *Beta vulgaris. Am. J. Bot.* **31**, 378–387.

Currier, H. B. (1951). Herbicidal properties of benzene and certain methyl derivatives. *Hilgardia* **20**, 383–406.

Curtis, L. C. (1944). The influence of guttation fluids on pesticides. *Phytopathology* **34**, 196–205.

Dainty, J. (1963a). Water relations of plant cells. *Adv. Bot. Res.* **1**, 279–326.

Dainty, J. (1963b). The polar permeability of plant cell membranes to water. *Protoplasma* **57**, 220–228.

Dainty, J. (1965). Osmotic flow. *Symp. Soc. exp. Biol.* **19**, 75–85.

Dainty, J. and Hope, A. B. (1959). The water permeability of cells of *Chara Australis* R.Br. *Aust. J. biol. Sci.* **12**, 136–145.

Dainty, J. and Ginzburg, B. Z. (1964a). The reflection coefficient of plant cell membranes for certain solutes. *Biochim. biophys. Acta* **79**, 129–137.

Dainty, J. and Ginzburg, B. Z. (1964b). The measurement of hydraulic conductivity (osmotic permeability to water) or internodal characean cells by means of transcellular osmosis. *Biochim. biophys. Acta* **79**, 102–111.

Dainty, J., Croghan, P. C. and Fensom, D. S. (1963). Electro-osmosis, with some applications to plant physiology. *Can. J. Bot.* **41**, 953–966.

Danielli, J. F. (1954). Morphological and molecular aspects of active transport. *Symp. Soc. exp. Biol.* **8**, 502–516.

Danielli, J. F. and Davson, H. (1952). The structure of the plasma membrane. *In*, "The Permeability of Natural Membranes" (H. Davson and J. F. Danielli, eds.) pp. 57–71. Cambridge University Press, Cambridge.

Daniels, F. and Alberty, R. A. (1961). "Physical Chemistry". John Wiley and Sons, New York.

Daughters, M. R. and Glenn, D. S. (1946). The role of water in freezing foods. *Refrig. Engng* **52**, 137–140.

Davies, J. T. and Rideal, E. K. (1961). "Interfacial Phenomena". Academic Press, New York and London.

Davis, C. H. (1940). Absorption of soil moisture by maize roots. *Bot. Gaz.* **101**, 791–805.

Davson, H. and Danielli, J. F. (1952). "The Permeability of Natural Membranes". 2nd Ed. Cambridge University Press, Cambridge.

Day, P. R. (1947). The moisture potential of soils. *Soil Sci.* **54**, 391–400.

Deacon, E. L. (1949). Vertical diffusion in the lowest layers of the atmosphere. *Q. Jl. R. met. Soc.* **75**, 89–103.

Deacon, E. L. (1950). The measurement and recording of the heat flux into the soil. *Q. Jl. R. met. Soc.* **76**, 479–483.

Deacon, E. L. and Swinbank, W. C. (1958). Comparison between momentum and water vapour transfer. *UNESCO Arid Zone Res.* **11**, 38–47.

Decker, J. P. and Wien, J. D. (1960). Transpirational surges in Tamarix and Eucalyptus as measured with an infrared gas analyzer. *Pl. Physiol., Lancaster* **35**, 340–343.

Decker, J. P., Gaylor, W. G. and Cole, F. D. (1962). Measuring transpiration of undisturbed Tamarisk shrubs. *Pl. Physiol., Lancaster* **37**, 393–397.

Denbigh, K. G. (1951). "The Thermodynamics of the Steady State". Methuen, London.

Denmead, O. T. (1964). Evaporation sources and apparent diffusivities in a forest canopy. *J. Appl. Meteorol.* **3**, 383–389.

Denmead, O. T. and Shaw, R. H. (1960). The effects of soil moisture stress at different stages of growth on the development of yield of corn. *Agron. J.* **52**, 272–274.

Denmead, O. T. and Shaw, R. H. (1962). Availability of soil water to plants as affected by soil moisture content and meteorological conditions. *Agron. J.* **54**, 385–390.

Dick, D. A. T. (1959). Osmotic properties of living cells. *Int. Rev. Cytol.* **8**, 387–448.

Diebold, C. H. (1954). Effect of tillage practices upon intake rates, run off and soil losses of dry farm-land soils. *Proc. Soil Sci. Soc. Am.* **18**, 88–91.

Dittmer, H. J. (1937). A quantitative study of the roots and root hairs of a winter rye plant (*Secale cereale*). *Am. J. Bot.* **24**, 417–420.

Dixon, H. H. (1914). "Transpiration and the Ascent of Sap in Plants". Macmillan, London.

van Duin, R. H. A. (1955). Tillage in relation to rainfall intensity and infiltration capacity of soils. *Neth. J. agric. Sci.* **3**, 182–191.

Durbin, R. P., Frank, H. and Solomon, A. K. (1956). Water flow through frog gastric mucosa. *J. gen. Physiol.* **39**, 535–551.

Duvdevani, S. (1953). Dew gradients in relation to climate, soil and topography. *In*, "Desert Research", pp. 136–152. Research Council of Israel, Spec. Pub. 2.

Duvdevani, S. (1957). Dew research for arid agriculture. *Discovery, Lond.* **18**, 330–334.

Dyer, A. J. (1961). Measurements of evaporation and heat transfer in the lower atmosphere by an automatic eddy correlation technique. *Q. Jl. R. met. Soc.* **87**, 401–412.

Dzerdzeevskii, B. L. (1958). On some climatological problems and microclimatological studies of arid and semi-arid regions in U.S.S.R. *UNESCO Arid Zone Res.* **16**, 315–325.

Eaton, F. M. (1927). The water-requirement and cell-sap concentration of Australian saltbush and wheat as related to the salinity of the soil. *Am. J. Bot.* **14**, 212–226.

Eaton, F. M. (1941). Water uptake and root growth as influenced by inequalities in the concentration of the substrate. *Pl. Physiol., Lancaster* **16**, 545–564.

Eaton, F. M. (1942). Toxicity and accumulation of chloride and sulfate salts in plants. *J. agric. Res.* **64**, 357–399.

Eaton, F. M. (1943). The osmotic and vitalistic interpretations of exudation. *Am. J. Bot.* **30**, 663–674.

Eaton, F. M. (1955). Physiology of the cotton plant. *A. Rev. Pl. Physiol.* **6**, 299–328.

Eckardt, F. E. (1960). Eco-physiological measuring techniques applied to research on water relations of plants in arid and semi-arid regions. *UNESCO Arid Zone Res.* **15**, 139–171.

Edlefsen, N. E. (1941). Some thermodynamic aspects of the use of soil-moisture by plants. *Trans. Amer. geophys. Un.* **22**, 917–940.

Edlefsen, N. E. and Anderson, A. B. C. (1943). Thermodynamics of soil moisture. *Hilgardia* **15**, 31–298.

Edlefsen, N. E. and Bodman, G. B. (1941). Field measurement of water movement through a silt loam soil. *J. Am. Soc. Agron.* **33**, 713–731.

Ehlig, C. F. (1960). Effect of salinity on four varieties of table grapes grown in sand culture. *Proc. Am. Soc. hort. Sci.* **76**, 323–335.

Ehlig, C. F. (1962). Measurement of energy status of water in plants with a thermocouple psychrometer. *Pl. Physiol., Lancaster* **37**, 288–290.

Ehlig, C. F. and Gardner, W. R. (1964). Relationship between transpiration and the internal water relations of plants. *Agron. J.* **56**, 127–130.

Ehrler, W. L., Nakayama, F. S. and van Bavel, C. H. M. (1965). Cyclic changes in water balance and transpiration of cotton leaves in a steady environment. *Physiologia Pl.* **18**, 766–775.

Eicke, R. (1954). Beitrag zur Frage des Hoftüpfelbaues der Koniferen. *Ber. dt. bot. Ges.* **67**, 213–217.

Eidmann, F. E. (1959). Die Interception in Buchen- und Fichtenbeständen. *Proc. int. Ass. Hydrology* (Hannover Symp.) **1**, 5–25.

Einsgraber, A. (1954). Über den Einfluss der Antrocknung auf die Assimilation und Atmung von Moosen und Flechten. *Flora, Jena* **141**, 432–475.

Eisenzopf, R. (1952). Ionenwirkungen auf die kutikuläre Wasseraufnahme von Koniferen. *Phyton, B. Aires* **4**, 149–159.

Elazari-Volcani, T. (1936). The influence of a partial interruption of the transpiration stream by root pruning and stem incisions on the turgor of citrus trees. *Palest. J. Bot. hort Sci.* **1**, 94–96.

Elrick, D. E. and Tanner, C. B. (1955). Influence of sample pretreatment on soil moisture retention. *Proc. Soil Sci. Soc. Am.* **19**, 279–282.

El-Sharkawy, M. A. and Hesketh, J. D. (1964). Effects of temperature and water deficit on leaf photosynthetic rates of different species. *Crop Sci.* **4**, 514–518.

Emerson, W. W. (1954). Water conduction by severed grass roots. *J. agric. Sci., Camb.* **45**, 241–245.

Epstein, E. (1956a). Mineral nutrition of plants: mechanism of uptake and transport. *A. Rev. Pl. Physiol.* **7**, 1–24.

Epstein, E. (1956b). Ion transport in plants. *Science, N.Y.* **124**, 937.

Esau, K. (1960) "Anatomy of Seed Plants". John Wiley and Sons, New York.

Esau, K. (1965). "Plant Anatomy". John Wiley and Sons, New York.

Evans. H. (1938). Studies on the absorbing surface of sugarcane root systems. I. Method of study with some preliminary results. *Ann. Bot.* (N.S.) **2**, 159–182.

Everett, D. H. and Stone, F. S. (1958). "The Structure and Properties of Porous Materials". Butterworths, London.

Fireman, M. (1957). Salinity and alkali problems in relation to high water tables in soils. *In*, "Drainage of Agricultural Lands" (J. N. Luthin, ed.) pp. 505–513. American Society of Agronomy, Madison, Wisconsin.

Fischbach, P. E. and Duley, F. L. (1950). Intake of water by claypan soils. *Proc. Soil Sci. Soc. Am.* **15**, 404–408.

Fitts, D. D. (1962). "Nonequilibrium Thermodynamics". McGraw-Hill, New York.

Fitzgerald, P. D. and Rickard, D. S. (1960). A comparison of Penman's and Thornthwaite's method of determining soil moisture deficits. *N.Z. Jl. agric. Res.* **3**, 106–112.

Fogg, G. E. (1947). Quantitative studies on the wetting of leaves by water. *Proc. R. Soc.* B **134**, 503–522.

Frank, H. S. (1958). Covalency in the hydrogen bond and the properties of water and ice. *Proc. R. Soc.* A **247**, 481–492.

Frank, H. S. (1963). Questions about water structure. *Proc. natn. Acad. Sci. U.S.A. Res. Council Publ.* **942**, 141–155.

Frank, H. S. and Evans, M. W. (1945). Free volume and entropy in condensed systems. III. Entropy in binary liquid mixtures; partial molal entropy in dilute solutions; structure and thermodynamics in aqueous electrolytes. *J. chem. Phys.* **13**, 507–532.

Frank, H. S. and Wen, W. (1957). Structural aspects of ion-solvent interaction in aqueous solutions: a suggested picture of water structure. *Discuss. Faraday Soc.* **24**, 133–140.

Frey-Wyssling, A. (1946). Zur wasser permeabilität des Protoplasmas. *Experimentia* **2**, 132–137.

Frey-Wyssling, A. (1953). "Submicroscopic Morphology of Protoplasm". Elsevier, Amsterdam.

Fritschen, L. J. and van Wijk, W. R. (1959). Use of an economical thermal transducer as a net radiometer. *Bull. Am. met. Soc.* **40**, 291–294.

Funk, J. P. (1959). Improved polythene-shielded net radiometers. *J. scient. Instrum.* **36**, 267–270.

Furr, J. R. and Reeve, J. O. (1945). The range of soil-moisture percentages through which plants undergo permanent wilting in some soils from semi-arid irrigated areas. *J. agric. Res.* **71**, 149–170.

Gaastra, P. (1959). Photosynthesis of crop plants as influenced by light, carbon dioxide, temperature and stomatal diffusion resistance. *Meded. Landb. Hogesch. Wageningen* **59**, 1–68.

Gaff, D. F. and Carr, D. J. (1961). The quantity of water in the cell wall and its significance. *Aust. J. biol. Sci.* **14**, 299–311.

Gaff, D. F. and Carr, D. J. (1964). An examination of the refractometric method for determining the water potential of plant tissues. *Ann Bot.* (N.S.) **28**, 351–368.

Gaff, D. F., Chambers, T. C. and Markus, K. (1964). Studies of extrafascicular movement of water in the leaf. *Aust. J. biol. Sci.* **17**, 581–586.

Gale, J. (1961). Studies on plant antitranspirants. *Physiologia Plant.* **14**, 777–786.

Gale, J. and Poljakoff-Mayber, A. (1965). Antitranspirants as a research tool for the study of the effects of water stress on plant behaviour. *UNESCO Arid Zone Res.* **25**, 269–274.

Gardner, W. R. (1958). Some steady state solutions of the unsaturated moisture flow equation with application to evaporation from a water table. *Soil. Sci.* **85**, 228–232.

Gardner, W. R. (1960a). Soil water relations in arid and semi-arid conditions. *UNESCO Arid Zone Res.* **15**, 37–61.

Gardner, W. R. (1960b). Dynamic aspects of water availability to plants. *Soil. Sci* **89**, 63–73.

Gardner, W. R. (1965). Dynamic aspects of soil water availability to plants. *A. Rev. Pl. Physiol.* **16**, 323–342.

Gardner, W. R. and Fireman, M. (1958). Laboratory studies of evaporation from soil columns in the presence of a water table. *Soil Sci.* **85**, 244–249.

Gardner, W. R. and Mayhugh, M. S. (1958). Solutions and tests of the diffusion equation for the movement of water in soil. *Proc. Soil Sci. Soc. Am.* **22**, 197–201.

Gardner, W. R. and Ehlig, C. F. (1962). Some observations on the movement of water to plant roots. *Agron. J.* **54**, 453–456.

Gardner, W. R. and Ehlig, C. F. (1963). The influence of soil water on transpiration by plants. *J. geophys. Res.* **68**, 5719–5724.

Gardner, W. R. and Nieman, R. H. (1964). Lower limit of water availability to plants. *Science, N.Y.* **143**, 1460–1462.

Gardner, W. R., Mayhugh, M. S., Goertzen, J. O. and Bower, C. A. (1959). Effect of electrolyte concentration and exchangeable-sodium percentage on diffusivity of water in soils. *Soil Sci.* **88**, 270–274.

Gates, C. T. (1955a). The response of the young tomato plant to a brief period of water shortage. I. The whole plant and its principal parts. *Aust. J. biol. Sci.* **8**, 196–214.

Gates, C. T. (1955b). The response of the young tomato plant to a brief period of water shortage. II. The individual leaves. *Aust. J. biol. Sci.* **8**, 215–230.

Gates, C. T. (1957). The response of the young tomato plant to a brief period of water shortage. III. Drafts in nitrogen and phosphorus. *Aust. J. biol. Sci.* **10**, 125–146.

Gates, C. T. (1964). The effect of water stress on plant growth. *J. Aust. Inst. agric. Sci.* **30**, 3–22.

Gates, C. T. and Bonner, J. (1959). The response of the young tomato plant to a brief period of water shortage. IV. Effects of water stress on the ribonucleic acid metabolism of tomato leaves. *Pl. Physiol., Lancaster* **34**, 49–55.

Gates, D. M. (1962). "Energy Exchange in the Biosphere". Harper and Row, New York.

Gates, D. M. (1963). Leaf temperature and energy exchange. *Arch. Met. Geophys. Bioklim.* B **12**, 321–336.

Gates, D. M. (1964). Leaf temperature and transpiration. *Agron. J.* **56**, 273–277.

Gates, D. M. and Tantraporn, W. (1952). The reflectivity of deciduous trees and herbaceous plants in the infrared to 25 microns. *Science, N.Y.* **115**, 613–616.

Gates, D. M., Tibbals, E. C. and Kreith, F. (1965). Radiation and convection for Ponderosa Pine. *Am. J. Bot.* **52**, 66–71.

Gates, D. M., Keegan, H. J., Schleter, J. C. and Weidner, V. R. (1965). Spectral properties of plants. *Appl. Optics* **4**, 11–20.

Gauch, H. G. and Wadleigh, C. H. (1944). Effects of high salt concentrations on growth of bean plants. *Bot. Gaz.* **105**, 379–387.

Geiger, R. (1957). "The Climate near the Ground". Harvard University Press, Cambridge, Mass.

Gessner, F. (1956). Die Wasseraufnahme durch Blätter und Samen. *In*, "Encyclopaedia of Plant Physiology", Vol. 3 (W. Ruhland, ed.) pp. 215–246. Springer-Verlag, Berlin.

Gibbs, R. D. (1958). Patterns in the seasonal water content of trees. *In*, "The Physiology of Forest Trees" (K. V. Thimann, ed.) pp 43–69. Ronald Press, New York.

Gier, J. L. and Dunkle, R. V. (1951). Total hemispherical radiometers. *Proc. Am. Inst. elect. Engrs.* **70**, 339–343.

Glasstone, J. (1947). "Thermodynamics for Chemists". van Nostrand, Princeton, New Jersey.

Glinka, Z. and Reinhold, L. (1962). Rapid changes in permeability of cell membranes to water brought about by carbon dioxide and oxygen. *Pl. Physiol., Lancaster* **37**, 481–486.

Glinka, Z. and Reinhold, L. (1964). Reversible changes in the hydraulic permeability of plant cell membranes. *Pl. Physiol., Lancaster* **39**, 1043–1050.

Goldacre, R. J. (1952). The folding and unfolding of protein molecules as a basis of osmotic work. *Int. Rev. Cytol.* **1**, 135–164.

Gortner, R. A., Lawrence, J. V. and Harris, J. A. (1916). The extraction of sap from plant tissues by pressure. *Biochem. Bull.* **5**, 139–142.

Granick, S. (1956). Plastid structure, development and inheritance *In*, "Encyclopaedia of Plant Physiology" Vol. 1, (W. Ruhland, ed.) pp. 507–564. Springer-Verlag, Berlin.

Greenfield, S. S. (1942). Inhibitory effects of inorganic compounds on photosynthesis in *Chlorella. Am. J. Bot.* **29**, 121–131.

Greenidge, K. N. (1957). Ascent of sap. *A. Rev. Pl. Physiol.* **8**, 237–256.

Greenway, H. (1962). Plant response to saline substrates. I. Growth and ion uptake of several varieties of *Hordeum* during and after sodium chloride treatment. *Aust. J. biol. Sci.* **15**, 16–38.

Greenway, H. and Thomas, D. A. (1965). Plant response to saline substrates. V. Chloride regulation in the individual organs of *Hordeum vulgare* during treatment with sodium chloride. *Aust. J. biol. Sci.* **18**, 505–524.

Gregory, F. G., Milthorpe, F. L., Pearse, H. L. and Spencer, E. J. (1950). Experimental studies of the factors controlling transpiration. II. The relation between transpiration rate and leaf water content. *J. exp. Bot.* **1**, 15–28.

Grieve, B. J. and Went, F. W. (1965). An electric hygrometer apparatus for measuring water-vapour loss from plants in the field. *UNESCO Arid Zone Res.* **25**, 247–257.

Groenewegen, H. and Mills, J. A. (1960). Uptake of mannitol into the shoots of intact barley plants. *Aust. J. biol. Sci.* **13**, 1–4.

de Groot, S. R. (1952). "Thermodynamics of Irreversible Processes". North-Holland, Amsterdam.

de Groot, S. R. and Mazur, P. (1962). "Non-equilibrium Thermodynamics". North-Holland, Amsterdam.

Grossenbacher, K. A. (1939). Autonomic cycle of rate of exudation of plants. *Am. J. Bot.* **26**, 107–109.

Guest, P. L. and Chapman, H. D. (1949). Investigations on the use of iron sprays, dusts, and soil applications to control iron chlorosis of citrus. *Proc. Am. Soc. hort. Sci.* **54**, 11–21.

Guggenheim, E. A. (1959). "Thermodynamics". North-Holland, Amsterdam.

Guilliermond, A. (1941). "The Cytoplasm of the Plant Cell". Chronica Botanica Co., Waltham, Mass.

Gurr, C. G. (1962). Use of gamma rays in measuring water content and permeability in unsaturated columns of soil. *Soil Sci.* **94**, 224–229.

Gurr, C. G., Marshall, T. J. and Hutton, J. T. (1952). Movement of water in soil due to a temperature gradient. *Soil Sci.* **74**, 335–345.

Hackett, D. P. (1952). The osmotic change during auxin-induced water uptake by potato tissue. *Pl. Physiol., Lancaster* **27**, 279–284.

Hackett, D. P. (1961). Effects of salts on DPNH oxidase activity and structure of sweet potato mitochondria. *Pl. Physiol., Lancaster* **36**, 445–452.

Hackett, D. P. and Thimann, K. V. (1952). The nature of the auxin-induced water uptake by potato tissue. *Am. J. Bot.* **39**, 553–560.

Hackett, D. P. and Thimann, K. V. (1953). The nature of auxin-induced water uptake by potato tissue. II. The relation between respiration and water absorption. *Am. J. Bot.* **40**, 183–188.

Hadley, W. A. and Eisenstadt, R. (1955). Thermally actuated moisture migration in granular media. *Trans. Am. geophys. Un.* **36**, 615–623.

Hagan, R. M. (1949). Autonomic diurnal cycles in the water relations of non-exuding detopped root systems. *Pl. Physiol., Lancaster* **24**, 441–454.

Hagan, R. M. and Vaadia, Y. (1960). Principles of irrigated cropping. *UNESCO Arid Zone Res.* **15**, 215–225.

Haines, F. M. (1952). The absorption of water by leaves in an atmosphere of high humidity. *J. exp. Bot.* **3**, 95–98.

Haines, F. M. (1953). The absorption of water by leaves in fogged air. *J. exp. Bot.* **4**, 106–107.

Haise, H. R. and Kelley, O. J. (1950). Causes of diurnal fluctuations of tensiometers. *Soil Sci.* **70**, 301–313.

Hall, D. M. and Jones, R. L. (1961). Physiological significance of surface wax on leaves. *Nature, Lond.* **191**, 95–96.

Hallaire, M. (1949). Profils hydriques envisagés comme la variation du potentiel capillaire avec la profondeur; nouvelle hypothese sur la diffusion capillaire de l'eau dans le sol. *C. r. hebd. Séanc. Acad. Sci., Paris* **229**, 1361–1363.

Hallaire, M. (1950). Profils hydriques en sols hétérogènes. *C. r. hebd. Séanc. Acad. Sci., Paris* **230**, 462–464.

Hallaire, M. (1955). The capillary movement of water in the soil and vertical variations of moisture under bare and cropped soils. *Sols afr.* **3**, 1–24.

Hamilton, E. L. and Rowe, P. B. (1949). "Rainfall interception by Chaparral vegetation in California". California Department of Natural Resources, Sacramento, California.

Handley, W. R. C. (1939). The effect of prolonged chilling on water movement and radial growth in trees. *Ann. Bot.* (N.S.) **3**, 803–813.

Harbeck, G. E., Kohler, M. A. and Coberg, G. E. (1958). Water loss investigations: Lake Mead studies. *U.S. Geol. Surv. Professional Paper* **298**.

Harrold, L. L. and Dreibelbis, F. R. (1951). Agricultural hydrology as evaluated by monolith lysimeters. *U.S. Dept. Agric. Tech. Bull.* **1050**.

Hayward, H. E. and Spurr, W. B. (1943). Effects of osmotic concentration of substrate on the entry of water into corn roots. *Bot. Gaz.* **105**, 152–164.

Hayward, H. E. and Spurr, W. B. (1944). Effects of iso-osmotic concentrations of inorganic and organic substrates on water entry into corn roots. *Bot. Gaz.* **106**, 131–139.

Hayward, H. E. and Wadleigh, C. H. (1949). Plant growth on saline and alkali soils. *Adv. Agron.* **1**, 1–38.

Hayward, H. E. and Bernstein, L. (1958). Plant-growth relationships on salt-affected soils. *Bot. Rev.* **24**, 584–635.

Hayward, H. E., Blair, W. M. and Skalling, P. E. (1942). Device for measuring entry of water into roots. *Bot. Gaz.* **104**, 152–160.

Heath, O. V. S. (1938). An experimental investigation of the mechanism of stomatal movement, with some preliminary observations upon the response of the guard cells to "shock". *New Phytol.* **37**, 385–395.

Heath, O. V. S. (1941). Experimental studies of the relation between carbon assimilation and stomatal movement. II. The use of the resistance porometer in estimating stomatal aperture and diffusive resistance. *Ann. Bot.* (N.S.) **5**, 455–500.

Heath, O. V. S. (1950). Studies in stomatal behaviour. V. The role of carbon dioxide in the light responses of stomata. Part 1. Investigation of the cause of abnormally wide stomatal opening within porometer cups. *J. exp. Bot.* **1**, 29–61.

Heath, O. V. S. (1959). The water relations of stomatal cells and the mechanisms of stomatal movement. *In*, "Plant Physiology" Vol. 2 (F. C. Steward, ed.) pp. 193–250. Academic Press, New York and London.

Heath, O. V. S. and Russell, J. (1954). Studies of stomatal behaviour. VI. An investigation of the light responses of wheat stomata with attempted elimination of control by the mesophyll. Part 1. Effects of light independent of carbon dioxide. *J. exp. Bot.* **5**, 1–15.

Heath, O. V. S. and Orchard, B. (1957). Mid-day closure of stomata. *Nature, Lond.* **180**, 180–181.

Heath, O. V. S. and Meidner, H. (1961). The influence of water strain on the minimum inter-cellular space carbon dioxide concentration, Γ, and stomatal movements in wheat leaves. *J. exp. Bot.* **12**, 226–242.

Heath, O. V. S. and Mansfield, T. A. (1962). A recording porometer with detachable cups operating on four separate leaves. *Proc. R. Soc.* A **156**, 1–13.

Hendrickson, A. H. and Veihmeyer, F. J. (1929). Irrigation experiments with peaches in California. *Calif. Agric. Exp. Sta. Bull.* **479**.

Hendrickson, A. H. and Veihmeyer, F. J. (1941). Some factors affecting the growth rate of pears. *Proc. Am. Soc. hort. Sci.* **39**, 1–7.

Hendrickson, A. H. and Veihmeyer, F. J. (1945). Permanent wilting percentages of soils obtained from field and laboratory trials. *Pl. Physiol., Lancaster* **20**, 517–539.

Hewlett, J. D. and Kramer, P. J. (1963). The measurement of water deficits in broadleaf plants. *Protoplasma* **57**, 381–391.

Hoagland, D. R. and Broyer, T. C. (1942). Accumulation of salt and permeability in plant cells. *J. gen. Physiol.* **25**, 865–880.

Hodges, T. K. and Vaadia, Y. (1964a). Uptake and transport of radiochloride and tritiated water by various zones of onion roots of different chloride status. *Pl. Physiol., Lancaster* **39**, 104–108.

Hodges, T. K. and Vaadia, Y. (1964b). Chloride uptake and transport in roots of different salt status. *Pl. Physiol., Lancaster* **39**, 109–114.

Hodgson, R. H. (1953). A study of the physiology of mycorrhizal roots on *Pinus taeda*, M.A. Thesis, Duke University, Durham, N.C.

Höfler, K. (1920). Ein Schema fur die osmotische Leistung der Pflanzenzelle. *Ber. dt. bot. Ges.* **38**, 288–298.

Hohn, K. (1954). Untersuchungen über das Wasserdampfaufnahme- und Wasserdampfabgabe- Vermögen höherer Landpflanzen. *Beitr. Biol. Pfl.* **30**, 159–178.

Holmes, J. W. (1955). Water sorption and swelling of clay blocks. *J. Soil Sci.* **6**, 200–208.

Holmes, J. W. (1956). Calibration and field use of the neutron scattering method of measuring soil water content. *Aust. J. appl. Sci.* **7**, 45–58.

Holmes, J. W. (1960). Water balance and the water-table in deep sandy soils of the upper south-east, South Australia. *Aust. J. agric. Res.* **11**, 970–988.

Holmes, J. W. and Turner, K. G. (1958). The measurement of water content of soils by neutron scattering. A portable apparatus for field use. *J. agric. Engng Res.* **3**, 199–204.

Holmes, J. W. and Jenkinson, A. F. (1959). Techniques for using the neutron moisture meter. *J. agric. Engng. Res.* **4**, 100–109.

Holmgren, P., Jarvis, P. G. and Jarvis, M. S. (1965). Resistance to carbon dicxide and water vapour transfer in leaves of different plant species. *Physiologia Plant.* **18**, 557–573.

van den Honert, T. H. (1948). Water transport in plants as a catenary process. *Discuss. Faraday Soc.* **3**, 146–153.

Horrocks, R. L. (1964). Wax and the water vapour permeability of apple cuticle. *Nature, Lond.* **203**, 547.

Horton, R. E. (1940). An approach toward a physical interpretation of infiltration-capacity. *Proc. Soil Sci. Soc. Am.* **5**, 399–417.

House, C. R. and Findlay, N. (1966). Water transport in isolated maize roots. *J. exp. Bot.* (In press)

House, G. J., Rider, N. E. and Tugwell, C. P. (1960). A surface energy-balance computer. *Q. Jl. R. met. Soc.* **86**, 215–231.

Hsu, S. T. (1963). "Engineering Heat Transfer". van Nostrand, Princeton, New Jersey.

Huber, B. (1932). Beobachtung und Messung pflanzlicher Sarströme. *Ber. dt. bot. Ges.* **50**, 89–109.

Huber, B. (1956). Die Gefäßleitung *In*, "Encyclopaedia of Plant Physiology" Vol. 3. (W. Ruhland, ed.) pp. 541–582. Springer-Verlag, Berlin.

Huber, B. and Höfler, K. (1930). Die Wasserpermeabilität des Protoplasmas. *Jb. wiss. Bot.* **73**, 351–511.

Huber, B. and Schmidt, E. (1936). Weitere thermoelektrische Untersuchungen über den Transpirationsstrom der Bäume. *Tharandt. forstl. Jb.* **87**, 369–412.

Huckenpahler, B. J. (1936). Amount and distribution of moisture in a living short-leaf pine. *J. For.* **34**, 399–401.

Hunt, F. M. (1951). Effects of flooded soil on growth of pine seedlings. *Pl. Physiol., Lancaster* **26**, 363–368.

Hylmö, B. (1953). Transpiration and ion absorption. *Physiologia Pl.* **6**, 333–405.

Hylmö, B. (1955). Passive components in the ion absorption of the plant, Part I. *Physiologia Pl.* **8**, 433–449.

Hylmö, B. (1958). Passive components in the ion absorption of the plant, Part II. *Physiologia Pl.* **11**, 382–400.

Itoh, M., Izawa, S. and Shibota, K. (1963). Shrinkage of whole chloroplasts upon illumination. *Biochim. biophys. Acta* **66**, 319–327.

Jackson, J. E. and Weatherley, P. E. (1962a). The effect of hydrostatic pressure gradients on the movement of potassium across the root cortex. *J. exp. Bot.* **37**, 128–143.

Jackson, J. E. and Weatherley, P. E. (1962b). The effect of hydrostatic pressure gradients on the movement of sodium and calcium across the root cortex. *J. exp. Bot.* **13**, 404–413.

Janes, B. E. (1954). Absorption and loss of water by tomato leaves in a saturated atmosphere. *Soil Sci.* **78**, 189–197.

Jarvis, P. G. and Jarvis, M. S. (1963a). Effects of several osmotic substrates on the growth of *Lupinus albus* seedlings. *Physiologia Pl.* **16**, 485–500.

Jarvis, P. G. and Jarvis, M. S. (1963b). The water relations of tree seedlings. III. Transpiration in relation to osmotic potential of the root medium. *Physologia Pl.* **16**, 269–275.

Jarvis, P. G. and Jarvis, M. S. (1963c). The water relations of tree seedlings. IV. Some aspects of the tissue water relations and drought resistance. *Physiologia Pl.* **16**, 501–516.

Jarvis, P. G. and Slatyer, R. O. (1966a). A controlled environment chamber for studies of gas exchange by each surface of a leaf. *CSIRO Div. Land Res. Tech. Paper* **29**.

Jarvis, P. G. and Slatyer, R. O. (1966b). Calibration of beta gauges for determining leaf water status. *Science, N.Y.* **153**, 78–79.

Jarvis, P. G., Rose, C. W. and Begg, J. E. (1966). An experimental and theoretical comparison of viscous and diffusive resistances to gas flow through amphistomatous leaves. *Agric. Met.* (in press).

Jensen, R. D. and Taylor, S. A. (1961). Effect of temperature on water transport through plants. *Pl. Physiol., Lancaster* **36**, 639–642.

Jensen, R. D., Taylor, S. A. and Wiebe, H. H. (1961). Negative transport and resistance to water flow through plants. *Pl. Physiol., Lancaster* **36**, 633–638.

Johnson, F. S. (1954). The solar constant. *J. Met.* **11**, 431–439.

Johnston, R. D. (1959). Control of water movement by stem chilling. *Aust. J. Bot.* **7**, 97–108.

Jones, H. E. and Kohnke, H. (1952). The influence of soil moisture tension on vapor movement of soil water. *Proc. Soil Sci. Soc. Am.* **16**, 245–248.

Kachinsky, N. A. (1936). Influence of the shape and size of flooded small plots on the permeability of soils. *Pochvovedenie* **1**, 62–78.

Kamiya, N. and Kuroda, K. (1956). Artificial modification of the osmotic pressure of the plant cell. *Protoplasma* **46**, 423–436.

Kamiya, N. and Tazawa, M. (1956). Studies of water permeability of a single plant cell by means of transcellular osmosis. *Protoplasma* **46**, 394–422.

Katchalsky, A. (1961). Membrane permeability and the thermodynamics of irreversible processes. *In*, "Membrane Transport and Metabolism" (A. Kleinzeller and A. Kotyk, eds.) pp. 69–86. Academic Press, New York and London.

Katchalsky, A. and Kedem, O. (1962). Thermodynamics of flow processes in biological systems. *Biophys. J.* **2**, 53–78.

Kauzmann, W. (1959). Some factors in the interpretation of protein denaturation. *Adv. Protein Chem.* **14**, 1–63.

Kavanau, J. L. (1964). "Water and Solute-Water Interactions". Holden-Day, Inc. San Francisco.

Kedem, O. (1965). Water flow in the presence of active transport. *Symp. Soc. exp. Biol.* **19**, 61–73.

Kedem, O. and Katchalsky, A. (1958). Thermodynamic analysis of the permeability of biological membranes to non-electrolytes. *Biochim. biophys. Acta* **27**, 229–246.

Kedem, O. and Katchalsky, A. (1961). A physical interpretation of the pheno-menological coefficients of membrane permeability. *J. gen Physiol.* **45**, 143–179.

Kedem, O. and Katchalsky, A. (1963a). Permeability of composite membranes. Part 1. Electric current, volume flow and flow of solute through membranes. *Trans. Faraday Soc.* **59**, 1918–1930.

Kedem, O. and Katchalsky, A. (1963b). Permeability of composite membranes. Part 2. Parallel elements. *Trans. Faraday Soc.* **59**, 1931–1940.

Kedem, O. and Katchalsky, A. (1963c). Permeability of composite membranes. Part 3. Series array of elements. *Trans. Faraday Soc.* **59**, 1941–1953.

Kelsey, K. E. (1957). The sorption of water vapour by wood. *Aust. J. appl. Sci.* **8**, 42–54.

Kessler, B. (1961). Nucleic acids as factors in drought resistance in higher plants. *Recent Adv. Bot.* **2**, 1153–1159.

Kessler, B., Engelberg, N., Chen, D. and Greenspan, H. (1964). Studies on physiological and biochemical problems of stress in higher plants. *Volcani Inst. Agric. Res. Spec. Bull.* **64**.

Ketellapper, H. J. (1963). Stomatal physiology. *A. Rev. Pl. Physiol.* **14**, 249–270.

King, K. M., Tanner, C. B. and Suomi, V. E. (1956). A floating lysimeter and its evaporation recorder. *Trans. Am. geophys. Un.* **37**, 738–742.

Kittredge, J. (1948). "Forest Influences". McGraw-Hill, New York.

Kleinzeller, A. and Kotyk, A. (1961). "Membrane Transport and Metabolism". Academic Press, New York and London.

Kleschnin, A. F. (1960). "Die Pflanze und das Licht". Akademische-Verlag, Berlin.

Klotz, I. M. (1958). Protein hydration and behaviour. *Science, N.Y.* **128**, 815–822.

Klotz, I. M. and Ayers, J. (1957). The hydrogen ion equilibria of a single group attached to serum albumin: Some implications as to the surface characteristics of protein molecules. *J. Am. chem. Soc.* **79**, 4078–4085.

Klute, A. (1952). A numerical method for solving the flow equation for water in unsaturated materials. *Soil Sci.* **73**, 105–116.

Knight, R. C. (1917). The interrelations of stomatal aperture, leaf water-content, and transpiration rate. *Ann. Bot.* **31**, 221–240.

Knoerr, K. R. and Gay, L. W. (1965). Tree leaf energy balance. *Ecology* **46**, 17–24.

Koller, D. and Samish, Y. (1964). A null-point compensating system for simul-taneous and continuous measurement of net photosynthesis and transpiration by controlled gas-stream analysis. *Bot. Gaz.* **125**, 81–88.

Korven, H. C. and Taylor, S. A. (1959). The Peltier effect and its use for deter-mining relative activity of soil water. *Can. J. Soil Sci.* **39**, 76–85.

Kostiakov, A. N. (1932). On the dynamics of the coefficient of water-percolation in soils, and on the necessity for studying it from a dynamic point of view for the purposes of amelioration. *Trans. Intern. Congr. Soil Sci.* **A6**, 17–21.

Kozlowski, T. T. (1943). Transpiration rates of some forest tree species during the dormant season. *Pl. Physiol., Lancaster* **18**, 252–260.

Kozlowski, T. T. (1964). "Water Metabolism in Plants". Harper and Row, New York.

Kramer, P. J. (1938). Root resistance as a cause of the absorption lag. *Am. J. Bot.* **25**, 110–113.

Kramer, P. J. (1940a). Root resistance as a cause of decreased water absorption by plants at low temperatures. *Pl. Physiol., Lancaster* **15**, 63–79.

Kramer, P. J. (1940b). Causes of decreased absorption of water by plants in poorly aerated media. *Am. J. Bot.* **27**, 216–220.

Kramer, P. J. (1942). Species differences with respect to water absorption at low soil temperatures. *Am. J. Bot.* **29**, 828–832.

Kramer, P. J. (1949). "Plant and Soil Water Relationships". McGraw-Hill, New York.

Kramer, P. J. (1950). Effects of wilting on the subsequent intake of water by plants. *Am. J. Bot.* **37**, 280–284.

Kramer, P. J. (1951). Causes of injury to plants resulting from flooding of the soil. *Pl. Physiol., Lancaster* **26**, 722–736.

Kramer, P. J. (1956a). Physical and physiological aspects of water absorption. *In*, "Encyclopedia of Plant Physiology" (W. Ruhland, ed.) Vol. 3, pp. 124–159. Springer-Verlag, Berlin.

Kramer, P. J. (1956b). Water content and water turnover in plant cells. *In*, "Encyclopedia of Plant Physiology" (W. Ruhland, ed.) Vol. 1, pp. 194–222. Springer-Verlag, Berlin.

Kramer, P. J. (1957). Outer space in plants. *Science, N.Y.* **125**, 633–635.

Kramer, P. J. (1959). Transpiration and the water economy of plants. *In*, "Plant Physiology" (F. C. Steward, ed.) Vol. 2, pp. 607–730. Academic Press, New York and London.

Kramer, P. J. and Wiebe, H. H. (1952). Longitudinal gradients of P^{32} absorption in roots. *Pl. Physiol., Lancaster* **27**, 661–674.

Kramer, P. J. and Jackson, W. T. (1954). Causes of injury to flooded tobacco plants. *Pl. Physiol., Lancaster* **29**, 241–245.

Kramer, P. J. and Kozlowski, T. T. (1960). "Physiology of Forest Trees". McGraw-Hill, New York.

Krause, H. (1935). Beiträge zur Kenntnis der Wasseraufnahme durch oberirdische Pflanzenorgane. *Öst. bot. Z.* **84**, 241–270.

Kreeb, K. (1963). Hydrature and plant production. *In*, "The Water Relations of Plants" (A. J. Rutter and F. H. Whitehead, eds.) pp. 272–288. Blackwell, London.

Kriedeman, P. E., Neales, T. F. and Ashton, D. H. (1965). Photosynthesis in relation to leaf orientation and light interception. *Aust. J. biol. Sci.* **17**, 591–600.

Kuiper, P. J. C. (1961). The effects of environmental factors on the transpiration of leaves, with special reference to stomatal light response. *Meded. Landb Hoogesch. Wageningen* **61**, 1–49.

Kuiper, P. J. C. (1963). Some considerations on water transport across living cell membranes. *In*, "Stomata and Water Relations in Plants" (I. Zelitch, ed.) pp. 59–68. Connecticut Agricultural Experiment Station, New Haven, Conn.

Kuiper, P. J. C. (1964a). Water uptake of higher plants as affected by root temperature. *Meded. Landb Hogesch. Wageningen* **63**, 1–11.

Kuiper, P. J. C. (1964b). Water transport across root cell membranes: Effect of alkenylsuccinic acids. *Science, N.Y.* **143**, 690–691.

Kuzmak, J. M. and Sereda, P. J. (1957). The mechanism by which water moves through a porous material subjected to a temperature gradient: 2. Salt tracer and streaming potential to detect flow in the liquid phase. *Soil Sci.* **84**, 419–422.

Laidler, K. J. and Shuler, K. E. (1949). The kinetics of membrane processes. I. The mechanism and the kinetic laws for diffusion through membranes. *J. chem. Phys.* **17**, 851–855.

Lambert, J. R. and van Schilfgaarde, J. (1965). A method of determining the water potential of intact plants. *Soil Sci.* **100**, 1–9.

Lang, A. R. G. and Barrs, H. D. (1965). An apparatus for measuring water potentials in the xylem of intact plants. *Aust. J. biol. Sci.* **18**, 487–497.

Langridge, J. (1963). Biochemical aspects of temperature response. *A. Rev. Pl. Physiol.* **14**, 441–462.

Laties, G. G. (1954). The osmotic inactivation *in situ* of plant mitochondrial enzymes. *J. exp. Bot.* **5**, 49–70.

Lebedeff, A. F. (1927). The movement of ground and soil waters. *Proc. Int. Congr. Soil Sci.* **1**, 459–494.

Lemon, E. R. (1963). Energy and water balance of plant communities. *In*, "Environmental Control of Plant Growth" (L. T. Evans, ed.) pp. 55–78. Academic Press, New York and London.

Lemon, E. R., Glaser, A. H. and Satterwhite, L. E. (1957). Some aspects of the relationship of soil, plant and meteorological factors to evapotranspiration. *Proc. Soil Sci. Soc. Am.* **21**, 464–468.

Lepeschkin, W. W. (1932). The influence of narcotics, mechanical agents and light upon the permeability of protoplasm. *Am. J. Bot.* **19**, 568–580.

Lettau, H. H. and Davidson, B. (1957). "Exploring the Atmosphere's First Mile", Vol. 1. Pergamon Press, London.

Levitt, J. (1947). The thermodynamics of active (non-osmotic) water absorption. *Pl. Physiol., Lancaster*, **22**, 514–525.

Levitt, J., Scarth, G. W. and Gibbs, R. D. (1936). Water permeability of isolated protoplasts in relation to volume change. *Protoplasma* **26**, 237–248.

Lewis, F. J. (1945). Physical condition of the surface of the mesophyll cell walls of the leaf. *Nature, Lond.* **156**, 407–490.

Lewis, F. J. (1948). Water movement in leaves. *Discuss. Faraday Soc.* **3**, 159–162.

Lewis, G. N. and Randall, M. (1961). (Revised by Pitzer, K. S. and Brown, L.). "Thermodynamics". McGraw-Hill, New York.

Lewis, M. R. (1937). The rate of infiltration of water in irrigation-practice. *Trans. Am. geophys. Un.* **18**, 361–368.

Liese, W. and Johann, I. (1954). Experimentelle Untersuchugen über die Feinstruktur der Hoftupfel bei den Koniferen. *Naturwissenschaften* **41**, 579.

Linsley, R. K., Kohler, M. A. and Paulhus, J. L. H. (1949). "Applied Hydrology". McGraw-Hill, New York.

List, R. J. (1963). "Smithsonian Meteorological Tables". Smithsonian Institution, Washington, D.C.

Livingston, B. E. (1906). The relation of desert plants to soil moisture and to evaporation. *Carnegie Inst. Washington Pub.* **50**.

Lloyd, F. E. (1908). The physiology of stomata. *Carnegie Inst. Washington Pub.* **82**.

Long, E. M. (1943). The effect of salt additions to the substrate on intake of water and nutrients by roots of approach-grafted tomato plants. *Am. J. Bot.* **30**, 594–601.

Loomis, W. E. (1934). Daily growth of maize. *Am. J. Bot.* **21**, 1–6.

Lopushinsky, W. and Kramer, P. J. (1961). Effect of water movement on salt movement through tomato roots. *Nature, Lond.* **192**, 994–995.

Loustalot, P. J. (1945). Influence of soil moisture conditions on apparent photosynthesis and transpiration of pecan leaves. *J. agric. Res.* **71**, 519–532.

Low, P. F. (1951). Force fields and chemical equilibrium in heterogeneous systems with special reference to soils. *Soil Sci.* **71**, 409–418.

Low, P. F. (1955). Effect of osmotic pressure on diffusion rate of water. *Soil Sci.* **80**, 95–100.

Low, P. F. (1961). Physical chemistry of clay-water interaction. *Adv. Agron.* **13**, 269–327.

Low, P. F. and Deming, J. M. (1953). Movement and equilibrium of water in heterogeneous systems with special reference to soils. *Soil. Sci* **75**, 187–202.

Lundegårdh, H. (1949). The effect of indole acetic acid on the bleeding of wheat roots. *Ark. Bot.* **1**, 295–299.

Luthin, J. N. (1957). "Drainage of Agricultural Lands". American Society of Agronomy, Madison, Wisconsin.

Lutz, H. J. (1944) Determination of certain physical properties of forest soils: I. Methods utilizing samples collected by metal cylinders. *Soil Sci.* **57**, 475–487.

Lybeck, B. R. (1959). Winter freezing in relation to the rise of sap in tall trees. *Pl. Physiol., Lancaster* **34**, 482–486.

Lyon, C. J. (1941). Osmotic pressure for the plant physiologist. *Science, N.Y.* **93**, 374–375.

McDermott, J. J. (1941). The effect of the method of cutting on the moisture content of samples from tree branches. *Am. J. Bot.* **28**, 506–508.

McIlroy, I. C. and Angus, D. E. (1964). Grass, water and soil evaporation at Aspendale. *Agric. Met.* **1**, 201–224.

McIlroy, I. C. and Sumner, C. J. (1961). A sensitive high-capacity balance for continuous automatic weighing in the field. *J. agric. Engng Res.* **6**, 252–258.

McIntyre, D. S. (1958). Permeability measurements of soil crusts formed by raindrop impact. *Soil Sci.* **85**, 185–189.

McQuilkin, W. E. (1935). Root development of pitch pine, with some comparative observations on shortleaf pine. *J. agric. Res.* **51**, 983–1016.

MacDougal, D. T., Overton, J. B. and Smith, G. B. (1929). The hydrostatic-pneumatic system of certain trees: movement of liquids and gases. *Carnegie Inst. Washington Pub.* **397**.

Macklon, A. E. S. and Weatherley, P. E. (1965). A vapour-pressure instrument for the measurement of leaf and soil water potential. *J. Exp. Botan.* **16**, 261–270.

MacRobbie, E. A. C. and Dainty, J. (1958). Ion transport in *Nitellopsis obtusa*. *J. gen. Physiol.* **42**, 335–353.

Magistad, O. C., Ayers, A. D., Wadleigh, C. H. and Gauch, H. G. (1943). Effect of salt concentration, kind of salt, and climate on plant growth in sand cultures. *Pl. Physiol,. Lancaster* **18**, 151–166.

Magness, J. R., Degman, E. S. and Furr, J. R. (1935). Soil moisture and irrigation investigations in eastern apple orchards. *U.S. Dep. Agric. Tech. Bull.* **491**.

Makkink, G. F. and van Heemst, H. D. J. (1956). The actual evapotranspiration as a function of the potential evapotranspiration and the soil moisture tension. *Neth. J. agric. Sci.* **4**, 67–72.

Marshall, T. J. (1958). A relation between permeability and size distribution of pores. *J. Soi Sci.* **9**, 1–8.

Marshall, T. J. (1959). "Relation between Water and Soil". Commonwealth Agricultural Bureaux, Farnham Royal, England.

Marshall, T. J. and Stirk, G. B. (1950). The effect of lateral movement of water in soil on infiltration measurements. *Aust. J. agric. Res.* **1**, 253–265.

Martin, E. (1943). Studies on evaporation and transpiration under controlled conditions. *Carnegie Inst. Washington Pub.* **550**.

Maskell, E. J. (1928). Experimental researches on vegetable assimilation and respiration. XVIII. The relation between stomatal opening and assimilation. A critical study of assimilation rates and porometer rates in leaves of cherry laurel. *Proc. R. Soc.* B **102**, 488–533.

Mason, T. G. and Phillis, E. (1939). Experiments on the extraction of sap from the vacuole of the leaf of the cotton plant and their bearing on the osmotic theory of water absorption of the cell. *Ann. Bot.* (*N.S.*) **3**, 531–544.

Mauro, A. (1957). Nature of solvent transfer in osmosis. *Science, N.Y.* **126**, 252–253.

Mauro, A. (1960). Some properties of ionic and non-ionic semipermeable membranes. *Circulation* **21**, 845–854.

Maximov, N. A. (1929). "The Plant in Relation to Water". Allen and Unwin, London.

Maximov, N. A. (1941). Influence of drought on physiological processes in plants. *In*, "Collection of papers on plant physiology in memory of K. A. Timiriazev" pp. 299–309. USSR Academy of Science, Moscow.

Mederski, H. J. (1961). Determination of internal water status of plants by beta ray gauging. *Soil Sci.* **92**, 143–146.

Mees, G. C. and Weatherley, P. E. (1957a). The mechanism of water absorption by roots. I. Preliminary studies on the effects of hydrostatic pressure gradients. *Proc. R. Soc.* B **147**, 367–380.

Mees, G. C. and Weatherley, P. E. (1957b). The mechanism of water absorption by roots. II. The role of hydrostatic pressure gradients across the cortex. *Proc. R. Soc.* B **147**, 381–391.

Meidner, H. (1954). Measurements of water intake from the atmosphere by leaves. *New Phytol.* **53**, 423–426.

Meidner, H. (1955). Changes in the resistance of the mesophyll tissue with changes in the leaf water content. *J. exp. Bot.* **6**, 94–99.

Meidner, H. (1965). Stomatal control of transpirational water loss. *Symp. Soc. exp. Biol.* **19**, 185–203.

Meidner, H. and Heath, O. V. S. (1959). Stomatal responses to temperature and carbon dioxide concentration in *Allium cepa* l. and their relevance to mid-day closure. *J. exp. Bot.* **10**, 206–219.

Meidner, H. and Heath, O. V. S. (1963). Rapid changes in transpiration in plants. *Nature, Lond.* **200**, 283–284.

Meidner, H. and Mansfield, T. A. (1965). Stomatal responses to illumination. *Biol. Rev.* **40**, 483–509.

Mellor, R. S., Salisbury, F. B. and Raschke, K. (1964). Leaf temperature in controlled environments. *Planta* **61**, 56–72.

la Mer, V. K. (1962). "Retardation of Evaporation by Monolayers". Academic Press, New York and London.

Mercer, F. V. (1955). The water relations of plant cells. *Proc. Linn. Soc. N.S.W.* **80**, 6–29.

Meyer, B. S. (1938). The water relations of plant cells. *Bot. Rev.* **4**, 531–547.

Meyer, B. S. (1945). A critical evaluation of the terminology of diffusion phenomena. *Pl. Physiol., Lancaster* **20**, 142–164.

Meyer, B. S. and Anderson, D. B. (1952). "Plant Physiology". Van Nostrand, New York.

Miller, E. C. (1938). "Plant Physiology". McGraw-Hill, New York.

Miller, E. J., Nielson, J. A. and Bandemer, S. L. (1937). Wax emulsions for spraying nursery stock and other plant materials. *Mich. Agric. Exp. Sta. Spec. Bull.* **282**.

Milthorpe, F. L. (1959). Transpiration from crop plants. *Fld Crop Abstr.* **12**, 1–9.

Milthorpe, F. L. (1960). The income and loss of water in arid and semi-arid zones. *UNESCO Arid Zone Res.* **15**, 9–36.

Milthorpe, F. L. and Spencer, E. J. (1957). Experimental studies of the factors controlling transpiration. *J. exp. Bot.* **8**, 413–437.

Monteith, J. L. (1957). Dew. *Q. Jl. R. met. Soc.* **83**, 322–341.

Monteith, J. L. (1958). The heat balance of soil beneath crops. *UNESCO Arid Zone Res.* **11**, 123–128.

Monteith, J. L. (1959). The reflection of short-wave radiation by vegetation. *Q. Jl. R. met. Soc.* **85**, 386–392.

Monteith, J. L. (1963a). Gas exchange in plant communities. *In*, "Environmental Control of Plant Growth" (L. T. Evans, ed.) pp. 95–112. Academic Press, New York and London.

Monteith, J. L. (1963b). Dew: Facts and Fallacies. *In*, "The Water Relations of Plants" (A. J. Rutter and F. H. Whitehead, eds.) pp. 37–56. Blackwell, London.

Monteith, J. L. (1965). Evaporation and environment. *Symp. Soc. exp. Biol.* **29**, 205–234.

Monteith, J. L. and Owen, P. C. (1958). A thermocouple method for measuring relative humidity in the range 95–100%. *J. scient. Instrum.* **35**, 443–446.

Moore, R. E. (1939). Water conduction from shallow water tables. *Hilgardia* **12**, 383–426.

Morgan, J. and Warren, B. E. (1938). X-ray analysis of the structure of water. *J. chem. Phys.* **6**, 666–673.

Morris, L. G. (1959). A recording weighing machine for the measurement of evapotranspiration and dewfall. *J. agric. Eng'ng Res.* **4**, 161–173.

Moss, D. N. (1966). Respiration of leaves in light and darkness. *Crop Sci.* **6**, 351–354.

Mothes, K. (1956). Der Einfluss des Wasserzustandes auf Fermentprozesse und Stoffumsatz. *In*, "Encyclopedia of Plant Physiology" (W. Ruhland, ed.) Vol. 3, pp. 656–664. Springer-Verlag, Berlin.

Mueller, L. E., Carr, P. H. and Loomis, W. E. (1954). The submicroscopic structure of plant surfaces. *Am. J. Bot.* **41**, 593–600.

Musgrave, G. W. (1955). How much of the rain enters the soil. *In*, "Water" (A. Stefferud, ed.) pp. 151–159. U.S. Dept. Agriculture, Washington, D.C.

Musgrave, R. B. and Moss, D. N. (1961). Photosynthesis under field conditions. I. A portable, closed system for determining net assimilation and respiration of corn. *Crop Sci.* **1**, 37–41.

Myers, G. M. P. (1951). The water permeability of unplasmolysed tissues. *J. Exp. Bot.* **2**, 129–144.

Nakayama, F. S. and Ehrler, W. L. (1964). Beta ray gauging technique for measuring leaf water content changes and moisture status of plants. *Pl. Physiol., Lancaster* **39**, 95–98.

Némethy, G. and Scheraga, H. A. (1962a). Structure of water and hydrophilic bonding in proteins. I. A model for the thermodynamic properties of liquid water. *J. chem. Phys.* **36**, 3382–3400.

Némethy, G. and Scheraga, H. A. (1962b). Structure of water and hydrophilic bonding in proteins. II. A model for the thermodynamic properties of aqueous solutions of hydro-carbon. *J. chem. Phys.* **36**, 3401–3417.

Nieman, R. H. (1962). Some effects of sodium chloride on growth, photosynthesis and respiration of twelve crop plants. *Bot. Gaz.* **123**, 279–285.

Nieman, R. H. and Poulson, L. L. (1962). The light dependence of nucleic acid and protein synthesis by isolated radish cotyledonary leaves. *Pl. Physiol., Lancaster* **37** (Supplement), 21–22.

Nishida, K. (1963). Studies on stomatal movement of Crassulacean plants in relation to the acid metabolism. *Physiologia, Pl.* **16**, 281–298.

Ogata, G., Richards, L. A. and Gardner, W. R. (1960). Transpiration of alfalfa determined from soil water content changes. *Soil Sci.* **89**, 179–182.

Onsager, L. (1931). Reciprocal relations in irreversible processes. I. *Phys. Rev.* **37**, 405–426.

Oppenheimer, H. R. and Mendel, K. (1939). On orange leaf transpiration under orchard conditions, Part I. Soil moisture high. *Palest. J. Bot. Rehovot Ser.* **2**, 171–250.

Ordin, L. (1958). The effect of water stress on the cell wall metabolism of plant tissue. *In*, "Radioisotopes in Scientific Research" Vol. 4, pp. 553–564. Pergamon Press, New York.

Ordin, L. (1960). Effect of water stress on cell wall metabolism of *Avena* coleoptile tissue. *Pl. Physiol., Lancaster* **35**, 443–450.

Ordin, L. and Bonner, J. (1956). Permeability of *Avena* coleoptile sections to water measured by diffusion of deuterium hydroxide. *Pl. Physiol., Lancaster* **31**, 53–57.

Ordin, L. and Gairon, S. (1961). Diffusion of tritiated water into roots as influenced by water status of tissue. *Pl. Physiol., Lancaster* **36**, 331–335.

Ordin, L., Applewhite, T. H. and Bonner, J. (1956). Auxin-induced water uptake by *Avena* coleoptile sections. *Pl. Physiol., Lancaster* **31**, 44–53.

Orlob, G. T. and Radhakrishna, G. N. (1958). The effects of entrapped gases on the hydraulic characteristics of porous media. *Trans. Am. geophys. Un.* **39**, 648–659.

van Overbeek, J. (1942). Water uptake by excised root systems of the tomato due to non-osmotic forces. *Am. J. Bot.* **29**, 677–682.

van Overbeek, J. (1944). Auxin water uptake and osmotic pressure in potato tissue. *Am. J. Bot.* **29**, 677–682.

van Overbeek, J. and Blondeau, R. (1954). Mode of action of phototoxic oils. *Weeds* **3**, 55–65.

Ovington, J. D. (1963). Quantitative ecology and the woodland ecosystem concept. *Adv. Ecol. Res.* **1**, 103–192.

Owen, P. C. (1953). The relation of germination of wheat to water potential. *J. exp. Bot.* **3**, 188–203.

Owen, P. C. (1958). The effects of short periods of water stress on the growth of sugar beet in pots. *New Phytol.* **57**, 318–325.

Packer, L., Siegenthaler, P. A. and Nobel, P. S. (1965). Light-induced high amplitude swelling of spinach chloroplasts. *Biochem. biophys. Res. Commun.* **18**, 474–477.

Pappenheimer, J. R. (1953). Passage of molecules through capillary walls. *Physologia Rev.* **33**, 387–423.

Parker, J. (1950). The effects of flooding on the transpiration and survival of some south-eastern forest tree species. *Pl. Physiol.*, *Lancaster* **25**, 453–460.

Parr, J. F. and Bertrand, A. R. (1960). Water infiltration into soils. *Adv. Agron.* **12**, 311–363.

Pasquill, F. (1949). Eddy diffusion of water and heat near the ground. *Proc. R Soc.* **A198**, 116–140.

Pauling, L. (1960). "The Nature of the Chemical Bond". Cornell University Press, Ithaca, New York.

Paulovassilis, A. (1962). Hysteresis of pore water, an application of the concept of independent domains. *Soil Sci.* **93**, 405–412.

Peck, A. J. (1960). Change of moisture tension with temperature and air pressure: theoretical. *Soil Sci.* **89**, 303–310.

Penman, H. L. (1940). Gas and vapour movements in the soil. I. The diffusion of vapours through porous solids. *J. agric. Sci.* **30**, 437–462.

Penman, H. L. (1948). Natural evaporation from open water, bare soil and grass. *Proc. R. Soc.* **A193**, 120–145.

Penman, H. L. (1949). The dependence of transpiration on weather and soil conditions. *J. Soil Sci.* **1**, 74–89.

Penman, H. L. (1955). "Humidity". Institute of Physics, London.

Penman, H. L. (1956). Evaporation: An introductory survey. *Neth. J. agric. Sci.* **4**, 9–29.

Penman, H. L. (1963). "Vegetation and Hydrology". Commonwealth Agricultural Bureaux, Farnham Royal, England.

Penman, H. L. and Schofield, R. K. (1951). Some physical aspects of assimilation and transpiration. *Symp. Soc. exp. Biol.* **5**, 115–129.

Penman, H. L. and Long, I. (1960). Weather in wheat—an essay in micrometeorology. *Q. Jl. R. met. Soc.* **86**, 16–50.

Pennell, G. A. and Weatherley, P. E. (1958). On the mechanism of sugar uptake by floating leaf discs. *New Phytol.* **57**, 326–339.

Perutz, M. F. (1946). The composition and swelling properties of haemoglobin crystals. *Trans. Faraday Soc.* B **42**, 187–195.

Petrie, A. H. K. and Wood, J. G. (1938a). Studies on the nitrogen metabolism of plants. I. The relation between the content of proteins, amino acids and water in the leaves. *Ann. Bot. (N.S.)* **2**, 33–60.

Petrie, A. H. K. and Wood, J. G. (1938b). Studies on the nitrogen metabolism of plants. III. On the effect of water content on the relationship between proteins and amino acids. *Ann. Bot. (N.S.)* **2**, 887–898.

Petrie, A. H. K. and Arthur, J. I. (1943). Physiological ontogeny in the tobacco plant. The effect of varying water supply on the drifts in dry weight and leaf area and on various components of the leaves. *Aust. J. exp. Biol. med. Sci.* **21**, 191–200.

Philip, J. R. (1954). An infiltration equation with physical significance. *Soil Sci.* **77**, 153–157.

Philip, J. R. (1955). The concept of diffusion applied to soil water. *Proc. natn. Acad. Sci. India* **24**, 93–104.

Philip, J. R. (1957a). The physical principles of soil water movement during the irrigation cycle. *Proc. Int. Congr. Irrig. Drain.* **8**, 125–154.

Philip, J. R. (1957b). Evaporation, and moisture and heat fields in the soil. *J. Met.* **14**, 354—366.

Philip, J. R. (1957c). The theory of infiltration: 5. The influence of the initial moisture content. *Soil Sci.* **84**, 329–339.

Philip, J. R. (1957d). The theory of infiltration: 1. The infiltration equation and its solution. *Soil Sci.* **83**, 345–357.

Philip, J. R. (1957e). Remarks on the analytical derivation of the Darcy equation. *Trans. Am. geophys. Un.* **38**, 782–784.

Philip, J. R. (1958a). The theory of infiltration: 6. Effect of water depth over soil. *Soil Sci.* **85**, 278–286.

Philip, J. R. (1958b). The theory of infiltration: 7. *Soil Sci.* **85**, 333–337.

Philip, J. R. (1958c). The osmotic cell, solute diffusibility, and the plant water economy. *Pl. Physiol., Lancaster* **33**, 264–271.

Philip, J. R. (1958d). Osmosis and diffusion in tissue: half-times and internal gradients. *Pl. Physiol., Lancaster* **33**, 275–278.

Philip, J. R. (1959). The theory of local advection. *J. Met.* **16**, 535–547.

Philip, J. R. (1961). The theory of heat flux meters. *J. geophys. Res.* **66**, 571–579.

Philip, J. R. (1964a). Sources and transfer processes in the air layers occupied by vegetation. *J. appl. Meteorol.* **3**, 390–395.

Philip, J. R. (1964b). Similarity hypothesis for capillary hysteresis in porous materials. *J. geophys. Res.* **69**, 1553–1562.

Philip, J. R. (1964c). The gain, transfer and loss of soil-water. *In*, "Water Resources, Use and Management" (E. S. Hills, ed.) pp. 257–275. Melbourne University Press, Melbourne.

Philip, J. R. (1966). Plant water relations: Some physical aspects. *A. Rev. Pl. Physiol.* **17**, 245–268.

Philip, J. R. and de Vries, D. A. (1957). Moisture movement in porous materials under temperature gradients. *Trans. Am geophys. Un.* **38**, 222–232.

Pillsbury, A. F. (1950). Effects of particle size and temperature on the permeability of sand to water. *Soil Sci.* **70**, 299–300.

Pillsbury, A. F. and Richards, S. J. (1954). Some factors affecting rates of irrigation water entry into Ramona sandy loam soil. *Soil Sci.* **78**, 211–217.

Pitman, M. G. (1963). The determination of the salt relations of the cytoplasmic phase in cells of beetroot tissue. *Aust. J. biol. Sci.* **16**, 647–668.

Plaut, Z. and Ordin, L. (1961). Effect of soil moisture content on the cell wall metabolism of sunflower and almond leaves. *Physiologia Pl.* **14**, 646–658.

Postlethwait, S. N. and Rogers, B. (1958). Tracing the path of the transpiration stream in trees by the use of radio-active isotopes. *Am. J. Bot.* **45**, 753–757.

Prescott, D. M. and Zeuthen, E. (1953). Comparison of water diffusion and water filtration across cell surfaces. *Acta physiol. scand.* **28**, 77–94.

Prescott, J. A. (1958). Climatic indices in relation to the water balance. *UNESCO Arid Zone Res.* **16**, 48–51.

Preston, R. D. (1952). Movement of water in higher plants. *In*, "Deformation and Flow in Biological Systems" (A. Frey-Wyssling, ed.) pp. 257–321. North Holland, Amsterdam.

Preston, R. D. (1958). The ascent of sap and the movement of soluble carbohydrates in stems of higher plants. *In*, "The Structure and Properties of Porous Materials" (D. H. Everett and F. S. Stone, eds.) pp. 366–382. Butterworths, London.

Preston, R. D. and Wardrop, A. B. (1949). The sub-microscopic organisation of the walls of conifer cambium. *Biochim. biophys. Acta* **3**, 549–559.

Preston, R. D., Nicolai, E., Reed, R. and Millard, A. (1948). An electron microscope study of cellulose in the wall of *Valonia ventricosa*. *Nature, Lond.* **162**, 665–667.

Prigogine, I. (1961). "Introduction to Thermodynamics of Irreversible Processes". John Wiley and Sons, New York.

Prigogine, I. and Defay, R. (Translated by D. H. Everett) (1954). "Chemical Thermodynamics". John Wiley and Sons, New York.

Privalov, P. J. (1958). The state and role of water in biological systems. *Biofizika* **3**, 738–743.

Probine, M. C. and Preston, R. D. (1962). Cell growth and the structure and mechanical properties of the wall in internodal cells of *Nitella opaca*. II. Mechanical properties of the walls. *J. exp. Bot.* **13**, 111–127.

Pruitt, W. O. and Angus, D. E. (1960). Large weighing lysimeter for measuring evapotranspiration. *Trans. Am. Soc. agric. Engrs* **3**, 13–18.

Quirk, J. P. and Schofield, R. K. (1955). The effect of electrolyte concentration on soil permeability. *J. Soil Sci.* **6**, 163–178.

Raber, O. (1937). Water utilization by trees, with special reference to the economic forest species of the north temperate zone. *U.S. Dept. Agric. Misc. Publ.* **257**.

Rabinowitch, E. I. (1945). "Photosynthesis" Vol. 1. Interscience, New York.

Radler, F. (1965). Reduction of loss of moisture by the cuticle wax components of grapes. *Nature, Lond.* **207**, 1002–1003.

Raney, F. and Vaadia, Y. (1965a). Movement and distribution of THO in tissue water and vapor transpired by shoots of *Helianthus* and *Nicotiana*. *Pl. Physiol., Lancaster* **40**, 383–388.

Raney, F. and Vaadia, Y. (1965b). Movement of tritiated water in the root system of *Helianthus annuus* in the presence and absence of transpiration. *Pl. Physiol., Lancaster* **40**, 378–382.

Raney, F. and Vaadia, Y. (1965c). Dispersion of THO and ^{36}Cl uptake by sunflower root systems. *Physiologia Pl.* **18**, 8–14.

Ranson, S. L. and Thomas, M. (1960). Crassulacean acid metabolism. *A. Rev. Pl. Physiol.* **11**, 81–110.

Raschke, K. (1956). Über die physikalischen Beziehungen zwischen Wärmeübergangszahl, Strahlungsaustausch, Temperatur und Transpiration eines Blattes. *Planta* **48**, 200–238.

Raschke, K. (1958). Über den Einflusz der Diffusionswiderstände auf die Transpiration und die Temperatur eines Blattes. *Flora, Jena* **146**, 546–578.

Raschke, K. (1960). Heat transfer between the plant and the environment. *A. Rev. Pl. Physiol.* **11**, 111–126.

Raschke, K. (1965a). Das Seifenblasenporometer (Zur Messung der Stomaweite an amphistomatischen Blättern). *Planta* **66**, 113–120.

Raschke, K. (1965b). Die Stomata als Glieder eines schwingungsfähigen CO_2-Regelsystems Experimenteller Nachweis an *Zea mays* L. *Z. Naturf.* **20**, 1261–1270.

Rawlins, S. L. (1963). Resistance to water flow in the transpiration stream. *In*, "Stomata and Water Relations in Plants" (I. Zelitch, ed.) pp. 69–85. Agricultural Experiment Station, New Haven, Connecticut.

Rawlins, S. L. (1964). A systematic error in leaf water potential measurements with a thermocouple psychrometer. *Science, N.Y.* **146**, 644–646.

Ray, P. M. (1960). On the theory of osmotic water movement. *Pl. Physiol., Lancaster* **35**, 783–795.

Ray, P. M. and Ruesink, W. (1963). Osmotic behaviour of oat coleoptile tissue in relation to growth. *J. gen. Physiol.* **47**, 83–101.

Reams, W. M. (1953). The occurrence and ontogeny of hydathodes in *Hygrophilia polysperma* T. Anders. *New Phytol.* **52**, 8–13.

Reed, J. F. (1939). Root and shoot growth of shortleaf and loblolly pines in relation to certain environmental conditions. *Duke University School of Forestry Bull.* **4**.

Reinders, D. E. (1938). The process of water-intake by discs of potato tuber tissue. *Proc. K.ned Akad. Wet.* C **41**, 820–831.

Reinders, D. E. (1942). Intake of water by parenchymatic tissue. *Recl. Trav. bot. néerl.* **39**, 1–140.

Reinhart, K. G. and Taylor, R. E. (1954). Infiltration and available water storage·capacity in the soil. *Trans. Am. geophys. Un.* **35**, 791–795.

Renner, O. (1910). Beiträge zur Physik der Transpiration. *Flora, Jena* **100**, 451–457.

Renner, O. (1915). Theoretisches und Experimentelles zur Kohasionstheorie der Wasserbewegung. *Jb. wiss. Bot.* **56**, 617–667.

Renner, O. (1929). Verfuche zur Bestimmung des Filtrations Widerstandes der Wurzen. *Jb. wiss. Bot.* **70**, 805–838.

Richards, B. G. (1965). Thermistor hygrometer for determining the free energy of moisture in unsaturated soils. *Nature, Lond.* **208**, 608–609.

Richards, L. A. (1931). Capillary conduction of liquids through porous mediums. *Physics* **1**. 318–333.

Richards, L. A. (1949). Methods of measuring soil moisture tension. *Soil Sci.* **68**, 95–112.

Richards, L. A. (1952). Report of the subcommittee on permeability and infiltration, Committee on Terminology, Soil Science Society of America. *Proc. Soil Sci. Soc. Am.* **16**, 85–88.

Richards, L. A. (1954). Diagnosis and improvement of saline and alkali soils. *U.S. Dept. Agric. Handbook* **60**.

Richards, L. A. and Weaver, L. R. (1944). Fifteen atmosphere percentage as related to the permanent wilting percentage. *Soil Sci.* **56**, 331–339.

Richards, L. A. and Wadleigh, C. H. (1952). Soil water and plant growth. *In,* "Soil Physical Conditions and Plant Growth" (B. T. Shaw, ed.) Academic Press, New York and London.

Richards, L. A. and Ogata, G. (1958). Thermocouple for vapour pressure measurement in .biological and soil systems at high humidity. *Science, N.Y.* **128**, 1089–1090.

Richards, L. A. and Richards, P. L. (1962). Radial-flow cell for soil-water measurements. *Proc. Soil Sci. Soc. Am.* **26**, 515–518.

Richards, L. A., Gardner, W. R. and Ogata, G. (1956). Physical processes determining water loss from soil. *Proc. Soil Sci. Soc. Am.* **20**, 311–314.

Rider, N. E. (1954). Evaporation from an oat field. *Q. Jl. R. met. Soc.* **80**, 198–212.

Rider, N. E. (1960). A system for recording and integrating physical measurements. *Aust. J. Phys.* **13**, 742–749.

Rijtema, P. E. (1965). An analysis of actual evapotranspiration. *Wageningen agric. res. Rep.* **659**.

Rim, M. (1954). Inverse moisture distribution in soil profiles: A manifestation of the dependence of soil moisture characteristics on compressive stress. *Trans. Int. Congr. Soil Sci.* **2**, 69–73.

Robbins, E. and Mauro, A. (1960). Experimental study of the independence of diffusion and hydrodynamic permeability coeficients in collodion membranes. *J. gen. Physiol.* **43**, 523–532.

Roberts, O. and Styles, S. A. (1939). An apparent connection between the presence of colloids and the osmotic pressures of conifer leaves, *Scient. Proc. R. Dubl. Soc.* **22**, 119–125.

Roberts, W. J. (1961). Reducing transpiration from plants. *J. geophys. Res.* **66**, 3309–3312.

Robertson, R. N. and Turner, J. F. (1951). The physiology of growth in apple fruits. II. Respiratory and other metabolic activities as functions of cell number and cell size in fruit development. *Aust. J. Sci. Res.* B **4**, 92–107.

Robins, J. S. and Domingo, C. E. (1953). Some effects of severe soil moisture deficits at specific growth stages of corn. *Agron. J.* **45**, 618–621.

Robinson, R. A. and Stokes, R. H. (1959). "Electrolyte Solutions". Butterworths, London.

Rollins, R. L., Spangler, M. G. and Kirkham, D. (1954). Movement of soil moisture under a thermal gradient. *Proc. Highw. Res. Bd.* **33**, 492–508.

Rose, C. W. and Stern, W. R. (1965). The drainage component of the water balance equation. *Aust. J. Soil Res.* **3**, 95–100.

Rose, C. W., Byrne, G. F. and Begg, J. E. (1966). An accurate hydraulic lysimeter with remote weight recording. *CSIRO Div. Land Res. Tech. Paper* **27**.

Rosene, H. F. (1943). Quantitative measurement of the velocity of water absorption in individual root hairs by a microtechnique. *Pl. Physiol., Lancaster* **18**, 588–607.

Rosene, H. F. (1944). Effect of cyanide on rate of exudation in excised onion roots. *Am. J. Bot.* **31**, 172–174.

Rosene, H. F. (1950). Effect of anoxia on water exchange and oxygen consumption of onion root tissues. *J. cell. comp. Physiol.* **35**, 179–193.

Rosene, H. F. (1954). The water absorptive capacity of root hairs. *Proc. Int. bot. Congr.* **11**, 217–218.

Rosene, H. F. and Bartlett, L. E. (1950). Effect of anoxia on water influx of individual radish root hair cells. *J. cell. comp. Physiol.* **36**, 83–96.

Rosene, H. F. and Walthall, A. M. J. (1954). Comparison of the velocities of water influx into young and old root hairs of wheat seedlings. *Physiologia Pl.* **7**, 190–194.

la Rue, C. D. (1952). Root-grafting in tropical trees. *Science, N.Y.* **115**, 296.

Ruhland, W. (1956a). "Encyclopedia of Plant Physiology" Vol. I. Springer-Verlag, Berlin.

Ruhland, W. (1956b). "Encyclopedia of Plant Physiology" Vol. II. Springer-Verlag, Berlin.

Ruhland, W. (1956c). "Encyclopedia of Plant Physiology" Vol. III. Springer-Verlag, Berlin.

Russell, E. W. (1961). "Soil Conditions and Plant Growth". Longmans, London.

Russell, M. B. (1952). Soil aeration and plant growth. *In*, "Soil Physical Conditions and Plant Growth" (B. T. Shaw, ed.) pp. 253–301. Academic Press, New York and London.

Russell, M. B. and Danielson, R. E. (1956). Time and depth patterns of water use by corn. *Agron. J.* **48**, 163–165.

Russell, M. B. and Woolley, J. T. (1961). Transport processes in the soil-plant system. *In*, "Growth in Living Systems" (M. X. Zarrow, ed.) pp. 695–722. Basic Books, New York.

Russell, R. S. and Shorrocks, V. M. (1959). The relationship between transpiration and the absorption of inorganic ions by intact plants. *J. Exp. Bot.* **10**, 301–316.

Russell, R. S. and Barber, D. A. (1960). The relationship between salt uptake and the absorption of water by intact plants. *A. Rev. Pl. Physiol.* **11**, 127–140.

Salter, P. J. (1954). The effects of different water-regimes on the growth of plants under glass. 1. Experiments with tomatoes (*Lycopersicum esculentum Mill.*) *J. hort. Sci.* **24**, 258–267.

Salter, P. J. (1958). The effects of different water-regimes on the growth of plants under glass. IV. Vegetative growth and fruit development in the tomato. *J. hort. Sci.* **33**, 1–12.

Savitz, D., Sidel, V. W. and Solomon, A. K. (1964). Osmotic properties of human red cells. *J. gen. Physiol.* **48**, 79–94.

Sayre, J. D. (1926). Physiology of stomata of *Rumex patientia*. *Ohio J. Sci.* **26**, 233–266.

Scarth, G. W. (1929). The influence of H-ion concentration on the turgor and movement of plant cells with special reference to stomatal behaviour. *Proc. Int. Congr. Pl. Sci.* **1**, 1151–1162.

Schiff, L. (1953). The effect of surface head on infiltration rates based on the performance of ring infiltrometers and ponds. *Trans. Am. geophys. Un.* **34**, 257–286.

Schneider, G. W. and Childers, N. F. (1941). Influence of soil moisture on photosynthesis, respiration and transpiration of apple leaves. *Pl. Physiol., Lancaster* **16**, 565–583.

Scholander, P. F. (1958). The rise of sap in lianas. *In*, "The Physiology of Forest Trees" (K. V. Thimann, ed.) pp. 3–17. Ronald Press, New York.

Scholander, P. F., Love, W. E. and Kanwisher, J. W. (1955). The rise of sap in tall grapevines. *Pl. Physiol., Lancaster* **30**, 93–104.

Scholander, P. F., Ruud, B. and Leivestad, H. (1957). The rise of sap in a tropical liana. *Pl. Physiol., Lancaster* **32**, 1–6.

Scholander, P. F., Hemmingsen, E. and Garey, W. (1961). Cohesive lift of sap in the rattan vine. *Science, N.Y.* **134**, 1835–1838.

Scholander, P. F., Hammel, H. T., Hemmingsen, E. and Garey, W. (1962). Salt balance in mangroves. *Pl. Physiol., Lancaster* **37**, 722–729.

Scholander, P. F., Hammel, H. T., Bradstreet, D. and Hemmingsen, E. A. (1965). Sap pressure in vascular plants. *Science, N.Y.* **148**, 339–346.

Scholte-Ubing, D. W. (1959). Über stralingsmetingen de warmtebalans en de verdamping van gras. *Meded. Landb Hogesch. Wageningen* **59**, 1–93.

Schroeder, G. L., Kraner, H. W. and Evans, R. D. (1965). Diffusion of radon in several naturally occuring soil types. *J. geophys. Res.* **70**, 471–474.

Scott, F. M. (1950). Internal suberization of tissues. *Bot. Gaz.* **111**, 378–394.

Scott, F. M., Schroeder, M. P. and Turrell, F. M. (1948). Development, cell shape, suberization of internal surface and abscission in the leaf of the Valencia orange, *Citrus sinensis*. *Bot. Gaz.* **109**, 381–411.

Seifriz, W. (1956). Microscopic and submicroscopic structure of cytoplasm. *In*, "Encyclopedia of Plant Physiol" (W. Ruhland, ed.) Vol. 1, pp. 301–339. Springer-Verlag, Berlin.

Shaw, B. T. (1952). "Soil Physical Conditions and Plant Growth". Academic Press, New York and London.

Shaw, B. T. and Baver, L. D. (1939). An electrothermal method for following the moisture changes of the soil *in situ. Proc. Soil Sci. Soc. Am.* **4**, 78–83.

Shaw, M. and Maclachlan, G. A. (1954a). Chlorophyll content and carbon dioxide uptake of stomatal cells. *Nature, Lond.* **173**, 29–30.

Shaw, M. and Maclachlan, G. A. (1954b). The physiology of stomata. I. Carbon dioxide fixation of guard cells. *Can. J. Bot.* **32**, 784–794.

Shimshi, D. (1963a). Effect of chemical closure of stomata on transpiration in varied soil and atmospheric environments. *Pl. Physiol., Lancaster* **38**, 709–712.

Shimshi, D. (1963b). Effect of soil moisture and phenylmercuric acetate upon stomatal aperture, transpiration, and photosynthesis. *Pl. Physiol., Lancaster* **38**, 713–721.

Shull, C. A. (1939). Atmospheric humidity and temperature in relation to the water system of plants and soils. *Pl. Physiol., Lancaster* **14**, 401–422.

Siegenthaler, P. and Packer, L. (1965). Light-dependent volume changes and reactions in chloroplasts. I. Action of alkenylsuccinic acids and phenylmercuric acetate and possible relation to mechanisms of stomatal control. *Pl. Physiol., Lancaster* **40**, 785–791.

Sierp, H. and Brewig, A. (1935). Quantitative Untersucchungen über die Wasserabsorptionzone der Wurzeln. *Jb. wiss. Bot.* **82**, 99–122.

Silva Fernandes, A. M., Baker, E. A. and Martin, J. T. (1964). Studies on plant cuticle. VI. The isolation and fractionation of cuticular waxes. *Ann. appl. Biol.* **53**, 43–58.

Singer, S. J. (1962). The properties of protein in non-aqueous solvents. *Adv. Protein Chem.* **17**, 1–68.

Skoog, F., Broyer, T. C. and Grossenbacher, K. A. (1938). Effects of auxin on rates, periodicity, and osmotic relations in exudation. *Am. J. Bot.* **25**, 749–759.

Slater, C. S. (1957). Cylinder infiltrometers for determining rates of irrigation. *Proc. Soil Sci. Soc. Am.* **21**, 457–460.

Slatyer, R. O. (1955). Studies of the water relations of crop plants grown under natural rainfall in northern Australia. *Aust. J. agric. Res.* **6**, 365–377.

Slatyer, R. O. (1956). Absorption of water from atmosphere of different humidity and its transport through plants. *Aust. J. biol. Sci.* **9**, 552–558.

Slatyer, R. O. (1957a). Significance of the permanent wilting percentage in studies of plant and soil water relations. *Bot. Rev.* **23**, 585–636.

Slatyer, R. O. (1957b). The influence of progressive increases in total soil moisture stress, on transpiration, growth, and internal water relationships of plants. *Aust. J. biol. Sci.* **10**, 320–336.

Slatyer, R. O. (1958). The measurement of diffusion pressure deficit in plants by a method of vapour equilibration. *Aust. J. biol. Sci.* **11**, 349–365.

Slatyer, R. O. (1960a). Aspects of the tissue water relationships of an important arid zone species (*Acacia aneura* F. Muell) in comparison with two mesophytes. *Bull. Res. Coun. Israel* 8D. 159–168.

Slatyer, R. O. (1960b). Agricultural Climatology of the Katherine area, N.T. *CSIRO Div. Land Res. Tech. Paper* **13**.

Slatyer, R. O. (1960c). Absorption of water by plants. *Bot. Rev.* **26**, 331–392.

Slatyer, R. O. (1961). Effects of several osmotic substrates on the water relations of tomato. *Aust. J. biol. Sci.* **14**, 519–540.

Slatyer, R. O. (1962a). Methodology of a water balance study conducted on a desert woodland (*Acacia aneura* F. Muell) community in central Australia. *UNESCO Arid Zone Res.* **16**, 15–26.

Slatyer, R. O. (1962b). Internal water relations of higher plants. *A. Rev. Pl. Physiol.* **13**, 351–378.

Slatyer, R. O. (1962c). Internal water balance of *Acacia aneura* F. Muell in relation to environmental conditions. *UNESCO Arid Zone Res.* **16**, 137–146.

Slatyer, R. O. (1963). Climatic control of plant water relations. *In*, "Environmental Control of Plant Growth" (L. T. Evans, ed.) pp. 33–54. Academic Press, New York and London.

Slatyer, R. O. (1965). Measurements of precipitation interception by an arid zone plant community (*Acacia aneura* F. Muell). *UNESCO Arid Zone Res.* **25**, 181–192.

Slatyer, R. O. (1966a). Some physical aspects of non-stomatal control of leaf transpiration. *Agric. Met.* **3**, 281-292

Slatyer, R. O. (1966b). An underlying cause of measurement discrepancies in determinations of osmotic characteristics in plant cells and tissues. *Protoplasma* **62**, 34–43.

Slatyer, R. O. (1966c). Terminology for cell and tissue water relations. *Zeit. Pflanzenphysiol.* (in press).

Slatyer, R. O. and Taylor, S. A. (1960). Terminology in plant-soil-water relations. *Nature, Lond.* **187**, 922–924.

Slatyer, R. O. and McIlroy, I. C. (1961). "Practical Microclimatology". UNESCO, Paris.

Slatyer, R. O. and Bierhuizen, J. F. (1964a). Transpiration from cotton leaves under a range of environmental conditions in relation to internal and external diffusive resistances. *Aust. J. biol. Sci.* **17**, 115–130.

Slatyer, R. O. and Bierhuizen, J. F. (1964b). The influence of several transpiration suppressants on transpiration, photosynthesis, and water use efficiency of cotton leaves. *Aust. J. biol. Sci.* **17**, 131–146.

Slatyer, R. O. and Bierhuizen, J. F. (1964c). The effect of several foliar sprays on transpiration and water use efficiency of cotton plants. *Agric. Met.* **1**, 42–53.

Slatyer, R. O. and Denmead, O. T. (1964). Water movement through the soil-plant-atmosphere system. *In*, "National Symposium on Water Resources, Use and Management" (E. S. Hills, ed.) pp. 276–289. Melbourne University Press, Melbourne.

Slatyer, R. O. and Barrs, H. D. (1965). Modifications to the relative turgidity technique with notes on its significance as an index of the internal water status of leaves. *UNESCO Arid Zone Res.* **25**, 331–342.

Slatyer, R. O. and Gardner, W. R. (1965). Overall aspects of water movement in plants and soils. *Symp. Soc. exp. Biol.* **19**, 113–129.

Slatyer, R. O. and Jarvis, P. G. (1966). A gaseous-diffusion porometer for continuous measurement of diffusive resistance of leaves. *Science, N.Y.* **151**, 574–576.

Slavik, B. (1958). The influence of water deficit on transpiration. *Physiologia Pl.* **11**, 524–536.

Slavik, B. (1963a). On the problem of the relationship between hydration of leaf tissue and intensity of photosynthesis and respiration. *In*, "The Water Relations of Plants" (A. J. Rutter and F. H. Whitehead, eds.) pp. 225–234. Blackwell, London.

Slavik, B. (1963b). The distribution pattern of transpiration rate, water saturation deficit, stomata number and size, photosynthetic and respiration rate in the area of the tobacco leaf blade. *Biol. Plant. Acad. Sci. Bohemoslov* **5**, 143–153.

Slavik, B. (1963c). Relationship between the osmotic potential of the cell sap and the water saturation deficit during the wilting of leaf tissue. *Biol. Plant. Acad. Sci. Bohemoslov.* **5**, 258–264.

Slavik, B. (1965). "Water Stress in Plants". Czechoslovakia Academy of Sciences, Prague.

Smith, R. M. and Browning, D. R. (1946). Some suggested laboratory standards of subsoil permeability. *Proc. Soil Sci. Soc. Am.* **11**, 21–26.

Smith, W. O. (1943). Thermal transfer of moisture in soils. *Trans. Am. geophys. Un.* **24**, 511–523.

Solomon, A. K. (1960). Red cell membrane, structure and ion transport. *J. gen. Physiol.* **43**, (Supplement) 1–15.

Spanner, D. C. (1951). The Peltier effect and its use in the measurement of suction pressure. *J. exp. Bot.* **11**, 145–168.

Spanner, D. C. (1953). On a new method for measuring the stomatal aperture of leaves. *J. exp. Bot.* **4**, 283–295.

Spanner, D. C. (1964). "Introduction to Thermodynamics". Academic Press, London and New York.

Specht, R. L. (1957). The water relationships in heath vegetation and pastures on the Makin sand. *Aust. J. Bot.* **5**, 151–172.

Spoehr, H. A. and Milner, H. W. (1939). Starch dissolution and amylolytic activity of leaves. *Proc. Am. phil. Soc.* **81**, 37–78.

Stadelmann, E. (1963). Vergleich und Umrechnung von Permeabilitäts-konstanten fur Wasser. *Protoplasma*, **57**, 660–678.

Stålfelt, M. G. (1955). The stomata as a hydrophotic regulator of the water deficit of the plant. *Physiologia Pl.* **8**, 572–593.

Stålfelt, M. G. (1956). Die stomatäre transpiration und die physiologie der spaltoffnungen. *In*, "Encyclopedia of Plant Physiology" (W. Ruhland, ed.) Vol. 3, pp. 351–426. Springer-Verlag, Berlin.

Stålfelt, M. G. (1961). The effect of the water deficit on the stomatal movements in a carbon dioxide-free atmosphere. *Physiologia Pl.* **14**, 826–843.

Stålfelt, M. G. (1962). The effect of temperature on opening of the stomatal cells. *Physiologia Pl.* **15**, 772–779.

Stamm, A. J. (1944). Surface properties of cellulosic materials. *In*, "Wood Chemistry" (L. E. Wide, ed.) pp. 449–550. Reinhold, New York.

Stanhill, G. (1961). A comparison of methods of calculating potential evapotranspiration from climatic data. *Ktavim.* **11**, 159–171.

Staple, W. J. and Lehane, J. J. (1954). Movement of moisture in unsaturated soils. *Can. J. agric. Sci.* **34**, 329–342.

Staple, W. J. and Lehane, J. J. (1962). Variability in soil moisture sampling. *Can. J. Soil Sci.* **42**, 157–164.

Staverman, A. J. (1951). The theory of measurement of osmotic pressure. *Recl. Trav. chim. Pays-Bas Belg.* **70**, 344–352.

Staverman, A. J. (1952). Apparent osmotic pressure of solutions of heterodisperse polymers. *Recl. Trav. chim. Pays-Bas Belg.* **71**, 623–633.

Stern, W. R. and Donald, C. M. (1962). The influence of leaf area and radiation on the growth of clover in swards. *Aust. J. agric. Res.* **13**, 615–623.

Stevens, A. B. P. (1956). The structure and development of hydathodes of *Caltha palustris* L. *New Phytol.* **55**, 339–345.

Steward, F. C. (1959). "Plant Physiology—A Treatise" Vol. 2. Academic Press, New York and London.

Stiles, W. (1922). The suction pressure of the plant cell. *Biochem. J.* **16**, 727–728.

Stocker, O. (1929). Eine Feldmethode zur Bestimmung der momentanen Transpirations- und Evaporationsgroesse. *Ber. dt. bot. Ges.* **47**, 126–129, 130–136.

Stocker, O. (1960). Physiological and morphological changes in plants due to water deficiency. *UNESCO Arid Zone Res.* **15**, 63–104.

Stocker, O. and Holdheide, W. (1937). Die Assimilation Helgoländer Gezeitenalgen während der Ebbezeit. *Z. Bot.* **32**, 1–59.

Stocker, O. and Ross, H. (1956). Reaktions- und Restitutionsphase der Plasmaviskosität bei Dürre- und Schuttelreizen. *Naturwissenschaften* **43**, 283–284.

Stocking, C. R. (1945). The calculation of tensions in *Cucurbita pepo*. *Am. J. Bot.* **32**, 126–134.

Stocking, C. R. (1956a). Hydration and cell physiology. *In*, "Encyclopedia of Plant Physiology" (W. Ruhland, ed.) Vol. 2, pp. 23–37. Springer-Verlag, Berlin.

Stocking, C. R. (1956b). Guttation and bleeding. *In*, "Encyclopedia of Plant Physiology" (W. Ruhland, ed.) Vol. 3, pp. 489–502. Springer-Verlag, Berlin.

Stocking, C. R. (1956c). Excretion by glandular organs. *In*, "Encyclopedia of Plant Physiology" (W. Ruhland, ed.) Vol. 3, pp. 503–510. Springer-Verlag, Berlin.

Stone, E. C. (1957a). Dew as an ecological factor. I. A review of the literature. *Ecology* **38**, 407–413.

Stone, E. C. (1957b). Dew as an ecological factor. II. The effect of artificial dew on the survival of *Pinus ponderosa* and associated species. *Ecology* **38**, 414–422.

Stone, E. C. and Fowells, H. A. (1955). Survival value of dew as determined under laboratory conditions with *Pinus ponderosa*. *Forest Sci.* **1**, 183–188.

Stone, E. C., Went, F. W. and Young, C. L. (1950). Water absorption from the atmosphere by plants growing in dry soil. *Science, N.Y.* **111**, 546–548.

Stone, J. F., Kirkham, D. and Read, A. A. (1955). Soil moisture determination by a portable neutron scatter meter. *Proc. Soil Sci. Soc. Am.* **19**, 418–423.

Strugger, S. (1949). "Praktikum der Zell- und Gewebephysiologie der Pflanze". Springer-Verlag, Berlin.

Strugger, S. and Peveling, E. (1961). Die elecktronenmikroskopische Analyse der extrafaszikulären Komponente des Transpirationsstromes mit Hilfe von Edelmetallsuspensoiden adäquater Dispersität. *Ber. dt. bot. Ges.* **74**, 300–304.

Sutton, O. G. (1953). "Micrometeorology". McGraw-Hill, New York.

Swinbank, W. C. (1951). The measurement of vertical transfer of heat and water vapour by eddies in the lower atmosphere. *J. Met.* **8**, 135–145.

Tanford, C. (1963). The structure of water and of aqueous solutions. *In*, "Temperature—Its Measurement and Control" (C. M. Herzfeld ed.) Vol. 3, pp. 123–129. Reinhold, New York.

Tanner, C. B. (1960). Energy balance approach to evapotranspiration from crops. *Proc. Soil Sci. Soc. Am.* **24**, 1–9.

Tanner, C. B. (1963), Energy relations in plant communities. *In*, "Environmental Control of Plant Growth" (L. T. Evans, ed.) pp. 141–148. Academic Press, New York and London.

Tanner, C. B. and Pelton, W. L. (1960). Potential evapotranspiration estimates by the approximate energy balance method of Penman. *J. geophys. Res.* **65**, 3391–3413.

Tanner, C. B., Peterson, A. E. and Love, J. R. (1960). Radiant energy exchange in a corn field. *Agron. J.* **52**, 373–379.

Taylor, C. A. and Furr, J. R. (1937). Use of soil moisture and fruit growth records for checking irrigation practices in citrus orchards. *U.S. Dept. Agric. Circ.* **426**.

Taylor, S. A. (1962). The influence of temperature upon the transfer of water in soil systems. *Meded. Landb Hogesch. Opzoek Stns Gent* **27**, 535–551.

Taylor, S. A. and Heuser, N. C. (1953). Water entry and downward movement in undisturbed soil cores. *Proc. Soil Sci. Soc. Am.* **17**, 195–201.

Taylor, S. A. and Cavazza, L. (1954). The movement of soil moisture in response to temperature gradients. *Proc. Soil Sci. Soc. Am.* **18**, 351–358.

Taylor, S. A. and Slatyer, R. O. (1961). Water-soil-plant relations terminology. *Proc. Int. Congr. Soil Sci.* **1**, 394–403.

Taylor, S. A. and Slatyer, R. O. (1962). Proposals for a unified terminology in studies of plant-soil-water relations. *UNESCO Arid Zone Res.* **16**, 339–349.

Taylor, S. A. and Cary, J. W. (1964). Linear equations for the simultaneous flow of matter and energy in a continuous soil system. *Proc. Soil Sci. Soc. Am.* **28**, 167–172.

Taylor, S. A., Evans, D. D. and Kemper, W. D. (1961). Evaluating soil water. *Utah Agric. Exp. Sta. Bull.* **426**.

Thames, J. L. (1961). Effects of wax coatings on leaf temperatures and field survival of *Pinus taeda* seedlings. *Pl. Physiol., Lancaster* **36**, 180–182.

Thoday, D. (1918). On turgescence and the absorption of water by the cells of plants. *New Phytol.* **17**, 108–113.

Thomas, M. D. (1955). Effect of ecological factors on photosynthesis. *A. Rev. Pl. Physiol.* **6**, 135–156.

Thomas, M. D. and Hill, G. R. (1949). Photosynthesis under field conditions. *In*, "Photosynthesis in Plants" (J. Franck and W. E. Loomis, eds.) pp. 19–52. Iowa State College Press, Ames, Iowa.

Thompson, L. M. (1957). "Soils and Soil Fertility". McGraw-Hill, New York.

Thornthwaite, C. W. (1948). An approach toward a rational classification of climate. *Geogrl. Rev.* **38**, 55–94.

Thornthwaite, C. W. (1954). A re-examination of the concept and measurement of potential evapotranspiration. *Johns Hopkins University Climatology Publication* No. 7.

Thornthwaite, C. W. and Holzman, B. (1939). The determination of evaporation from land and water surfaces. *Mon. Weath. Rev.* **67**, 4–11.

Tibbals, E. C., Carr, E. K., Gates, D. M. and Kreith, F. (1964). Radiation and convection in conifers. *Am. J. Bot.* **51**, 529–538.

Ting, I. P. and Loomis, W. E. (1963). Diffusion through stomates. *Am. J. Bot.* **50**, 866–872.

Tisdall, A. L. (1951). Antecedent soil moisture and its relation to infiltration. *Aust. J. agric. Res.* **2**, 342–354.

Turrell, F. M. (1936). The area of the internal exposed surface of dicotyledon leaves. *Am. J. Bot.* **23**, 255–264.

Turrell, F. M. (1942). A quantatitive morphological analysis of large and small leaves of alfalfa with special reference to internal surface. *Am. J. Bot.* **29**, 400–415.

UNESCO (1958a). "Climatology—Reviews of Research". Arid Zone Research Vol. 10. UNESCO, Paris.

UNESCO (1958b). "Climatology and Microclimatology". Arid Zone Research Vol. 11. UNESCO, Paris.

UNESCO (1960). "Reviews of Research on Plant-Water Relationships in Arid Conditions". Arid Zone Research Vol. 15. UNESCO, Paris.

UNESCO (1962). "Plant-water Relationships in Arid and Semi-Arid Conditions". Arid Zone Research Vol. 16. UNESCO, Paris.

Ursprung, A. (1915). Über die Kohasion des Wassers im Farnannulus. *Ber. dt. bot. Ges.* **33**, 153–162.

Ursprung, A. and Blum, G. (1916). Uber die Verteilung des osmotischen Wertes in der Pflanze. *Ber. dt. bot. Ges.* **34**, 88–104.

Ussing, H. H. (1952). Some aspects of the application of tracers in permeability studies. *Adv. Enzymol.* **13**, 21–65.

Vaadia, Y. and Marr, A. G. (1961). Rapid cryoscopic technique for measuring osmotic properties of drop size samples. *Pl. Physiol., Lanceaster* **36**, 677–680.

Vaadia, Y. and Waisel, Y. (1963). Water absorption of the aerial organs of plants. *Physiologia Pl.* **16**, 44–51.

Vaadia, Y., Raney, F. C. and Hagan, R. M. (1961). Plant water deficits and physiological processes. *A. Rev. Pl. Physiol.* **12**, 265–292.

Vartapetyan, B. B. (1960a). Further investigation of water exchange in plants using H_2O^{18} (English translation). *Fiziologiya Rast.* **7**, 395–397.

Vartapetyan, B. B. (1960b). Participation of H_2O^{18} in the metabolism of photosynthetic tissues (English translation). *Fiziologiya Rast.* **7**, 414–418.

Vartapetyan, B. B. and Kursanov, A. L. (1959). A study of water metabolism of plants using water containing heavy oxygen, H_2O^{18} (English translation). *Fiziologiya Rast.* **6**, 144–150.

Vasil'eva, N. G. and Burkina, Z. S. (1960). Water conditions in cell organelles (English translation). *Fiziologiya Rast.* **7**, 401–406.

Veihmeyer, F. J. (1956). Soil moisture. *In*, "Encyclopedia of Plant Physiology" (W. Ruhland, ed.) Vol. 3, pp. 64–123. Springer-Verlag, Berlin.

Veihmeyer, F. J. and Hendrickson, A. H. (1927). Soil moisture conditions in relation to plant growth. *Pl. Physiol., Lancaster* **2**, 71–82.

Veihmeyer, F. J. and Hendrickson, A. H. (1949). Methods of measuring field capacity and permanent wilting percentage of soils. *Soil Sci.* **68**, 75–94.

Veihmeyer, F. J. and Hendrickson, A. H. (1950). Soil moisture in relation to plant growth. *A. Rev. Pl. Physiol.* **1**, 285–304.

Verduin, J. and Loomis, W. E. (1944). Absorption of carbon dioxide by maize. *Pl. Physiol., Lancaster* **19**, 278–293.

Villegas, R., Barton, T. C. and Solomon, A. K. (1958). The entrance of water into beef and dog red cells. *J. gen. Physiol.* **42**, 355–369.

Virgin, H. I. (1953). Physical properties of protoplasm. *A. Rev. Pl. Physiol.* **4**, 363–382.

Visser, W. C. (1964). Moisture requirements of crops and rate of moisture depletion of the soil. *Inst. Land Water Manag. Res. Tech. Bull.* **32**.

Vomocil, J. A. (1954). *In situ* measurement of soil bulk density. *Agric. Engng* **35**, 651–654.

de Vries, D. A. (1952a). Het warmtegeleidingsvermogen van grond. *Meded. Landb Hogesch. Wageningen* **52**, 1–73.

de Vries, D. A. (1952b). The thermal conductivity of granular materials. *Annexe* 1952–1 *Bul. Inst. Intern. du Froid*, 115–131.

de Vries, D. A. (1958). Simultaneous transfer of heat and moisture in porous media. *Trans. Am. geophys. Un.* **39**, 909–916.

de Vries, D. A. and Peck, A. J. (1958). On the cylindrical probe method of measuring thermal conductivity with special reference to soils. I. Extension of theory and discussion of probe characteristics. *Aust. J. Phys.* **11**, 255–271.

Wadleigh, C. H. (1946). The integrated soil moisture stress upon a root system in a large container of saline soil. *Soil Sci.* **61**, 225–238.

Wadleigh, C. H. and Ayers, A. D. (1945). Growth and biochemical composition of bean plants as conditioned by soil moisture tension and salt concentration. *Pl. Physiol., Lancaster* **20**, 106–132.

Wadleigh, C. H. and Gauch, H. G. (1948). Rate of leaf elongation as affected by the intensity of the total soil moisture stress. *Pl. Physiol., Lancaster* **23**, 485–495.

Wadleigh, C. H., Gauch, H. G. and Magistad, O. C. (1946). Growth and rubber accumulation in guayule as conditioned by soil salinity and irrigation regime. *U. S. Dept. Agric. Tech. Bull.* **925**.

Waggoner, P. E. (1965). Calibration of a porometer in terms of diffusive resistance. *Agric. met.* **2**, 317–329.

Wakeshima, H. (1961). On the theory of the fracture of liquids. *J. phys. Soc. Japan.* **16**, 6–14.

Walker, D. A. and Zelitch, I. (1963). Some effects of metabolic inhibitors, temperature, and anaerobic conditions on stomatal movement. *Pl. Physiol., Lancaster* **38**, 390–396.

Wallihan, E. F. (1964). Modification and use of an electric hygrometer for estimating relative stomatal apertures. *Pl. Physiol., Lancaster* **39**, 86–90.

Walter, H. (1929). Plasmaquellung und assimilation. *Protoplasma* **6**, 113–156.

Walter, H. (1931). "Die Hydratur der Pflanze". Gustav Fischer, Jena.

Walter, H. (1936). Die ökologischen Verhältnisse in der Namib-Nebelwüste. (Südwestafrika). *Jb. wiss. Bot.* **84**, 58–222.

Walter, H. (1955). The water economy and hydrature of plants. *A. Rev. Pl. Physiol.* **6**, 239–252.

Walter, H. (1963). Zur Klärung des spezifischen Wasserzustandes im Plasma und in der Zellwand bei höheren Pflanzen und seine Bestimmung. *Ber. dt. bot. Ges.* **76**, 40–53.

Wassink, E. C. (1959). Efficiency of light energy conversion in plant growth. *Pl. Physiol.* **34**, 356–361.

Watson, D. J. (1952). The physiological basis of variation in yield. *Adv. Agron.* **4**, 101–145.

Watson, D. J. (1963). Climate, weather and plant yield. *In*, "Environmental Control of Plant Growth" (L. T. Evans, ed.) pp. 337–350. Academic Press, New York and London.

Weatherley, P. E. (1950). Studies in the water relations of the cotton plant. I. The field measurements of water deficits in leaves. *New Phytol.* **49**, 81–87.

Weatherley, P. E. (1951). Studies in the water relations of the cotton plant. II. Diurnal and seasonal fluctuations and environmental factors. *New Phytol.* **50**, 36–51.

Weatherley, P. E. (1954). Preliminary investigations into the uptake of sugars by floating leaf disks. *New Phytol.* **53**, 204–216.

Weatherley, P. E. (1955). On the uptake of sucrose and water by floating leaf disks under aerobic and anaerobic conditions. *New Phytol.* **54**, 13–28.

Weatherley, P. E. (1961). A new micro-osmometer, *J. exp. Bot.* **11**, 258–268.

Weatherley, P. E. (1962). The mechanism of sieve tube translocation: Observation, experiment and theory. *Advan. Sci.* **18**, 571–577.

Weatherley, P. E. (1963). The pathway of water movement across the root cortex and leaf mesophyll of transpiring plants. *In*, "The Water Relations of Plants" (A. J. Rutter and F. H. Whitehead, eds.) pp. 85–100. Blackwell, London.

Weatherley, P. E. and Slatyer, R. O. (1957). Relationship between relative turgidity and diffusion pressure deficit in plants. *Nature, Lond.* **179**, 1085–1086.

Webb, E. K. (1960). An investigation of the evaporation from Lake Eucumbene. *CSIRO Div. Meteorol. Physics Tech. Paper* **10**.

Weinmann, H. and le Roux M. (1946). A critical study of the torsion balance of measuring transpiration. *S. Afr. J. Sci.* **42**, 147–153.

Werner, H. O. (1954). Influence of atmospheric and soil moisture conditions on diurnal variations in relative turgidity of potato leaves. *Neb. Agric. Exp. Sta. Res. Bull.* **176**.

Wesseling, J. and van Wijk, W. R. (1957). Soil physical conditions in relation to drain depth. *In*, "Drainage of Agricultural Lands" (J. N. Luthin, ed.) pp. 461–504. American Society of Agronomy, Madison, Wisconsin.

Whiteman, P. C. and Wilson, C. L. (1963). Estimation of diffusion pressure deficit by correlation with relative turgidity and beta radiation absorption. *Aust. J. Biol. Sci.* **16**, 140–146.

Whiteman, P. C. and Koller, D. (1964). Saturation deficit of the mesophyll evaporating surfaces in a desert halophyte. *Science, N.Y.* **146**, 1320–1321.

Whitney, J. B. (1942). Effects of the composition of the soil atmosphere on the absorption of water by plants. Ohio State University, Ph.D. Thesis.

Wiersma, D. and Veihmeyer, F. J. (1954). Absence of water exudation from roots of plants grown in an atmosphere of high humidity. *Soil Sci.* **78**, 33–36.

van Wijk, W. R. (1963). "Physics of Plant Environment". North-Holland Amsterdam.

Wilcox, J. C. (1960). Rate of soil drainage following an irrigation. II. Effects on determination of rate of consumption use. *Can. J. Soil Sci.* **40**, 15–27.

Wilson, A. M. and McKell, C. M. (1961). Effect of soil moisture stress on absorption and translocation of phosphorus applied to leaves of sunflower. *Pl. Physiol., Lancaster* **36**, 762–765.

Wilson, C. C. (1948). Diurnal fluctuations in growth in length of tomato stem. *Pl. Physiol., Lancaster* **23**, 156–157.

Wilson, C. W., Boggess, W. R. and Kramer, P. J. (1953). Diurnal fluctuations in the moisture content of some herbaceous plants. *Am. J. Bot.* **40**, 97–100.

Wind, G. P. (1955a). A field experiment concerning capillary rise of moisture in a heavy clay soil. *Neth. J. agric. Sci.* **3**, 60–69.

Wind, G. P. (1955b). Flow of water through plant roots. *Neth. J. agric. Sci.* **3**, 259–264.

Wind, G. P. (1960). Capillary rise and some applications of the theory of moisture movement in unsaturated soils. *Versl. Meded.* **5**, 1–15.

Withrow, R. B. and Withrow, A. P. (1956). Generation, control and measurement of visible and non visible radiant energy. *In*, "Radiation Biology" (A. Hollaender, ed.) Vol. 3, pp. 125–258. McGraw-Hill, New York.

Van't Woudt, B. D. and Hagan, R. M. (1957). Crop responses at excessively high soil moisture levels. *In*, "Drainage of Agricultural Lands" (J. N. Luthin, ed.) pp. 514–578. American Society of Agronomy, Madison, Wisconsin.

Wyllie, M. R. J. and Rose, W. D. (1950). Application of the Kozeny equation to consolidated media. *Nature, Lond.* **165**, 972.

Yamada, Y., Tamai, S. and Miyaguchi, T. (1958). Measurement of the thickness of leaves using S^{35}. *In*, "Proceedings 2nd Japanese Conference on Radio-isotopes" pp. 1692–1700.

Youngs, E. G. (1957). Moisture profiles during vertical infiltration. *Soil Sci.* **84**, 283–290.

Zamfirescu, N. (1931). Cercetari asupra Absorptiunii apei prin organele aeriene ale plantelor. *Bull. Min. Agr. Domenilor* **3**, 3–6.

Zelitch, I. (1961). Biochemical control of stomatal opening in leaves. *Proc. natn. Acad. Sci. U.S.A.* **47**, 1423–1433.

Zelitch, I. (1963). The control and mechanisms of stomatal movement. *In*, "Stomata and Water Relations in Plants" (I. Zelitch, ed.) pp. 18–42. Connecticut Agricultural Experiment Station, New Haven, Conn.

Zelitch, I. (1964). Reduction of transpiration of leaves through stomatal closure by alkenyl-succinic acids. *Science, N.Y.* **143**, 692–693.

Zelitch, I. (1965). Environmental and biochemical control of stomatal movement in leaves. *Biol. Rev.* **40**, 463–482.

Zelitch, I. and Waggoner, P. E. (1962a). Effect of chemical control of stomata on transpiration and photosynthesis. *Proc. natn. Acad. Sci. U.S.A.* **48**, 1101–1108.

Zelitch, I. and Waggoner, P. E. (1962b). Effect of chemical control of stomata on transpiration of intact plants. *Proc. natn. Acad. Sci. U.S.A.* **48**, 1297–1299.

Zimmermann, M. H. (1964a). Effect of low temperature on ascent of sap in trees. *Pl. Physiol., Lancaster* **39**, 568–572.

Zimmermann, M. H. (1964b). Sap movements in trees. *Biorheology* **2**, 15–27.

Zimmermann, M. H. (1965). Water movement in stems of tall plants. *Symp. Soc. exp. Biol.* **19**, 151–155.

Author Index

The numbers in italics refer to the page in the Bibliography in which the reference appears.

Subject Index

A

Absorption of water by plants, *see* Water absorption

Activation energy
and heat of transfer of soil water, 118
and Q_{10}, 180
factors affecting, 180–181
of water transport through membranes, 180

Active water uptake, 174–175, 195–197

Activity of water, 20–21
activity coefficient, 20
relationship to chemical potential, 20–21

Advection of energy,
effect on evaporation, 39–40
effect on surface energy balance, 36–40

Aeration
effect on membrane permeability, 182–183
effect on water uptake by roots, 206–207

Albedo
of leaves, 238
of plant communities, 32

Anatomy
of cells, 127–130
of leaves, 215–219
of roots, 198–203
of stems, 211–212

Availability of soil water
and development of internal water deficits, 275–282
and permanent wilting percentage, 77, 277–279
effect of salinity on, 301–308
for plant growth, 77–78, 299–301
for transpiration, 78, 221–230

B

Bound water, 137
and protoplasmic hydration, 285–287

Boundary layers
at cell surfaces, 190–193
at leaf surfaces, 241–244

Bowen ratio
for leaves, 246–247
for plant community surfaces, 36–38

C

Capillary phenomena
in cell walls, 130–132
in soil water retention, 69–70
in water flow from a water table, 118–121
relationship to vapour pressure, 13

Cells
forces holding water in, 130, 137
permeability of, 127–130
structure of, 127–130
volume changes in, 175–178
water and solute transfer in, 161–171
water content of, 127–128, 136–137
water distribution in, 127–130
water potential of, 143–150

Cell wall
hydration, 132–133
permeability, 176, 219–220
structure, 130–131

Chemical potential, 17–21
alternative expressions for, 18–21
and activity, 20
and free energy, 18
and vapour pressure, 20–21
dependence on composition, 18–19
dependence on pressure, 18–19
dependence on temperature, 18–19
of water in plants, 143–150
of water in soils, 78–84
of water in solutions, 19–21

Cohesion theory for sap ascent, 212–214

Colloid—water structures, 8
and protoplasmic hydration, 285–287
and water exchangeability, 137
in cytoplasm and cell walls, 130–135